BAYESIAN MODEL COMPARISON

ADVANCES IN ECONOMETRICS

Series Editors: Juan Carlos Escanciano,
Thomas B. Fomby,
R. Carter Hill, Eric Hillebrand,
David Jacho-Chávez, Ivan Jeliazkov,
Daniel L. Millimet and
Rodney Strachan

Recent Volumes:

Volume 23: Bayesian Econometrics – Edited by Siddhartha Chib, Gary Koop, Bill Griffiths and Dek Terrell

Volume 24: Measurement Error: Consequences, Applications and Solutions – Edited by Jane Binner, David Edgerton and Thomas Elger

Volume 25: Nonparametric Econometric Methods – Edited by Qi Li and Jeffrey S. Racine

Volume 26: Maximum Simulated Likelihood Methods and Applications – Edited by R. Carter Hill and William Greene

Volume 27A: Missing Data Methods: Cross-Sectional Methods and Applications – Edited by David M. Drukker

Volume 27B: Missing Data Methods: Time-Series Methods and Applications – Edited by David M. Drukker

Volume 28: DSGE Models in Macroeconomics: Estimation, Evaluation and New Developments – Edited by Nathan Balke, Fabio Canova, Fabio Milani and Mark Wynne

Volume 29: Essays in Honor of Jerry Hausman – Edited by Badi H. Baltagi, Whitney Newey, Hal White and R. Carter Hill

Volume 30: 30th Anniversary Edition – Edited by Dek Terrell and Daniel Millmet

Volume 31: Structural Econometric Models – Edited by Eugene Choo and Matthew Shum

Volume 32: VAR Models in Macroeconomics — New Developments and Applications: Essays in Honor of Christopher A. Sims – Edited by Thomas B. Fomby, Lutz Kilian and Anthony Murphy

Volume 33: Essays in Honor of Peter C. B. Phillips – Edited by Thomas B. Fomby, Yoosoon Chang and Joon Y. Park

ADVANCES IN ECONOMETRICS VOLUME 34

BAYESIAN MODEL COMPARISON

EDITED BY

IVAN JELIAZKOV
Department of Economics, University of California – Irvine, Irvine, CA, USA

DALE J. POIRIER
Department of Economics, University of California – Irvine, Irvine, CA, USA

United Kingdom – North America – Japan
India – Malaysia – China

Emerald Group Publishing Limited
Howard House, Wagon Lane, Bingley BD16 1WA, UK

First edition 2014

British Library Cataloguing in Publication Data
A catalogue record for this book is available from the British Library

ISBN: 978-1-78441-185-5
ISSN: 0731-9053 (Series)

ISOQAR REGISTERED
Certificate Number 1985
ISO 14001

ISOQAR certified Management System, awarded to Emerald for adherence to Environmental standard ISO 14001:2004.

INVESTOR IN PEOPLE

CONTENTS

LIST OF CONTRIBUTORS

Daniel Felix Ahelegbey	Graduate School of Economics and Management, University of Venice, Venice, Italy
Miguel Belmonte	Department of Economics, University of Strathclyde, Glasgow, UK
Gail Blattenberger	Department of Economics, University of Utah, Salt Lake City, UT, USA
Martin Burda	Department of Economics, University of Toronto, Toronto, Canada
Garland Durham	Orfalea College of Business, California State Polytechnic Institute, San Luis Obispo, CA, USA
Richard Fowles	Department of Economics, University of Utah, Salt Lake City, UT, USA
John Geweke	School of Business, University of Technology Sydney, Sydney, Australia; Erasmus University, Rotterdam, The Netherlands; Colorado State University, Fort Collins, CO, USA
Benjamin J. Gillen	Department of Economics, California Institute of Technology, Pasadena, CA, USA
Paolo Giudici	Department of Economics and Management, University of Pavia, Pavia, Italy
Esther Hee Lee	EViews, IHS Global Inc., Irvine, CA, USA
Ivan Jeliazkov	Department of Economics, University of California – Irvine, Irvine, CA, USA

Gary Koop	Department of Economics, University of Strathclyde, Glasgow, UK
Peter D. Loeb	Department of Economics, Rutgers University, Newark, NJ, USA
Enrique Martínez-García	Research Department, Federal Reserve Bank of Dallas, Dallas, TX, USA
Elías Moreno	Department of Statistics, University of Granada, Granada, Spain
Luís Raúl Pericchi	Department of Mathematics, University of Puerto Rico, Rio Piedras, San Juan, Puerto Rico, USA
Dale J. Poirier	Department of Economics, University of California – Irvine, Irvine, CA, USA
Hyungsik Roger Moon	Department of Economics, University of Southern California, Los Angeles, CA, USA
Matthew Shum	Department of Economics, California Institute of Technology, Pasadena, CA, USA
Angela Vossmeyer	Department of Economics, University of California – Irvine, Irvine, CA, USA
Guillaume Weisang	Graduate School of Management, Clark University, Worcester, MA, USA
Mark A. Wynne	Research Department, Federal Reserve Bank of Dallas, Dallas, TX, USA

PREFACE

Addressing model uncertainty is of fundamental importance in statistical data analysis. In practice, researchers rarely know the "correct model" and instead have to sort through a variety of possible specifications that differ along multiple margins. For instance, one has to decide what explanatory variables should be included, determine the functional specification through which they enter the model, and examine alternative distributional assumptions, dependence structures, and so on. Important advances in modeling have guided the richness of potential choices, whereas major recent developments in estimation methodology, simulation, and computation have facilitated their practical implementation.

The Bayesian approach offers a fully general probabilistic framework for simultaneously addressing both model and parameter uncertainty. Given data y and a set of competing models $\{M_1,\ldots, M_J\}$, each specified by a sampling density $f(y|\theta_j,M_j)$, prior $p(\theta_j|M_j)$ on the model parameters θ_j, and prior model probability $p(M_j)$, the posterior model probabilities are given by

$$p(M_j|y) = \frac{p(y|M_j)p(M_j)}{\sum_k p(y|M_k)p(M_k)},$$

where the marginal likelihood $p(y|M_j)$ is given by

$$p(y|M_j) = \int f(y|\theta_j, M_j)p(\theta_j|M_j)d\theta_j.$$

The posterior model probabilities can then be used for model comparison, model selection, and model averaging, depending on the inferential task at hand. In addition, any pair of competing models M_i and M_j can be compared on the basis of their posterior odds

$$\frac{p(M_i|y)}{p(M_j|y)} = \frac{p(y|M_i)}{p(y|M_j)} \times \frac{p(M_i)}{p(M_j)},$$

given by product of the Bayes factor (the ratio of marginal likelihoods) and the prior odds. If the goal is to select a single model from the set of alternatives, then one can naturally consider the model with the highest posterior model probability. On the other hand, if the goal is to produce a forecast and there are multiple models with nontrivial probabilities, these models and their respective probabilities can be used in model averaging. The idea behind model averaging is that instead of forming a predictive density for some future outcome y_f

$$p(y_f|y, M_j) = \int f(y_f|y, \theta_j, M_j)p(\theta_j|y, M_j)d\theta_j.$$

based on a single model, one can account for model uncertainty by considering a weighted predictive distribution given by

$$p(y_f|y) = \sum\nolimits_j p(y_f|y, M_j)p(M_j|y).$$

The conceptual simplicity and intellectual appeal of this framework are nicely complemented by the proven ability of modern Bayesian computing to handle previously intractable problems. Every aspect of the Bayesian model comparison framework offers ample opportunities for interesting research. Important computational issues arise in evaluating marginal likelihoods and Bayes factors and in simulating the model parameters from their respective posterior distributions. Additional topics of interest include the suitability of information criteria to adequately approximate marginal likelihoods and Bayes factors in small samples. Specification of the primitives of this framework, $f(y|\theta_j,M_j)$ and $p(\theta_j|M_j)$, has also received much attention. In the case of $f(y|\theta_j,M_j)$, modeling advances have led to the development of methods for estimating and comparison of parametric and more flexible semiparametric and nonparametric models. Research examining the priors $p(\theta_j|M_j)$, on the other hand, has led to considerable innovations and the development of important types of prior distributions, such as intrinsic priors, that have desirable practical properties and have been of much interest in objective Bayesian analysis.

The aforementioned computational, modeling, and theoretical developments have been reflected throughout this volume of Advances in Econometrics. The volume begins with a set of papers that provide important computational contributions to the literature and study key contemporary questions in macroeconomics and finance. The methodological issues examined in these papers include massively parallel computation, dynamic model selection in time-varying parameter models, small sample model comparison of structural macroeconomic models, Bayesian thresholding methods in hierarchical graphical models, Hamiltonian Monte Carlo and parallelization in multi-core environments, and adaptive reversible jump MCMC for variable selection.

The second half of the volume is devoted to papers that offer innovations in theory and methodology, and study applications in microeconomics, industrial organization, banking, and transportation. Among the topics considered in these papers are the modeling of endogeneity in discrete data models with applications in banking and labor economics, the implementation of parameter-based and non-parameter-based approaches to variable selection, survey of key results in objective Bayesian model selection methodology, multistep LASSO estimators for structural models of demand, and the use of parameter expansion algorithms for the estimation of copula models for correlated counts with an application to technology patents.

The content of this volume of Advances in Econometrics is a manifestation of the major improvements in model building and evaluation that have been achieved in the Bayesian paradigm in recent years, which should appeal to readers with computational, modeling, theoretical, and applied inclinations. The volume contains contributions from leading international scholars, summarizes central topics and developments in the literature, and provides new state-of-the-art techniques, methodology, and findings that we hope will serve as a basis for important further research in years to come.

In closing, we wish to express our gratitude to the authors and referees for their hard work; we also wish to thank the production team at Emerald for their patience and professionalism during the preparation of the volume. Last but not least, we wish to thank the sponsors whose generous support made possible a research conference on Bayesian model comparison that was held on the campus of the University of California, Irvine, in February 2014:

- EViews/HIS;
- Office of Research and Council on Research, Computing and Libraries, UCI;
- Donald Bren School of Information and Computer Science, UCI;
- School of Social Sciences, UCI;
- Department of Economics, UCI.

Ivan Jeliazkov
Dale J. Poirier
Editors

ADAPTIVE SEQUENTIAL POSTERIOR SIMULATORS FOR MASSIVELY PARALLEL COMPUTING ENVIRONMENTS

Garland Durham[a] and John Geweke[b]

[a]*Orfalea College of Business, California State Polytechnic Institute, San Luis Obispo, CA, USA*
[b]*School of Business, University of Technology Sydney, Sydney, Australia; Erasmus University, Rotterdam, The Netherlands; Colorado State University, Fort Collins, CO, USA*

ABSTRACT

Massively parallel desktop computing capabilities now well within the reach of individual academics modify the environment for posterior simulation in fundamental and potentially quite advantageous ways. But to fully exploit these benefits algorithms that conform to parallel computing environments are needed. This paper presents a sequential posterior simulator designed to operate efficiently in this context. The simulator makes fewer analytical and programming demands on investigators, and is faster, more reliable, and more complete than conventional posterior simulators. The paper extends existing sequential Monte Carlo methods

Bayesian Model Comparison
Advances in Econometrics, Volume 34, 1–44

ISSN: 0731-9053/doi:10.1108/S0731-905320140000034003

and theory to provide a thorough and practical foundation for sequential posterior simulation that is well suited to massively parallel computing environments. It provides detailed recommendations on implementation, yielding an algorithm that requires only code for simulation from the prior and evaluation of prior and data densities and works well in a variety of applications representative of serious empirical work in economics and finance. The algorithm facilitates Bayesian model comparison by producing marginal likelihood approximations of unprecedented accuracy as an incidental by-product, is robust to pathological posterior distributions, and provides estimates of numerical standard error and relative numerical efficiency intrinsically. The paper concludes with an application that illustrates the potential of these simulators for applied Bayesian inference.

Keywords: Bayesian model comparison; marginal likelihood; graphics processing unit; particle filter; posterior simulation; sequential Monte Carlo

JEL classifications: Primary, C11; Secondary, C630

1. INTRODUCTION

Bayesian approaches have inherent advantages in solving inference and decision problems, but practical applications pose challenges for computation. As these challenges have been met Bayesian approaches have proliferated and contributed to the solution of applied problems. McGrayne (2011) has recently conveyed these facts to a wide audience.

The evolution of Bayesian computation over the past half-century has conformed with exponential increases in speed and decreases in the cost of computing. The influence of computational considerations on algorithms, models, and the way that substantive problems are formulated for statistical inference can be subtle but is hard to overstate. Successful and innovative basic and applied research recognizes the characteristics of the tools of implementation from the outset and tailors approaches to those tools.

Recent innovations in hardware (graphics processing units, or GPUs) provide individual investigators with massively parallel desktop processing at reasonable cost. Corresponding developments in software (extensions of the C programming language and mathematical applications software) make these attractive platforms for scientific computing. This paper extends and applies existing sequential Monte Carlo (SMC) methods for posterior simulation in this context. The extensions fill gaps in the existing theory to provide a thorough and practical foundation for sequential posterior simulation, using approaches suited to massively parallel computing environments. The application produces generic posterior simulators that make substantially fewer analytical and programming demands on investigators implementing new models and provides faster, more reliable, and more complete posterior simulation than do existing methods inspired by conventional predominantly serial computational methods.

Sequential posterior simulation grows out of SMC methods developed over the past 20 years and applied primarily to state space models ("particle filters"). Seminal contributions include Baker (1985, 1987), Gordon, Salmond, and Smith (1993), Kong, Liu, and Wong (1994), Liu and Chen (1995, 1998), Chopin (2002, 2004), Del Moral, Doucet, and Jasra (2006, 2011), Andrieu, Doucet, and Holenstein (2010), and Chopin and Jacob (2010).

The posterior simulator proposed in this paper, which builds largely on Chopin (2002, 2004), has attractive properties.

- It is highly generic (easily adaptable to new models). All that is required of the user is code to generate from the prior and evaluate the prior and data densities.
- It is computationally efficient. This is in large part due to the specifics of our implementation of the simulator, which makes effective use of low-cost massively parallel computing hardware.
- Since the simulator provides a sample from the posterior density at each observation date conditional on information available at that time, it is straightforward to compute moments of arbitrary functions of interest conditioning on relevant information sets. Marginal likelihood and predictive scores, which are fundamental to Bayesian model comparison, are immediately available. Also immediately available is the probability integral transform at each observation date, which provides a powerful diagnostic tool for model exploration ("generalized residuals"; e.g., Diebold, Gunther, & Tay, 1998).
- Estimates of numerical standard error and relative numerical efficiency (RNE) are provided as an intrinsic part of the simulator. This is important but often neglected information for the researcher interested in generating dependable and reliable results.

- The simulator is robust to irregular posteriors (e.g., multimodality), as has been pointed out by Jasra, Stephens, and Holmes (2007) and others.
- The simulator has well-founded convergence properties.

But, although the basic idea of sequential posterior simulation goes back at least to Chopin (2002, 2004), and despite its considerable appeal, the idea has seen essentially no penetration in mainstream applied econometrics. Applications have been limited to relatively simple illustrative examples (although this has begun to change very recently; see Chopin, Jacob, & Papaspiliopoulis, 2013; Fulop & Li, 2011; Herbst & Schorfheide, 2012). This is in stark contrast to applications of SMC methods to state space filtering ("particle filters"), which have seen widespread use.

There has also been recent work on developing new Markov chain Monte Carlo (MCMC) methods that address some of the issues noted above. Roberts and Rosenthal (2007) develop adaptive MCMC methods, providing a mechanism whereby proposal densities can be tuned automatically based on their performance in the process of running the algorithm. Jasra et al. (2007) and others have suggested the idea of running a large number of parallel MCMC chains. And, Yang and Dunson (2013) propose a sequential MCMC algorithm designed to alleviate the particle degeneracy problem associated with SMC methods.

Relative to MCMC, which has become a mainstay of applied work, sequential posterior simulators are computationally costly when applied in a conventional serial computing environment. Our interest in sequential posterior simulation is largely motivated by the recent availability of low cost hardware supporting massively parallel computation, for which the method is ideally suited.

The massively parallel hardware device used in this paper is a commodity graphical processing unit (GPU), which provides hundreds of cores at a cost of well under one dollar (US) each. But in order to realize the huge potential gains in computing speed made possible by such hardware, algorithms that conform to the single-instruction multiple-data (SIMD) paradigm are needed. The simulators presented in this paper conform to the SIMD paradigm by design and realize the attendant increases in computing speed in practical application.

But there are also some central issues regarding properties of sequential posterior simulators that have not been resolved in the literature. Since reliable applied work depends on the existence of solid theoretical underpinnings, we address these as well.

Our main contributions are as follows.

1. *Theoretical basis.* Whereas the theory for sequential posterior simulation as originally formulated by Chopin (2002, 2004) assumes that key elements of the algorithm are fixed and known in advance, practical applications demand algorithms that are adaptive, with these elements constructed based on the information provided by particles generated in the course of running the algorithm.

   While some progress has been made in developing a theory that applies to such adaptive algorithms (e.g., Douc & Moulines, 2008), we expect that a solid theory that applies to the kind of algorithms that are used and needed in practice is likely to remain unattainable in the near future.

   In Section 3 we provide an approach that addresses this problem in a different way, providing a posterior simulator that is highly adapted to the models and data at hand, while satisfying the relatively straightforward conditions elaborated in the original work of Chopin (2002, 2004).
2. *Numerical accuracy.* While Chopin (2004) provides a critical central limit theorem, the extant literature does not provide any means of estimating the variance, which is essential for assessing the numerical accuracy and RNE of moment approximations (Geweke, 1989). This problem has proved difficult in the case of MCMC (Flegal & Jones, 2010) and appears to have been largely ignored in the SMC literature. In Section 2.3 we propose an approach to resolving this issue which entails no additional computational effort and is natural in the context of the massively parallel environment as well as key in making efficient use of it.
3. *Marginal likelihood.* While the approach to assessing the asymptotic variance of moment approximations noted in the previous point is useful, it does not apply directly to marginal likelihood, which is critical to Bayesian model comparison. We address this issue in Section 4.
4. *Parallel implementation.* It has been noted that sequential posterior simulation is highly amenable to parallelization going back to at least Chopin (2002), and there has been some work toward exploring parallel implementations (e.g., Fulop & Li, 2011; Lee, Yau, Giles, Doucet, & Homes, 2010). Building on this work, we provide a software package that includes a full GPU implementation of the simulator, with specific details of the algorithm tuned to the massively parallel environment as outlined in Section 3 and elsewhere in this paper. The software is fully object-oriented, highly modular, and easily extensible. New models are

easily added, providing the full benefits of GPU computing with little effort on the part of the user. Geweke, Durham, and Xu (2013) provide an example of such an implementation for the logit model. In future work, we intend to build a library of models and illustrative applications utilizing this framework which will be freely available. Section 3.4 provides details about this software.

5. *Specific recommendations.* The real test of the simulator is in its application to problems that are characteristic of the scale and complexity of serious disciplinary work. In Section 5, we provide one such application to illustrate key ideas. In other ongoing work, including Geweke et al. (2013) and a longer working paper written in the process of this research (Durham & Geweke, 2011), we provide several additional substantive applications. As part of this work, we have come up with some specific recommendations for aspects of the algorithm that we have found to work well in a wide range of practical applications. These are described in Section 3.3 and implemented fully in the software package we are making available with this paper.

2. POSTERIOR SIMULATION IN A MASSIVELY PARALLEL COMPUTING ENVIRONMENT

The sequential simulator proposed in this paper is based on ideas that go back to Chopin (2002, 2004), with even earlier antecedents including Gilks and Berzuini (2001) and Fearnhead (1998). The simulator begins with a sample of parameter vectors ("particles") from the prior distribution. Data is introduced in batches, with the steps involved in processing a single batch of data referred to as a cycle. At the beginning of each cycle, the particles represent a sample from the posterior conditional on information available up to that observation date. As data is introduced sequentially, the posterior is updated using importance sampling (Kloek & van Dijk, 1978), the appropriately weighted particles representing an importance sample from the posterior at each step. As more data is introduced, the importance weights tend to become "unbalanced" (a few particles have most of the weight, while many others have little weight), and importance sampling becomes increasingly inefficient. When some threshold is reached, importance sampling stops and the cycle comes to an end. At this point, a resampling step is undertaken, wherein particles are independently resampled in proportion to their importance weights. After this step, there will be many copies of particles with high weights, while particles with low weights will tend to drop out of the sample. Finally, a sequence of Metropolis steps is undertaken in order to rejuvenate the diversity of the particle sample. At the conclusion of the cycle, the collection of particles once again represents a sample from the posterior, now incorporating the information accrued from the data newly introduced.

This sequential introduction of data is natural in a time-series setting, but also applicable to cross-sectional data. In the latter case, the sequential ordering of the data is arbitrary, though some orderings may be more useful than others.

Much of the appeal of this simulator is due to its amenability to implementation using massively parallel hardware. Each particle can be handled in a distinct thread, with all threads updated concurrently at each step. Communication costs are low. In applications where computational cost is an important factor, nearly all of the cost is incurred in evaluating densities of data conditional on a candidate parameter vector and thus isolated to individual threads. Communication costs in this case are a nearly negligible fraction of the total computational burden.

The key to efficient utilization of the massively parallel hardware is that the workload be divided over many SIMD threads. For the GPU hardware used in this paper optimal usage involves tens of thousands of threads. In our implementation, particles are organized in a relatively small number of groups each with a relatively large number of particles (in the application in Section 5 there are 2^6 groups of 2^{10} particles each). This organization of particles in groups is fundamental to the supporting theory. Estimates of numerical error and RNE are generated as an intrinsic part of the algorithm with no additional computational cost, while the reliability of these estimates is supported by a sound theoretical foundation. This structure is natural in a massively parallel computing environment, as well critical in making efficient use of it.

The remainder of this section provides a more detailed discussion of key issues involved in posterior simulation in a massively parallel environment. It begins in Section 2.1 with a discussion of the relevant features of the hardware and software used. Section 2.2 sets up a generic model for Bayesian inference along with conditions used in deriving the analytical properties of various sequential posterior simulators in Section 3. Section 2.3 stipulates a key convergence condition for posterior simulators, and then shows that if this condition is met there are attractive generic methods for approximating the standard error of numerical approximation in massively parallel computing environments. Section 3 then develops sequential posterior simulators that satisfy this condition.

2.1. Computing Environment

The particular device that motivates this work is the GPU. As a practical matter several GPUs can be incorporated in a single server (the "host") with no significant complications, and desktop computers that can accommodate up to eight GPUs are readily available. The single- and multiple-GPU environments are equivalent for our purposes. A single GPU consists of several multiprocessors, each with several cores. The GPU has global memory shared by its multiprocessors, typically one to several gigabytes (GB) in size, and local memory specific to each multiprocessor, typically on the order of 50–100 kilobytes (KB) per multiprocessor. (For example, this research uses a single Nvidia GTX 570 GPU with 15 multiprocessors, each with 32 cores. The GPU has 1.36 GB of local memory, and each multiprocessor has 49 KB of memory and 32 KB of registers shared by its cores.) The bus that transfers data between GPU global and local memory is significantly faster than the bus that transfers data between host and device, and accessing local memory on the multiprocessor is faster yet. For greater technical detail on GPUs, see, for example, Hendeby, Karlsson, and Gustafsson (2010), Lee et al. (2010), and Souchard et al. (2010).

This hardware has become attractive for scientific computing with the extension of scientific programming languages to allocate the execution of instructions between host and device and facilitate data transfer between them. Of these the most significant has been the compute unified device architecture (CUDA) extension of the C programming language (Nvidia, 2013). CUDA abstracts the host–device communication in a way that is convenient to the programmer yet faithful to the aspects of the hardware important for writing efficient code.

Code executed on the device is contained in special functions called kernels that are invoked by the host code. Specific CUDA instructions move the data on which the code operates from host to device memory and instruct the device to organize the execution of the code into a certain number of blocks with a certain number of threads each. The allocation into blocks and threads is the virtual analogue of the organization of a GPU into multiprocessors and cores.

While the most flexible way to develop applications that make use of GPU parallelization is through C/C++ code with direct calls to the vendor-supplied interface functions, it is also possible to work at a higher level of abstraction. For example, a growing number of mathematical libraries have been ported to GPU hardware (e.g., Innovative Computing Laboratory, 2013). Such libraries are easily called from standard scientific programming languages and can yield substantial increases in performance for some applications. In addition, Matlab (2013) provides a library of kernels, interfaces for calling user-written kernels, and functions for host–device data transfer from within the Matlab workspace.

2.2. *Models and Conditions*

We augment standard notation for data, parameters, and models. The relevant observable random vectors are Y_t $(t=1,\ldots,T)$ and $Y_{t_1:t_2}$ denotes the collection $\{Y_{t_1},\ldots,Y_{t_2}\}$. The observation of Y_t is y_t, $y_{t_1:t_2}$ denotes the collection $\{y_{t_1},\ldots,y_{t_2}\}$, and therefore $y_{1:T}$ denotes the data. This notation assumes ordered observations, which is natural for time series. If $\{Y_t\}$ is independent and identically distributed the ordering is arbitrary.

A model for Bayesian inference specifies a $k\times 1$ unobservable parameter vector $\theta\in\Theta$ and a conditional density

$$p(Y_{1:T}|\theta)=\prod_{t=1}^{T} p(Y_t|Y_{1:t-1},\theta) \tag{1}$$

with respect to an appropriate measure for $Y_{1:T}$. The model also specifies a prior density $p(\theta)$ with respect to a measure ν on Θ. The posterior density $p(\theta|y_{1:T})$ follows in the usual way from Eq. (1) and $p(\theta)$.

The objects of Bayesian inference can often be written as posterior moments of the form $E[g(\theta)|y_{1:T}]$, and going forward we use $g(\theta)$ to refer to such a generic function of interest. Evaluation of $g(\theta)$ may require simulation, for example, $g(\theta)=E[h(Y_{T+1:T+f})|\theta]$, and conventional approaches based on the posterior simulation sample of parameters (e.g., Geweke, 2005, Section 1.4) apply in this case. The marginal likelihood $p(y_{1:T})$ famously does not take this form, and Section 4 takes up its approximation with sequential posterior simulators and, more specifically, in the context of the methods developed in this paper.

Several conditions come into play in the balance of the paper.

Condition 1. (*Prior distribution*). *The model specifies a proper prior distribution. The prior density kernel can be evaluated with SIMD-compatible code. Simulation from the prior distribution must be practical but need not be SIMD-compatible.*

It is well understood that a model must take a stance on the distribution of outcomes $Y_{1:T}$ a priori if it is to have any standing in formal Bayesian model comparison, and that this requires a proper prior distribution. This requirement is fundamentally related to the generic structure of sequential posterior simulators, including those developed in Section 3, because they require a distribution of θ before the first observation is introduced. Given a proper prior distribution the evaluation and simulation conditions are weak. Simulation from the prior typically involves minimal computational cost and thus we do not require it to be SIMD-compatible. However, in practice it will often be so.

Condition 2. (*Likelihood function evaluation*). *The sequence of conditional densities*

$$p(y_t|y_{1:t-1}, \theta) \quad (t = 1, \ldots, T) \tag{2}$$

can be evaluated with SIMD-compatible code for all $\theta \in \Theta$.

Evaluation with SIMD-compatible code is important to computational efficiency because evaluation of Eq. (2) constitutes almost all of the floating point operations in typical applications for the algorithms developed in this paper. Condition 2 excludes situations in which unbiased simulation of Eq. (2) is possible but exact evaluation is not, which is often the case in nonlinear state space models, samples with missing data, and in general any case in which a closed form analytical expression for Eq. (2) is not available. A subsequent paper will take up this extension.

Condition 3. (*Bounded likelihood*). *The data density $p(y_{1:T}|\theta)$ is bounded above by $\overline{p} < \infty$ for all $\theta \in \Theta$.*

This is one of two sufficient conditions for the central limit theorem invoked in Section 3.1. It is commonly but not universally satisfied. When it is not, it can often be attained by minor and innocuous modification of the likelihood function; Section 5 provides an example.

Condition 4. (*Existence of prior moments*). *If the algorithm is used to approximate* $E[g(\theta)|y_{1:T}]$, *then* $E[g(\theta)^{2+\delta}]<\infty$ *for some* $\delta>0$.

In any careful implementation of posterior simulation the existence of relevant posterior moments must be verified analytically. This condition, together with Condition 3, is sufficient for the existence of $\mathrm{E}[g(\theta)|y_{1:T}]$ and $\mathrm{var}[g(\theta)|y_{1:T}]$. Condition 4 also comes into play in establishing a central limit theorem.

2.3. Assessing Numerical Accuracy and Relative Numerical Efficiency

Consider the implementation of any posterior simulator in a parallel computing environment like the one described in Section 2.1. Following the conventional approach, we focus on the posterior simulation approximation of

$$\overline{g} = \mathrm{E}[g(\theta)|y_{1:T}] \tag{3}$$

The posterior simulator operates on parameter vectors, or particles, θ_{jn} organized in J groups of N particles each; define $\mathcal{J}=\{1,\ldots,J\}$ and $\mathcal{N}=\{1,\ldots,N\}$. This organization is fundamental to the rest of the paper.

Each particle is associated with a core of the device. The CUDA software described in Section 2.1 copes with the case in which JN exceeds the number of cores in an efficient and transparent fashion. Ideally the posterior simulator is SIMD-compatible, with identical instructions executed on all particles. Some algorithms, like the sequential posterior simulators developed in Section 3, come quite close to this ideal. Others, for example, Gibbs samplers with Metropolis steps, may not. The theory in this section applies regardless. However, the advantages of implementing a posterior simulator in a parallel computing environment are driven by the extent to which the algorithm is SIMD-compatible.

Denote the evaluation of the function of interest at particle θ_{jn} by $g_{jn}=g(\theta_{jn})$ and within-group means by $\overline{g}_j^N = N^{-1}\sum_{n=1}^N g_{jn}$ $(j\in\mathcal{J})$. The posterior simulation approximation of $\overline{g}$ is the grand mean

$$\overline{g}^{(J,N)} = J^{-1}\sum_{j=1}^{J}\overline{g}_j^N = (JN)^{-1}\sum_{j=1}^{J}\sum_{n=1}^{N} g_{jn} \tag{4}$$

In general we seek posterior simulators with the following properties.

Condition 5. (*Asymptotic normality of posterior moment approximation*). *The random variables* $\overline{g}_j^N$ $(j \in \mathcal{J})$ *are independently and identically distributed. There exists* $v > 0$ *for which*

$$N^{1/2}(\overline{g}_j^N - \overline{g}) \xrightarrow{d} N(0,v) \quad (j \in \mathcal{J}) \tag{5}$$

as $N \to \infty$.

Going forward, v will denote a generic variance in a central limit theorem. Convergence (5) is part of the rigorous foundation of posterior simulators, for example, Geweke (1989) for importance sampling, Tierney (1994) for MCMC, and Chopin (2004) for a sequential posterior simulator. For importance sampling $\overline{g}_j^N = \sum_{n=1}^{N} \omega(\theta_{jn}) g_{jn} / \sum_{n=1}^{N} \omega(\theta_{jn})$ $(j \in \mathcal{J})$, where $\omega(\theta)$ is the ratio of the posterior density kernel to the source density kernel.

For Eq. (5) to be of any practical use in assessing numerical accuracy there must also be a simulation-consistent approximation of v. Such approximations are immediate for importance sampling. They have proven more elusive for MCMC; for example, see Flegal and Jones (2010) for discussion and an important contribution. To our knowledge there is no demonstrated simulation-consistent approximation of v for sequential posterior simulators in the existing literature.

However, for any simulator satisfying Condition 5 this is straightforward. Given Condition 5, it is immediate that

$$(JN)^{1/2}(\overline{g}^{(J,N)} - \overline{g}) \xrightarrow{d} N(0,v) \tag{6}$$

Define the estimated posterior simulation variance

$$\hat{v}^{(J,N)}(g) = [N/(J-1)] \sum_{j=1}^{J} (\overline{g}_j^N - \overline{g}^{(J,N)})^2 \tag{7}$$

and the numerical standard error (NSE)

$$\text{NSE}^{(J,N)}(g) = [\hat{v}^{(J,N)}/JN]^{1/2} = \left\{ [J(J-1)]^{-1} \sum_{j=1}^{J} (\overline{g}_j^N - \overline{g}^{(J,N)})^2 \right\}^{1/2} \tag{8}$$

Note the different scaling conventions in Eqs. (7) and (8): in Eq. (7) the scaling is selected so that $\hat{v}^{(J,N)}(g)$ approximates v in Eq. (6) because we will use this expression mainly for methodological developments; in Eq. (8) the scaling is selected to make it easy to appraise the reliability of numerical approximations of posterior moments like those reported in Section 5.

One conventional assessment of the efficiency of a posterior simulator is its RNE (Geweke, 1989),

$$\text{RNE}^{(J,N)}(g) = v\hat{a}r^{(J,N)}(g)/\hat{v}^{(J,N)}(g)$$

where $v\hat{a}r^{(J,N)}(g) = (NJ-1)^{-1}\sum_{j=1}^{J}\sum_{n=1}^{N}(g_{jn} - \overline{g}^{(J,N)})^2$, the simulation approximation of $\text{var}[g(\theta)|y_{1:T}]$.

Proposition 1. *Condition 5 implies*

$$(J-1)\hat{v}^{(J,N)}(g)/v \xrightarrow{d} \chi^2(J-1) \tag{9}$$

and

$$(JN)^{1/2}(\overline{g}^{(J,N)} - \overline{g})/[\hat{v}^{(J,N)}(g)]^{1/2} \xrightarrow{d} t(J-1) \tag{10}$$

both as $N \to \infty$. Conditions 1 through 4 imply

$$E[\hat{v}^{(J,N)}(g)] = var[\overline{g}^{(J,N)}] \tag{11}$$

Proof. From Condition 5,

$$N\sum_{j=1}^{J}(\overline{g}_j^N - \overline{g}^{(J,N)})^2/v \xrightarrow{d} \chi^2(J-1) \tag{12}$$

as $N \to \infty$. Substituting Eq. (7) in Eq. (12) yields Eq. (9). Condition 5 also implies

$$\left(\frac{NJ}{v}\right)^{1/2}(\overline{g}^{(J,N)} - \overline{g}) \xrightarrow{d} N(0,1) \tag{13}$$

as $N \to \infty$. Since Eqs. (12) and (13) are independent in the limiting distribution, we have Eq. (10).

Conditions 1 through 4 imply the existence of the first two moments of g_{jn}. Then Eq. (11) follows from the fact that the approximations $\overline{g}_j$ are independent and identically distributed across $j = 1, \ldots, J$. ■

3. PARALLEL SEQUENTIAL POSTERIOR SIMULATORS

We seek a posterior simulator that is generic, requiring little or no intervention by the investigator in adapting it to new models beyond the provision of the software implicit in Conditions 1 and 2. It should reliably assess the numerical accuracy of posterior moment approximations, and this should be achieved in substantially less execution time than would be required using the same or an alternative posterior simulator for the same model in a serial computing environment.

Such simulators are necessarily adaptive: they must use the features of the evolving posterior distribution, revealed in the particles $\theta_{jn} (j \in \mathcal{J}, n \in \mathcal{N})$, to design the next steps in the algorithm. This practical requirement has presented a methodological conundrum in the SMC literature, because the mathematical complications introduced by even mild adaptation lead to analytically intractable situations in which demonstration of the central limit theorem in Condition 5 is precluded. This section solves this problem and sets forth a particular adaptive sequential posterior simulator that has been successful in applications we have undertaken with a wide variety of models. Section 5 details one such application.

Section 3.1 places a nonadaptive sequential posterior simulator (Chopin, 2004) that satisfies Condition 5 into the context developed in the previous section. Section 3.2 introduces a technique that overcomes the analytical difficulties associated with adaptive simulators. It is generic, simple, and imposes little additional overhead in a parallel computing environment. Section 3.3 provides details on a particular variant of the algorithm that we have found to be successful for a variety of models representative of empirical research frontiers in economics and finance. Section 3.4 discusses a software package implementing the algorithm that we are making available.

3.1. Nonadaptive Simulators

We rely on the following mild generalization of the SMC algorithm of Chopin (2004), cast in the parallel computing environment detailed in the previous section. In this algorithm, the observation dates at which each cycle terminates $(t_1, \ldots, t_L)$ and the parameters involved in specifying the Metropolis updates $(\lambda_1, \ldots, \lambda_L)$ are assumed to be fixed and known in advance, in conformance with the conditions specified by Chopin (2004).

Algorithm 1. (*Nonadaptive*). *Let* $t_0, \ldots, t_L$ *be fixed integers with* $0 = t_0 < t_1 < \cdots < t_L = T$ *and let* $\lambda_1, \ldots, \lambda_L$ *be fixed vectors.*

1. Initialize $\ell = 0$ and let $\theta_{jn}^{(\ell)} \sim^{iid} p(\theta) \quad (j \in \mathcal{J}, n \in \mathcal{N})$.
2. For $\ell = 1, \ldots, L$
   a. Correction (C) phase:
      i. $w_{jn}(t_{\ell-1}) = 1 \quad (j \in \mathcal{J}, n \in \mathcal{N})$
      ii. For $s = t_{\ell-1} + 1, \ldots, t_\ell$

$$w_{jn}(s) = w_{jn}(s-1) \cdot p(y_s | y_{1:s-1}, \theta_{jn}^{(\ell-1)}) \quad (j \in \mathcal{J}, n \in \mathcal{N}) \tag{14}$$

      iii. $w_{jn}^{(\ell-1)} := w_{jn}(t_\ell) \quad (j \in \mathcal{J}, n \in \mathcal{N})$

   b. Selection (S) phase, applied independently to each group $j \in \mathcal{J}$: Using multinomial or residual sampling based on $\{w_{jn}^{(\ell-1)} (n \in \mathcal{N})\}$, select

$$\{\theta_{jn}^{(\ell,0)} (n \in \mathcal{N})\} \text{ from } \{\theta_{jn}^{(\ell-1)} (n \in \mathcal{N})\}$$

   c. Mutation (M) phase, applied independently across $j \in \mathcal{J}, n \in \mathcal{N}$:

$$\theta_{jn}^{(\ell)} \sim p(\theta | y_{1:t_\ell}, \theta_{jn}^{(\ell,0)}, \lambda_\ell) \tag{15}$$

   where the drawings are independent and the p.d.f. (15) satisfies the invariance condition

$$\int_\Theta p(\theta | y_{1:t_\ell}, \theta^*, \lambda_\ell) p(\theta^* | y_{1:t_\ell}) d\nu(\theta^*) = p(\theta | y_{1:t_\ell}) \tag{16}$$

3. $\theta_{jn} := \theta_{jn}^{(L)} \quad (j \in \mathcal{J}, n \in \mathcal{N})$

The algorithm is nonadaptive because $t_0, \ldots, t_L$ and $\lambda_1, \ldots, \lambda_L$ are predetermined before the algorithm starts. Going forward it will be convenient to denote the cycle indices by $\mathcal{L} = \{1, \ldots, L\}$. At the conclusion of the algorithm, the simulation approximation of a generic posterior moment is Eq. (4).

Proposition 2. *If Conditions 1 through 4 are satisfied then Algorithm 1 satisfies Condition 5.*

Proof. The results follow from Chopin (2004), Theorem 1 (for multinomial resampling) and Theorem 2 (for residual resampling). The assumptions made in Theorems 1 and 2 are

1. $L = T$ and $t_\ell = \ell$ $(\ell = 0, \ldots, T)$,
2. The functions $p(\theta)p(y_{1:t} \mid \theta)(t = 1, \ldots, T)$ are integrable on Θ, and
3. The moments $E(g(\theta)^{2+\delta} \mid y_{1:t})(t = 1, \ldots, T)$ exist.

Assumption 1 is merely a change in notation; Conditions 1 through 3 imply assumption 2; and Conditions 1 through 4 imply assumption 3. Theorems 1 and 2 in Chopin (2004) are stated using the weighted sample at the end of the last C phase, but as that paper states, they also apply to the unweighted sample at the end of the following M phase. At the conclusion of each cycle ℓ, the J groups of N particles each, $\theta_{jn}^{(\ell)}$ $(n \in \mathcal{N})$, are mutually independent because the S phase is executed independently in each group. ■

At the end of each cycle ℓ, all particles $\theta_{jn}^{(\ell)}$ are identically distributed with common density $p(\theta \mid y_{1:t_\ell})$. Particles in different groups are independent, but particles within the same group are not. The amount of dependence within groups depends upon how well the M phases succeed in rejuvenating particle diversity.

3.2. *Adaptive Simulators*

In fact the sequences $\{t_\ell\}$ and $\{\lambda_\ell\}$ for which the algorithm is sufficiently efficient for actual application are specific to each problem. As a practical matter these sequences must be tailored to the problem based on the characteristics of the particles $\{\theta_{jn}^{(\ell)}\}$ produced by the algorithm itself. This leads to algorithms of the following type.

Algorithm 2. (*Adaptive*). *Algorithm 2 is the following generalization of Algorithm 1. In each cycle ℓ,*

1. t_ℓ may be random and depend on $\theta_{jn}^{(\ell-1)}$ $(j \in \mathcal{J}, n \in \mathcal{N})$;
2. The value of λ_ℓ may be random and depend on $\theta_{jn}^{(\ell,0)}$ $(j \in \mathcal{J}, n \in \mathcal{N})$.

While such algorithms can be effective and are what is needed for practical applied work, the theoretical foundation relevant to Algorithm 1 does not apply.

The groundwork for Algorithm 1 was laid by Chopin (2004). With respect to the first condition of Algorithm 2, in Chopin's development each cycle consists of a single observation, while in practice cycle lengths are often adaptive based on the effective sample size criterion (Liu & Chen, 1995). Progress has been made toward demonstrating Condition 5 in this case only recently (Del Moral et al., 2011; Douc & Moulines, 2008), and it appears that Conditions 1 through 4 are sufficient for the assumptions made in Douc and Moulines (2008). With respect to the second condition of Algorithm 2, demonstration of Condition 5 for the adaptations that are necessary to render sequential posterior simulators even minimally efficient appears to be well beyond current capabilities.

There are many specific adaptive algorithms, like the one described in Section 3.3 below, that will prove attractive to practitioners even in the absence of sound theory for appraising the accuracy of posterior moment approximations. In some cases Condition 5 will hold; in others, it will not but the effects will be harmless; and in still others Condition 5 will not hold and the relation of Eq. (4) to Eq. (3) will be an open question. Investigators with compelling scientific questions are unlikely to forego attractive posterior simulators awaiting the completion of their proper mathematical foundations.

The following algorithm resolves this dilemma, providing a basis for implementations that are both computationally effective and supported by established theory.

Algorithm 3. (*Hybrid*)

1. Execute Algorithm 2. Retain $\{t_\ell, \lambda_\ell\}$ $(\ell \in \mathcal{L})$ and discard θ_{jn} $(j \in \mathcal{J}, n \in \mathcal{N})$.
2. Execute Algorithm 1 using the retained $\{t_\ell, \lambda_\ell\}$ but with a new seed for the random number generator used in the S and M phases.

Proposition 3. C*If Conditions 1 through 4 are satisfied then Algorithm 3 satisfies Condition 5.*

Proof. Because the sequences $\{t_\ell\}$ and $\{\lambda_\ell\}$ produced in step 1 are fixed (predetermined) with respect to the random particles $\theta_{jn}(j \in \mathcal{J}, n \in \mathcal{N})$ generated in step 2, the conditions of Algorithm 1 are satisfied in step 2. From Proposition 2, Condition 5 is therefore satisfied. ■

3.3. A Specific Adaptive Simulator

Applications of Bayesian inference are most often directed by investigators who are not specialists in simulation methodology. A generic version of Algorithm 2 that works well for wide classes of existing and prospective models and data sets with minimal intervention by investigators is therefore attractive. While general descriptions of the algorithm such as provided in Sections 3.1–3.2 give a useful basis for discussion, what is really needed for practical use is a detailed specification of the actual steps that need to be executed. This section provides the complete specification of a particular variant of the algorithm that has worked well across a spectrum of models and data sets characteristic of current research in finance and economics. The software package that we are making available with this paper provides a full implementation of the algorithm presented here (see Section 3.4).

C phase termination. The C phase is terminated based on a rule assessing the effective sample size (Kong, Liu, & Wong, 1994; Liu & Chen, 1995). Following each update step s as described in Algorithm 1, part 2(a)ii, compute

$$\mathrm{ESS}(s) = \frac{\left[\sum_{j=1}^{J} \sum_{n=1}^{N} w_{jn}(s)\right]^2}{\sum_{j=1}^{J} \sum_{n=1}^{N} w_{jn}(s)^2}$$

It is convenient to work with the relative sample size, $\mathrm{RSS} = \mathrm{ESS}/(JN)$, where JN is the total number of particles. Terminate the C phase and proceed to the S phase if $\mathrm{RSS} < 0.5$.

We have found that the specific threshold value used has little effect on the performance of the algorithm. Higher thresholds imply that the C phase is terminated earlier, but fewer Metropolis steps are needed to obtain adequate diversification in the M phase (and inversely for lower thresholds). The default threshold in the software is 0.5. This can be easily modified by users, but we have found little reason to do so.

Resampling method. We make several resampling methods available. The results of Chopin (2004) apply to multinomial and residual resampling (Baker, 1985, 1987; Liu & Chen, 1998). Stratified (Kitagawa, 1996) and systematic (Carpenter, Clifford, & Fearnhead, 1999) samplers are also of potential interest. We use residual sampling as the default. It is substantially more efficient than multinomial at little additional computational cost. Stratified and systematic resamplers are yet more efficient as well as being less costly and simpler to implement; however, we are not aware of any available theory supporting their use and so recommend them only for experimental use at this point.

Metropolis steps. The default is to use a sequence of Gaussian random walk samplers operating on the entire parameter vector in a single block. At iteration r of the M phase in cycle ℓ, the variance of the sampler is obtained by $\sum_{\ell r} = h_{\ell r}^2 V_{\ell r}$ where $V_{\ell r}$ is the sample variance of the current population of particles and $h_{\ell r}$ is a "stepsize" scaling factor. The stepsize is initialized at 0.5 and incremented depending upon the Metroplis acceptance rate at each iteration. Using a target acceptance rate of 0.25 (Gelman, Roberts, & Gilks, 1996), the stepsize is incremented by 0.1 if the acceptance rate is greater than the target and decremented by 0.1 otherwise, respecting the constraint $0.1 \leq h_{\ell r} \leq 1.0$.

An alternative that has appeared often in the literature is to use an independent Gaussian sampler. This can give good performance for relatively simple models, and is thus useful for illustrative purposes using implementations that are less computationally efficient. But for more demanding applications, such as those that are likely to be encountered in practical work, the independent sampler lacks robustness. If the posterior is not well-represented by the Gaussian sampler, the process of particle diversification can be very slow. The difficulties are amplified in the case of high-dimensional parameter vectors. And the sampler fails spectacularly in the presence of irregular posteriors (e.g., multimodality).

The random walk sampler gives up some performance in the case of near Gaussian posteriors (though it still performs well in this situation), but is robust to irregular priors, a feature that we consider to be important. The independent sampler is available in the software for users wishing to experiment. Other alternatives for the Metropolis updates are also possible, and we intend to investigate some of these in future work.

M phase termination. The simplest rule would be to terminate the M phase after some fixed number of Metropolis iterations, R. But in practice, it sometimes occurs that the C phase terminates with a very low RSS. In this case, the M phase begins with relatively few distinct particles and we would like to use more iterations in order to better diversify the population of particles. A better rule is to end the M phase after $R_\ell = \kappa \cdot \overline{R}$ iterations if $\text{RSS} < D_2$ and after $R_\ell = \overline{R}$ iterations if $D_2 \leq \text{RSS} < D_1$ (the default settings in the software are $\overline{R} = 7$, $\kappa = 3$, $D_1 = 0.50$, and $D_2 = 0.20$). We refer to this as the "deterministic stopping rule."

But, the optimal settings, especially for $\overline{R}$, are highly application dependent. Setting $\overline{R}$ too low results in poor rejuvenation of particle diversity and the RNE associated with moments of functions of interest will typically be low. This problem can be addressed by either using more particles or a higher setting for $\overline{R}$. Typically, the latter course of action is more efficient if RNE drops below about 0.25; that is, finesse (better rejuvenation) is generally preferred over brute force (simply adding more particles). On the other hand, there is little point in continuing M phase iterations once RNE gets close to one (indicating that particle diversification is already good, with little dependence among the particles in each group). Continuing past this point is simply a waste of computing resources.

RNE thus provides a useful diagnostic of particle diversity. But it also provides the basis for a useful alternative termination criterion. The idea is to terminate M phase iterations when the average RNE of numerical approximations of selected test moments exceeds some threshold. We use a default threshold of $E_1 = 0.35$. It is often useful to compute moments of functions of interest at particular observation dates. While it is possible to do this directly using weighted samples obtained within the C phase, it is more efficient to terminate the C phase and execute S and M phases at these dates prior to evaluating moments. In such cases, it is useful to target a higher RNE threshold to assure more accurate approximations. We use a default of $E_2 = 0.90$ here. We also set an upper bound for the number of Metropolis iterations, at which point the M phase terminates regardless of RNE. The default is $R_{\max} = 100$.

While the deterministic rule is simpler, the RNE-based rule has important advantages: it is likely to be more efficient than the deterministic rule in practice; and it is fully automated, thus eliminating the need for user intervention. Some experimental results illustrating these issues are provided in Section 5.

An important point to note here is that after the S phase has executed, RSS no longer provides a useful measure of the diversity of the particle population (sincle the particles are equally-weighted). RNE is the only useful measure of which we are aware.

Parameterization. Bounded support of some parameters can impede the efficiency of the Gaussian random walk in the M phase, which has unbounded support. We have found that simple transformations, for example, a positive parameter σ to $\theta = \log(\sigma)$ or a parameter ρ defined on $(-1, 1)$ to $\theta = \text{atanh}(\rho)$, can greatly improve performance. This requires attention to the Jacobian of the transformation in the prior distribution, but it is often just as appealing simply to place the prior distribution directly on the transformed parameters. Section 5.1 provides an example.

The posterior simulator described in this section is adaptive (and thus a special case of Algorithm 2). That is, key elements of the simulator are not known in advance, but are determined on the basis of particles generated in the course of executing it. In order to use the simulator in the context of the hybrid method (Algorithm 3), these design elements must be saved for use in a second run. For the simulator described here, we need the observations $\{t_\ell\}$ at which each C phase termination occurs, the variance matrices $\{\Sigma_{\ell r}\}$ used in the Metropolis updates, and the number of Metropolis iterations executed in each M phase $\{R_\ell\}$. The hybrid method then involves running the simulator a second time, generating new particles but using the design elements saved from the first run. Since the design elements are now fixed and known in advance, this is a special case of Algorithm 1 and thus step 2 of Algorithm 3. And therefore, in particular, Condition 5 is satisfied and the results of Section 2.3 apply.

3.4. Software

The software package that we are making available with this paper (http://www.quantosanalytics.org/garland/mp-sps_1.1.zip) implements the algorithm developed in Section 3.3. The various settings described there are provided as defaults but can be easily changed by users. Switches are available allowing the user to choose whether the algorithm runs entirely on the CPU or using GPU acceleration, and whether to run the algorithm in adaptive or hybrid mode.

The software provides log marginal likelihood (and log score if a burn-in period is specified) together with estimates of NSE, as described in Section 4.2. Posterior mean and standard deviation of functions of interest at specified dates are provided along with corresponding estimates of RNE and NSE, as described in Section 2.3. RSS at each C phase update and

RNE of specified test moments at each M phase iteration are available for diagnostic purposes, and plots of these are generated as part of the standard output.

For transparency and accessibility to economists doing practical applied work, the software uses a Matlab shell (with extensive use of the parallel/GPU toolbox), with calls to functions coded in C/CUDA for the computationally intensive parts. For simple problems where computational cost is low to begin with, the overhead associated with using Matlab is noticeable. But, for problems where computational cost is actually important, the cost is nearly entirely concentrated in C/CUDA code. Very little efficiency is lost by using a Matlab shell, while the gain in transparency is substantial.

The software is fully object-oriented, highly modular, and easily extensible. It is straightforward for users to write plug-ins implementing new models, providing access to the full benefits of GPU acceleration with little programming effort. All that is needed is a Matlab class with methods that evaluate the data density and moments of any functions of interest. The requisite interfaces are all clearly specified by abstract classes.

To use a model for a particular application, data and a description of the prior must be supplied as inputs to the class constructor. Classes implementing simulation from and evaluation of some common priors are included with the software. Extending the package to implement additional prior specifications is simply a matter of providing classes with this functionality.

The software includes several models (including the application described in Section 5 of this paper), and we are in the process of extending the library of available models and illustrative applications using this framework in ongoing work. Geweke et al. (2013) provide an example of one such implementation for the logit model.

4. PREDICTIVE AND MARGINAL LIKELIHOOD

Marginal likelihoods are fundamental to the Bayesian comparison, averaging, and testing of models. Sequential posterior simulators provide simulation-consistent approximations of marginal likelihoods as a by-product – almost no additional computation is required. This happens because the marginal likelihood is the product of predictive likelihoods over the sample, and the C phase approximates these terms in the same way that classical importance sampling can be used to approximate marginal likelihood (Geweke, 2005, Section 8.2.2; Kloek & van Dijk, 1978). By contrast there is no such generic approach with MCMC posterior simulators.

The results for posterior moment approximation apply only indirectly to marginal likelihood approximation. This section describes a practical adaptation of the results to this case, and a theory of approximation that falls somewhat short of Condition 5 for posterior moments.

4.1. Predictive Likelihood

Any predictive likelihood can be cast as a posterior moment by defining

$$g(\theta; t, s) = p(y_{t+1:s} \mid y_{1:t}, \theta) \quad (s > t) \tag{17}$$

Then

$$p(y_{t+1:s} \mid y_{1:t}) = E[g(\theta; t, s) \mid y_{1:t}] := \overline{g}(t, s) \tag{18}$$

Proposition 4. *If Conditions 1, 2, and 3 are satisfied, then in Algorithms 1 and 3 Condition 5 is satisfied for the function of interest* (17) *and the sample* $y_{1:t}$.

Proof. It suffices to note that Conditions 1 through 3 imply that Condition 4 is satisfied for Eq. (17). ■

The utility of Proposition 4 stems from the fact that simulation approximations of many moments (18) are computed as by-products in Algorithm 1. Specifically, in the C phase at (14) for $t = t_{\ell-1}$ and $s = t_{\ell-1}+1, \ldots, t_\ell$

$$w_{jn}(s) = g(\theta_{jn}^{(\ell-1)}; t, s) \quad (j \in \mathcal{J}, n \in \mathcal{N}, \ell \in \mathcal{L}) \tag{19}$$

From Eq. (6) Proposition 4 applied over the sample $y_{1:t_{\ell-1}}$ implies for $t = t_{\ell-1}$ and $s > t_{\ell-1}$

$$\overline{g}^{(J,N)}(t, s) = (JN)^{-1} \sum_{j=1}^{J} \sum_{n=1}^{N} g(\theta_{jn}^{(\ell-1)}; t, s) \xrightarrow{p} p(y_{t+1:s} \mid y_{1:t}) \quad (\ell \in \mathcal{L})$$

Algorithms 1 and 3 therefore provide weakly consistent approximations to Eq. (18) expressed in terms of the form (19). The logarithm of this approximation is a weakly consistent approximation of the more commonly reported log predictive likelihood. The numerical standard error of $\overline{g}^{(J,N)}(t,s)$ is given by Eq. (8), and the delta method provides the numerical standard error for the log predictive likelihood.

The values of s for which this result is useful are limited because as s increases, $g(\theta;t,s)$ becomes increasingly concentrated and the computational efficiency of the approximations $\overline{g}^{(J,N)}(t,s)$ of $\overline{g}(t,s)$ declines. Indeed, this is why particle renewal (in the form of S and M phases) is undertaken when the effective sample size drops below a specified threshold.

The sequential posterior simulator provides access to the posterior distribution at each observation in the C phase using the particles with the weights computed at (14). If at this point one executes the auxiliary simulations $Y_{sjn}^{(\ell-1)} \sim p(Y_s|y_{1:s-1}, \theta_{jn}^{(\ell-1)})$ $(j \in \mathcal{J}, n \in \mathcal{N})$ then a simulation-consistent approximation of the cumulative distribution of a function $F(Y_s)$, evaluated at the observed value $F(y_s)$, is

$$\frac{\sum_{j=1}^{J}\sum_{n=1}^{N} w_{jn}(s-1) I_{[-\infty, F(y_s)]}[F(Y_{sjn}^{(\ell-1)})]}{\sum_{j=1}^{J}\sum_{n=1}^{N} w_{jn}(s-1)}$$

These evaluations are the essential element of a probability integral transform test of model specification (Berkowitz, 2001; Diebold, Gunther, & Tay, 1998; Geweke & Amisano, 2010; Rosenblatt, 1952; Smith, 1985).

Thus by accessing the weights $w_{jn}(s)$ from the C phase of the algorithm, the investigator can compute simulation consistent approximations of any set of predictive likelihoods (18) and can execute probability integral transform tests for any function of Y_t.

4.2. Marginal Likelihood

To develop a theory of sequential posterior simulation approximation to the marginal likelihood, some extension of the notation is useful. Denote

$$\overline{w}_j^N(\ell-1) = N^{-1}\sum_{n=1}^{N} w_{jn}^{(\ell-1)} \quad (\ell \in \mathcal{L}, j \in \mathcal{J})$$

$$\overline{w}_j^N = \prod_{\ell=1}^{L} \overline{w}_j^N(\ell - 1) \quad (j \in \mathcal{J})$$

$$\overline{w}^{(J,N)}(\ell - 1) = J^{-1} \sum_{j=1}^{J} \overline{w}_j^N(\ell - 1) \quad (\ell \in \mathcal{L})$$

and then

$$\overline{w}^{(J,N)} = J^{-1} \sum_{j=1}^{J} \overline{w}_j^N \quad \tilde{w}^{(J,N)} = \prod_{\ell=1}^{L} \overline{w}^{(J,N)}(\ell - 1)$$

It is also useful to introduce the following condition, which describes an ideal situation that is not likely to be attained in practical applications but is useful for expositional purposes.

Condition 6. *In the mutation phase of Algorithm 1*

$$p(\theta \mid y_{1:t_\ell}, \theta^*, \lambda) = p(\theta \mid y_{1:t_\ell}) \; \forall \, \theta^* \in \Theta$$

With the addition of Condition 6 it would follow that $v = \text{var}[g(\theta)|y_{1:T}]$ in Condition 5, and computed values of relative numerical efficiencies would be about 1 for all functions of interest $g(\theta)$. In general Condition 6 is unattainable in any interesting application of an sequential posterior simulator, for if it were the posterior distribution could be sampled by direct Monte Carlo.

Proposition 5. *If Conditions 1, 2, and 3 are satisfied then in Algorithms 1 and 3*

$$\overline{w}^{(J,N)} \xrightarrow{p} p(y_{1:T}), \quad \tilde{w}^{(J,N)} \xrightarrow{p} p(y_{1:T}) \tag{20}$$

as $N \to \infty$, *and*

$$\text{E}\left\{ [J(J-1)]^{-1} \sum_{j=1}^{J} (\overline{w}_j^N - \overline{w}^{(J,N)})^2 \right\} = \text{var}(\overline{w}^{(J,N)}) \tag{21}$$

If, in addition, Condition 6 is satisfied then

$$(JN)^{-1/2}[\overline{w}^{(J,N)} - p(y_{1:T})] \xrightarrow{d} N(0,v) \tag{22}$$

Proof. From Proposition 4, Condition 5 is satisfied for $g(\theta; t_{\ell-1}, t_\ell)(\ell \in \mathscr{L})$. Therefore $\overline{w}_j^N(\ell-1) \to^p p(y_{t_{\ell-1}+1:t_\ell})$ as $N \to \infty$ $(j \in \mathscr{J}; \ell \in \mathscr{L})$. The result (20) follows from the decomposition

$$p(y_{1:T}) = \prod_{\ell=1}^{L} p\left(y_{t_{\ell-1}+1:t_\ell} \mid y_{1:t_{\ell-1}}\right)$$

Condition 3 implies that moments of $\overline{w}_j^N$ of all orders exist. Then Eq. (21) follows from the mutual independence of $\overline{w}_1^N, \ldots, \overline{w}_J^N$.

Result (22) follows from the mutual independence and asymptotic normality of the JL terms $\overline{w}_j^N(\ell)$, and v follows from the usual asymptotic expansion of the product in $\overline{w}_j^N$. ■

Note that without Condition 6, $w_j^N(\ell)$and $w_j^N(\ell')$ are not independent and so Condition 5 does not provide an applicable central limit theorem for the product $\overline{w}_j^N$ (although Condition 5 does hold for each $\overline{w}_j^N(\ell-1)$ $(j \in \mathscr{J}, \ell \in \mathscr{L})$ individually, as shown in Section 4.1).

Let $v\hat{a}r(\overline{w}^{(J,N)})$ denote the term in braces in Eq. (21). Proposition 5 motivates the working approximations

$$\overline{w}^{(J,N)} \stackrel{\cdot}{\sim} N[p(y_{1:T}), \hat{\text{var}}(\overline{w}^{(J,N)})]$$

$$\log(\overline{w}^{(J,N)}) \stackrel{\cdot}{\sim} N\left[\log p(y_{1:T}), \frac{\hat{\text{var}}(\overline{w}^{(J,N)})}{(\overline{w}^{(J,N)})^2}\right] \tag{23}$$

Standard expansions of $\log \overline{w}^N$ and $\log \tilde{w}^N$ suggest the same asymptotic distribution, and therefore comparison of these values in relation to the working standard error from Eq. (23) provides one indication of the adequacy of the asymptotic approximation. Condition 6 suggests that these normal approximations will be more reliable as more Metropolis iterations are undertaken in the algorithm detailed in Section 3.3, and more generally, the more effective is the particle diversity generated by the M phase.

5. APPLICATION: EXPONENTIAL GENERALIZED AUTOREGRESSIVE CONDITIONAL HETEROSKEDASTICITY MODEL

The predictive distributions of returns to financial assets are central to the pricing of their derivatives like futures contracts and options. The literature modeling asset return sequences as stochastic processes is enormous and has been a focus and motivation for Bayesian modeling in general and application of SMC methods in particular. One of these models is the exponential generalized autoregressive conditional heteroskedasticity (EGARCH) model introduced by Nelson (1991). The example in this section works with a family of extensions developed in Durham and Geweke (2013) that is highly competitive with many stochastic volatility models.

In the context of this paper the EGARCH model is also of interest because its likelihood function is relatively intractable. The volatility in the model is the sum of several factors that are exchangeable in the posterior distribution. The return innovation is a mixture of normal distributions that are also exchangeable in the posterior distribution. Both features are essential to the superior performance of the model (Durham & Geweke, 2013). Permutations in the ordering of factors and mixture components induce multimodal distributions in larger samples. Models with these characteristics have been widely used as a drilling ground to assess the performance of simulation approaches to Bayesian inference with ill-conditioned posterior distributions (e.g., Jasra et al., 2007). The models studied in this section have up to $(4!)^2 = 576$ permutations, and potentially as many local modes. Although it would be possible to optimize the algorithm for these characteristics, we intentionally make no efforts to do so. Nonetheless, the irregularity of the posteriors turns out to pose no difficulties for the algorithm.

Most important, in our view, this example illustrates the potential large savings in development time and intellectual energy afforded by the algorithm presented in this paper compared with other approaches that might be taken. We believe that other existing approaches, including importance sampling and conventional variants on MCMC, would be substantially more difficult. At the very least they would require experimentation with tuning parameters by Bayesian statisticians with particular skills in these numerical methods, even after using the approach of Geweke (2007) to deal with dimensions of posterior intractability driven by exchangeable parameters in the posterior distribution. The algorithm replaces this effort with a systematic updating of the posterior density, thereby releasing the time of investigators for more productive and substantive efforts.

5.1. Model and Data

An EGARCH model for a sequence of asset returns $\{y_t\}$ has the form

$$v_{kt} = \alpha_k v_{k,t-1} + \beta_k(|\varepsilon_{t-1}| - (2/\pi)^{1/2}) + \gamma_k \varepsilon_{t-1} \quad (k = 1,\ldots,K) \tag{24}$$

$$y_t = \mu_Y + \sigma_Y \exp\left(\sum_{k=1}^{K} v_{kt}/2\right)\varepsilon_t \tag{25}$$

The return disturbance term ε_t is distributed as a mixture of I normal distributions,

$$p(\varepsilon_t) = \sum_{i=1}^{I} p_i \phi(\epsilon_t; \mu_i, \sigma_i^2) \tag{26}$$

where $\phi(\cdot; \mu, \sigma^2)$ is the Gaussian density with mean μ and variance σ^2, $p_i > 0$ $(i = 1,\ldots, I)$, and $\sum_{i=1}^{I} p_i = 1$. The parameters of the model are identified by the conditions $E(\varepsilon_t) = 0$ and $\mathrm{var}(\varepsilon_t) = 1$; equivalently,

$$\sum_{i=1}^{I} p_i \mu_i = 0, \quad \sum_{i=1}^{I} p_i(\mu_i^2 + \sigma_i^2) = 1 \tag{27}$$

The models are indexed by K, the number of volatility factors, and I, the number of components in the return disturbance normal mixture, and we refer to the specification (24)–(25) as `egarch KI`. The original form of the EGARCH model (Nelson, 1991) is Eqs. (24)–(25) with $I = K = 1$.

The algorithm operates on transformed parameters, as described in Section 3.3. The vector of transformed parameters is denoted θ. The prior distributions for the elements of θ are all Gaussian. Priors and transformations are detailed in Table 1. The intermediate parameters p_i^*, μ_i^*, and σ_i^* are employed to enforce the normalization (27) and undergo the further transformations

$$p_i = p_i^* / \sum_{i=1}^{I} p_i^*, \quad \mu_i^{**} = \mu_i^* - \sum_{i=1}^{I} p_i \mu_i^*, \quad c = \left[\sum_{i=1}^{I} p_i ((\mu_i^{**})^2 + (\sigma_i^*)^2) \right]^{-1/2} \tag{28}$$

$$\mu_i = c\mu_i^{**}, \quad \sigma_i = c\sigma_i^* \quad (i = 1,\ldots,I) \tag{29}$$

The truncation of the parameters θ_{8i} $(i = 1,\ldots,I)$ bounds the likelihood function above, thus satisfying Condition 3. The initial simulation from the prior distribution is trivial, as is evaluation of the prior density.

Evaluation of $p(y_t | y_{1:t-1}, \theta)$ entails the following steps, which can readily be expressed in SIMD-compatible code, satisfying Condition 2.

1. Transform the parameter vector θ to the parameters of the model (24)–(25) using the fourth column of Table 1 and Eqs. (28)–(29).
2. Compute v_{kt} $(k = 1,\ldots,K)$ using Eq. (24), noting that ε_{t-1} and $v_{k,t-1}$ $(k = 1,\ldots,K)$ are available from the evaluation of $p(y_{t-1} | y_{1:t-2}, \theta)$. As is conventional in these models the volatility states are initialized at $v_{k0} = 0$ $(k = 1,\ldots,K)$.

Table 1. Parameters and Prior Distributions for the EGARCH Models.

	Mean	SD	Transformation
θ_1	0	1	$\mu_Y = \theta_1 / 1000$
θ_2	$\log(0.01)$	1	$\sigma_Y = \exp(\theta_2)$
θ_{3k}	$\tanh^{-1}(0.95)$	1	$\alpha_k = \tanh(\theta_{3k})$
θ_{4k}	$\log(0.10)$	1	$\beta_k = \exp(\theta_{4k})$
θ_{5k}	0	0.2	$\gamma_k = \theta_{5k}$
θ_{6i}	0	1	$p_i^* = \tanh(\theta_{6i}) + 1$
θ_{7i}	0	1	$\mu_i^* = \theta_{7i}$
θ_{8i}	0	1	$\sigma_i^* = \exp(\theta_{8i})$

All parameters have Gaussian priors with means and standard deviations indicated above (the prior distribution of θ_{8i} is truncated above at -3.0). Indices i and k take on the values $i = 1,\ldots,I$ and $k = 1,\ldots,K$.

3. Compute $h_t = \sigma_Y \exp\left(\sum_{k=1}^{K} v_{kt}/2\right)$ and $\varepsilon_t = (y_t - \mu_Y)/h_t$.
4. Evaluate $p(y_t \mid y_{1:t-1}, \theta) = (2\pi)^{-1/2} h_t^{-1} \sum_{i=1}^{I} \left\{ p_i \frac{1}{\sigma_i} \exp\left[-(\varepsilon_t - \mu_i)^2 / 2\sigma_i^2\right] \right\}$.

The observed returns are $y_t = \log(p_t/p_{t-1})$ $(t = 1, \ldots, T)$ where p_t is the closing Standard and Poors 500 index on trading day t. We use returns beginning January 3, 1990 $(t=1)$ and ending March 31, 2010 $(t = T = 5,100)$.

5.2. *Performance*

All of the inference for the EGARCH models is based on $2^{16} = 65,536$ particles in $J = 2^6 = 64$ groups of $N = 2^{10} = 1,024$ particles each. Except where specified otherwise, we use the C phase stopping rule with $D_1 = 0.50$ and $D_2 = 0.20$, and the RNE-based M phase stopping rule with $E_1 = 0.35$, $E_2 = 0.90$, and $R_{\max} = 300$. Unless otherwise noted, reported results are obtained from step 2 of the hybrid algorithm. See Section 3.3 for details.

For the RNE-based rule, we use the following test functions: log volatility, $g_1(\theta, y_{1:t}) = \log\sigma_Y + \sum_{k=1}^{K} v_{kt}/2$; skewness of the mixture distribution (26), $g_2(\theta, y_{1:t}) = E(\varepsilon_{t+1}^3 \mid \theta)$; and 3% loss probability, $g_3(\theta, y_{1:t}) = P(Y_{t+1} < -0.03 \mid \theta, y_{1:t})$. Posterior means of these test functions are of potential interest in practice, and we return to a more detailed discussion of these applications in Section 5.3.

The exercise begins by comparing the log marginal likelihood of the 10 variants of this model indicated in column 1 of Table 2. The log marginal likelihood and its numerical standard error (NSE) are computed as described in Section 4.2. "Log score" is the log predictive likelihood $p(y_{505:5100} \mid y_{1:504})$; observation 505 is the return on the first trading day of 1992.

The table provides results in pairs corresponding to the two steps of the hybrid algorithm (Algorithm 3). In step 1, numerical approximations are based on particles generated using the simulator detailed in Section 3.3 run adaptively. In step 2, new particles are obtained by rerunning the algorithm of Section 3.3 (nonadaptively) using design elements retained from the first run.

There are two striking features in each pair of results: (1) numerical standard errors are mutually consistent given the reliability inherent with $J = 64$ groups of particles; (2) differences in log score or marginal likelihood are, in turn, consistent with these numerical standard errors. These findings are consistent with two conjectures: (1) the central limit theorem for posterior

Table 2. Bayesian Model Comparison.

Model	Hybrid Step	Compute Time (Seconds)	Cycles	Metropolis Steps	Log ML	NSE	Log Score	NSE
egarch_11	1	74	54	515	16,641.92	0.1242	15,009.20	0.1029
egarch_11	2	65	54	515	16,641.69	0.0541	15,009.08	0.0534
egarch_12	1	416	65	2,115	16,713.44	0.0814	15,075.82	0.0596
egarch_12	2	812	65	2,115	16,713.60	0.0799	15,075.91	0.0649
egarch_21	1	815	62	4,439	16,669.40	0.0705	15,038.40	0.0630
egarch_21	2	732	62	4,439	16,669.39	0.0929	15,038.41	0.0887
egarch_22	1	1,104	71	4,965	16,736.81	0.0704	15,100.17	0.0534
egarch_22	2	991	71	4,965	16,736.89	0.0864	15,100.25	0.0676
egarch_23	1	2,233	77	7,490	16,750.77	0.0683	15,114.24	0.0455
egarch_23	2	2,093	77	7,490	16,750.83	0.0869	15,114.21	0.0512
egarch_32	1	1,391	76	6,177	16,735.04	0.0870	15,099.91	0.0650
egarch_32	2	1,276	76	6,177	16,734.94	0.0735	15,099.90	0.0540
egarch_33	1	2,685	82	8,942	16,748.74	0.0703	15,113.62	0.0397
egarch_33	2	2,619	82	8,942	16,748.75	0.0646	15,113.68	0.0456
egarch_34	1	3,036	82	8,311	16,748.78	0.0671	15,113.76	0.0486
egarch_34	2	2,878	82	8,311	16,748.64	0.0716	15,113.64	0.0413
egarch_43	1	2,924	79	9,691	16,745.62	0.0732	15,112.36	0.0462
egarch_43	2	2,741	79	9,691	16,745.61	0.0725	15,112.41	0.0534
egarch_44	1	3,309	82	9,092	16,745.63	0.1025	15,112.33	0.0451
egarch_44	2	3,133	82	9,092	16,745.54	0.0643	15,112.31	0.0508

moments (Proposition 1) applies to the simulator detailed in Section 3.3 when run adaptively as well as when run nonadaptively in the context of the hybrid method; (2) conclusion (22) of Proposition 5 does not require Condition 6, which is ideal rather than practical, and is true under Conditions 1 through 3, which are weak and widely applicable.

We emphasize that these are working conjectures. They are not theorems that are likely to be proved any time in the near future, if ever. Our current recommendation for practical application, pending more extensive experience with these matters, is to run the simulator adaptively in the work of model development and modification; then check results using the full hybrid method (step 2 of Algorithm 3) at regular intervals to guard against unpleasant surprises; and to always do so before making results available publicly.

Bayes factors strongly favor the `egarch_23` model, as do log scores. In comparison with all the models that it nests, the Bayes factor is at least exp(15). In comparison with all the models that nest `egarch_23`, the Bayes factor ranges from exp(2) to over exp(5). This pattern is classic: the data provide strong diagnostics of underfitting, while the evidence of overfitting is weaker because it is driven primarily by the prior distribution. The 95% confidence intervals for log Bayes factors are generally shorter than 0.2. Going forward, all examples in this section utilize the `egarch_23` model.

Compute time increases substantially as a function of I and K in the model specification (24)–(26). There are three reasons for this: the larger models require more floating point operations to evaluate conditional data densities; the larger models exhibit faster reduction of effective sample size in the C phase, increasing the number of cycles L and thereby the number of passes through the M phase; and the larger models require more Metropolis iterations in each M phase to meet the RNE threshold for the stopping rule.

Table 3 shows the outcome of an exercise exploring the relationship between the number of Metropolis iterations undertaken in the M phases and the performance of the algorithm. The first eight rows of the table use the deterministic M phase stopping rule with $\overline{R} = 5, 8, 13, 21, 34, 55, 89$, and 144. The last row of the table, indicated by "*" in the $\overline{R}$ field, uses the RNE-based stopping rule with $E_1 = 0.35$, $E_2 = 0.90$, and $R_{max} = 300$. The last column of the table reports a measure of computational performance: precision (i.e., $1/NSE^2$) normalized by computational cost (time in seconds).

The ratio of precision to time is the relevant criterion for comparing two means of increasing numerical accuracy: more steps in the M phase versus

Table 3. Sensitivity to Number of Metropolis Steps in M Phase.

$\overline{R}$	Compute Time (Seconds)	Cycles	Metropolis Steps	Log Score	NSE	Precision /Time
5	91	77	395	15,113.13	0.4597	0.052
8	127	75	616	15,114.38	0.5441	0.027
13	205	78	1,040	15,114.02	0.2888	0.059
21	304	76	1,638	15,114.39	0.1880	0.093
34	482	76	2,652	15,114.27	0.1277	0.127
55	776	77	4,345	15,114.23	0.0849	0.179
89	1,245	77	7,031	15,114.16	0.0592	0.229
144	2,120	79	11,664	15,114.27	0.0395	0.302
*	2,329	77	8,531	15,114.24	0.0427	0.236

more particles. The ratio of precision to time is constant in the latter strategy. (In fact, on a GPU, it increases up to a point because the GPU works more efficiently with more threads, but the application here with 2^{16} particles achieves full efficiency.) Therefore adding more steps in the M phase is the more efficient strategy so long as the ratio of precision to time continues to increase. Table 3 shows that in this example, adding iterations is dominant at $\overline{R}=89$ and is likely dominant at $\overline{R}=144$.

Ideally, one would select $\overline{R}$ to maximize the performance measure reported in the last column of the table. The RNE-based stopping rule, shown in the last line of the table, does a good job of automatically picking an appropriate stopping point without requiring any user input or experimentation. The number of Metropolis iterations varies from one M phase to the next with this stopping rule, averaging just over 100 in this application. Although the RNE-based rule uses fewer Metropolis steps overall relative to $\overline{R}=144$, total execution time is greater. This is because the Metropolis iterations are more concentrated toward the end of the sample period with the RNE-based rule, where they are more computationally costly.

The fact that so many Metroplis iterations are needed to get good particle diversity is symptomatic of the extremely irregular posterior densities implied by this model (we return to this issue in Section 5.4). For models with posterior densities that are closer to Gaussian, many fewer Metropolis iterations will typically be needed.

5.3. Posterior Moments

Models for asset returns, like the EGARCH models considered here, are primarily of interest for their predictive distributions. We illustrate this application using the three functions of interest $g_i(\theta, y_{1:t})$ introduced in Section 5.2.

Moments are evaluated by Monte Carlo approximation over the posterior distribution of θ using the particles obtained at time t, as described in Section 2.3. The last observation of the sample, March 31, 2010 in this application ($t=5,100$), is typically of interest. For illustration, we specify the same date one year earlier, March 31, 2009 ($t=4,848$), as an additional observation of interest. Volatility is much higher on the earlier date than it is on the later date. At each date of interest, the C phase is terminated (regardless of the RSS) and S and M phases are executed.

Fig. 1 shows RNE for all three test functions $g_i(\theta, y_{1:t_\ell})$ at each Metropolis iteration of each cycle ℓ. The figure reports results for four

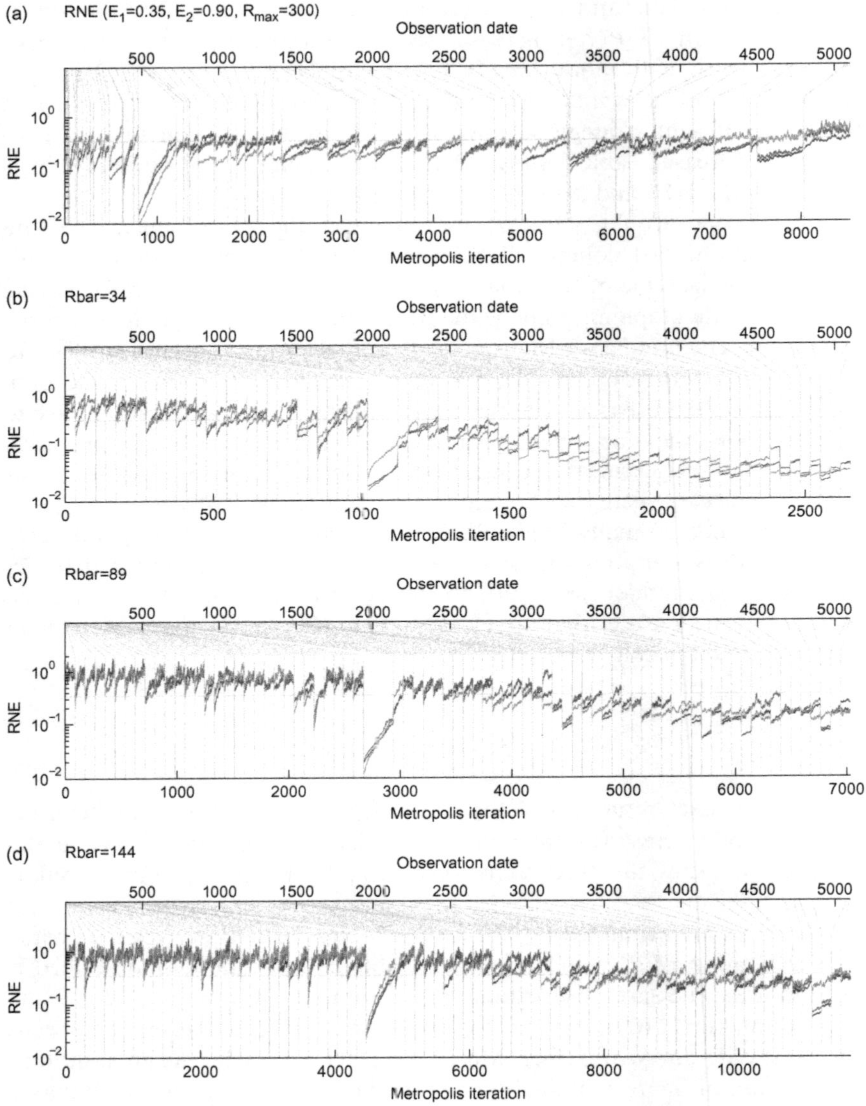

Fig. 1. RNE of Test Functions at Each Metropolis Iteration: (a) RNE-Based Stopping Rule; (b) $\overline{R}=34$; (c) $\overline{R}=89$; (d) $\overline{R}=144$.

different M phase stopping rules: the RNE-based rule (with $E_1 = 0.35$, $E_2 = 0.90$, and $R_{\max} = 300$) and the deterministic rule with $\overline{R}$ = 34, 89, and 144. The beginning of each M phase is indicated by a vertical line. The lower axis of each panel indicates the Metropolis iteration (cumulating across cycles); the top axis indicates the observation date t at which each M phase takes place; and the left axis indicates RNE. For reference, Fig. 1 includes a horizontal line indicating RNE = 0.35. This is the default target for the RNE-based stopping rule and serves as a convenient benchmark for the other rules as well.

In the early part of the sample, the deterministic rules execute many more Metropolis steps than needed to achieve the nominal target of RNE = 0.35. However, these require relatively little time because sample size t is small. As noted above, there is little point in undertaking additional Metropolis iterations once RNE approaches one, as happens toward the beginning of the sample for all three deterministic rules shown in the figure.

Toward the end of the sample, achieving any fixed RNE target in the M phase requires more iterations due to the extreme non-Gaussianity of the posterior (see Section 5.4 for a more detailed discussion of this issue). The RNE-based rule adapts automatically, performing iterations only as needed to meet the RNE target, implying more iterations as sample size increases in this application.

At observation $t = 474$, November 15, 1991, the C phase terminates with very low RSS (regardless of M phase stopping rule used), the result of a return that is highly unlikely conditional on the model and past history of returns. The deterministic rules undertake additional Metropolis iterations to compensate, as detailed in Section 3.3. The RNE-based rule also requires more iterations than usual to meet the relevant threshold; but in this case the number of additional iterations undertaken is determined algorithmically.

Table 4 reports details on the posterior mean approximations for the two dates of interest. Total computation time for running the simulator across the full sample is provided for reference. The last line in each panel, indicated by "*" in the $\overline{R}$ field, uses the RNE-based stopping rule. For both dates, the NSE of the approximation declines substantially as $\overline{R}$ increases. Comparison of the compute times reported in Table 4 again suggests that increasing $\overline{R}$ is more efficient for reducing NSE than would be increasing N, up to at least $\overline{R} = 89$.

Some bias in the moment approximations is evident with low values of $\overline{R}$. The issue is most apparent for the 3% loss probability on March 31,

Table 4. Moment Approximations.

$\overline{R}$	Compute Time	10 × Volatility State				1000 × (3% Loss Probability)				10 × Skewness			
		E	SD	NSE	RNE	*E*	SD	NSE	RNE	*E*	SD	NSE	RNE
March 31, 2009													
5	91	−37.405	0.373	0.027	0.003	95.877	7.254	0.513	0.003	−1.999	0.823	0.063	0.003
8	127	−37.388	0.390	0.029	0.003	96.457	7.543	0.531	0.003	−2.204	0.837	0.060	0.003
13	205	−37.347	0.366	0.020	0.005	97.311	7.167	0.361	0.006	−2.348	0.774	0.034	0.008
21	304	−37.382	0.352	0.014	0.009	96.767	6.938	0.255	0.011	−2.530	0.788	0.030	0.011
34	482	−37.339	0.349	0.011	0.015	97.582	6.913	0.211	0.016	−2.569	0.812	0.021	0.023
55	776	−37.340	0.353	0.007	0.034	97.591	6.970	0.142	0.037	−2.563	0.811	0.011	0.079
89	1245	−37.330	0.364	0.006	0.055	97.735	7.133	0.109	0.065	−2.574	0.812	0.010	0.105
144	2120	−37.332	0.355	0.004	0.141	97.743	6.996	0.066	0.172	−2.580	0.816	0.007	0.230
*	2329	−37.334	0.359	0.003	0.170	97.687	7.037	0.063	0.192	−2.587	0.818	0.005	0.406
March 31, 2010													
5	91	−50.563	0.309	0.014	0.007	0.596	0.253	0.013	0.006	−2.078	0.816	0.063	0.003
8	127	−50.514	0.309	0.011	0.012	0.662	0.276	0.013	0.007	−2.274	0.838	0.059	0.003
13	205	−50.517	0.309	0.009	0.016	0.696	0.284	0.010	0.013	−2.425	0.771	0.033	0.008
21	304	−50.512	0.310	0.007	0.031	0.736	0.295	0.008	0.021	−2.599	0.780	0.028	0.012
34	482	−50.492	0.309	0.006	0.044	0.754	0.301	0.006	0.045	−2.616	0.795	0.018	0.030
55	776	−50.491	0.310	0.004	0.097	0.755	0.302	0.004	0.103	−2.610	0.797	0.010	0.093
89	1245	−50.482	0.315	0.003	0.164	0.765	0.308	0.003	0.162	−2.640	0.794	0.008	0.145
144	2120	−50.483	0.314	0.002	0.381	0.765	0.308	0.002	0.240	−2.632	0.800	0.006	0.276
*	2329	−50.482	0.315	0.002	0.460	0.769	0.307	0.002	0.588	−2.643	0.796	0.005	0.452

2010. Since volatility is low on that date, the probability of realizing a 3% loss is tiny and arises from tails of the posterior distribution of θ, which is poorly represented in small samples. For example, with $\overline{R}=13$, RNE is 0.013, implying that each group of size $N=1,024$ has an effective sample size of only about 13 particles. There is no evidence of bias for $\overline{R}\geq 89$ or with the RNE-based rule.

Plots such as those shown in Fig. 1 provide useful diagnostics and are provided as a standard output of the software. For example, it is easy to see that with $\overline{R}=34$ (panel (b) of the figure) not enough iterations are performed in the M phases, resulting in low RNE toward the end of the sample. With lower values of $\overline{R}$ the degradation in performance is yet more dramatic. As $\overline{R}$ is increased to 89 and 144 in panels (c) and (d) of the figure, respectively, the algorithm is better able to maintain particle diversity through the entire sample. The RNE-based rule does a good job at choosing an appropriate number of iterations in each M phase and does so without the need for user input or experimentation.

5.4. Robustness to Irregular Posterior Distributions

In the `egarch_23` model there are two permutations of the factors v_{kt} and six permutations of the components of the normal mixture probability distribution function of ε_t. This presents a severe challenge for single-chain MCMC as discussed by Celeux, Hurn, and Robert (2000) and Jasra et al. (2007), and for similar reasons importance sampling is also problematic. The problem can be mitigated (Frühwirth-Schnatter, 2001) or avoided entirely (Geweke, 2007) by exploiting the special " mirror image" structure of the posterior distribution. But these models are still interesting as representatives of multimodal and ill-behaved posterior distributions in the context of generic posterior simulators. We focus here on the six permutations of the normal mixture in `egarch_23`.

Consider a 9×1 parameter subvector ψ with three distinct values of the triplets (p_s, μ_s, σ_s) $(s=A, B, C)$. There are six distinct ways in which these values could be assigned to components $i=1,2,3$ of the normal mixture (26) in the `egarch_23` model. These permutations define six points ψ_u $(u=1,\ldots,6)$. For all sample sizes t, the posterior densities $p(\psi_u|y_{1:t})$ at these six points are identical. Let ψ' be a different parameter vector with analogous permutations ψ'_u $(u=1,\ldots,6)$. As the sample adds evidence $p(\psi_u|y_{1:t})/p(\psi_{u'}|y_{1:t}) \rightarrow^{as} 0$ or $p(\psi_u|y_{1:t})/p(\psi_{u'}|y_{1:t}) \rightarrow^{as} \infty$. Thus, a specific triplet set of triplets (p_s, μ_s, σ_s) $(s = A, B, C)$ and its permutations will emerge as pseudo-true values of the parameters (Geweke, 2005, Section 3.4).

The marginal distributions will exhibit these properties as well. Consider the pair (σ_1, σ_2), which is the case portrayed in Fig. 2. The scatterplot is symmetric about the axis $\sigma_1 = \sigma_2$ in all cases. As sample size t increases six distinct and symmetric modes in the distribution gradually emerge. These reflect the full marginal posterior distribution for the normal mixture components of the `egarch_23` model (i.e., marginalizing on all other parameters) that is roughly centered on the components

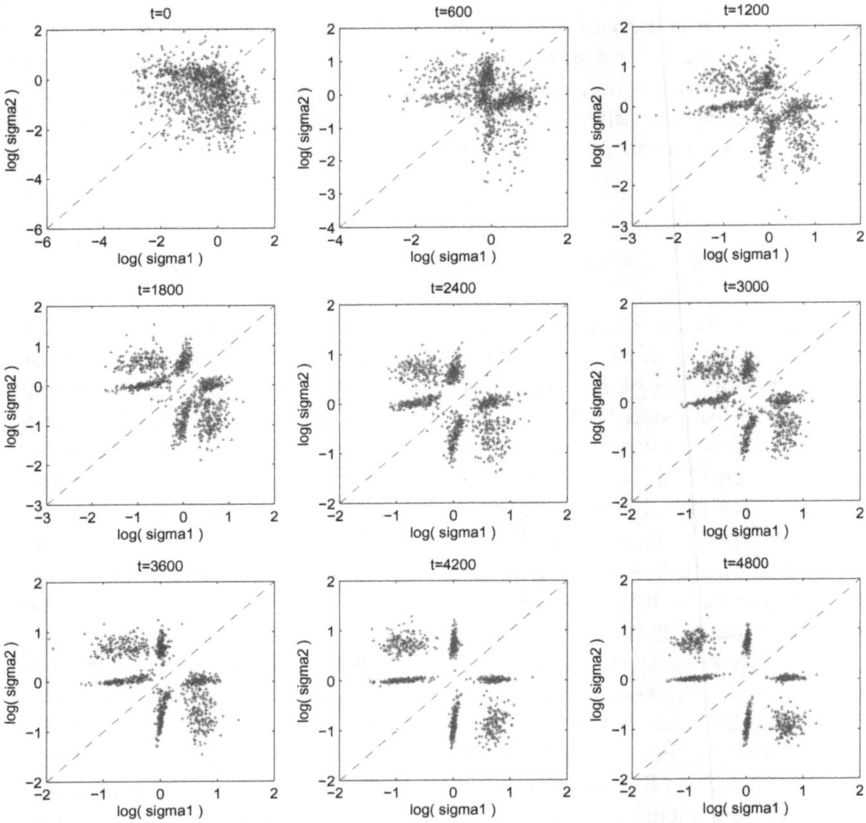

Fig. 2. Scatterplots of a Subset of Particles $(\log \sigma_1, \log \sigma_2)$ from Selected Posterior Distributions Conditional on $y_{1:t}$ $(J = 64, N = 4{,}096, D_1 = 0.5, D_2 = 0.2, \overline{R} = 34)$.

($p=0.17$, $\mu=0.16$, $\sigma=0.40$), ($p=0.85$, $\mu=0.01$, $\sigma=1.01$) and ($p=0.01$, $\mu=-1.36$, $\sigma=1.96$). The progressive decrease in entropy with increasing t illustrates how the algorithm copes with ill-behaved posterior distributions. Particles gradually migrate toward concentrations governed by the evidence in the sample. Unlike MCMC there is no need for particles to migrate between modes, and unlike importance sampling there is no need to sample over regions eliminated by the data (on the one hand) or to attempt to construct multimodal source distributions (on the other).

Similar phenomena are also evident for the other mixture parameters as well as for the parameters of the GARCH factors. The algorithm proposed in this paper adapts to these situations without specific intervention on a case-by-case basis.

5.5. Comparison with Markov chain Monte Carlo

To benchmark the performance of the algorithm against a more conventional approach, we constructed a straightforward Metropolis random walk MCMC algorithm, implemented in C code on a recent vintage CPU using the `egarch_23` model. The variance matrix was tuned manually based on preliminary simulations, which required several hours of investigator time and computing time. The algorithm required 12,947 seconds for 500,000 iterations. The numerical approximation of the moment of interest $1000 \cdot \mathrm{E}[P(Y_{t+1} < -0.03 | y_{1:t}, \theta)]$ for March 31, 2010, the same one addressed in Table 4, produced the result 0.763 and an NSE of 0.003. Taking the square of NSE to be inversely proportional to the number of MCMC iterations, an NSE of 0.002 (the result for the RNE-based stopping rule in Table 4) would require about 1,840,000 iterations and 47,600 seconds computing time. Thus posterior simulation for the full sample would require about 20 times as long using random walk Metropolis in a conventional serial computing environment.

Although straightforward, this analysis severely understates the advantage of the sequential posterior simulator developed in this paper relative to conventional approaches. The MCMC simulator does not traverse all six mirror images of the posterior density. This fact greatly complicates attempts to recover marginal likelihood from the MCMC simulator output (see Celeux et al. 2000). To our knowledge the only reliable way to attack this problem using MCMC is to compute predictive likelihoods from the prior distribution and the posterior densities $p(\theta | y_{1:t})$ $(t=1,\ldots,T-1)$. Recalling that $T=5{,}100$, and taking computation time to be proportional to sample size (actually, it is somewhat greater) yields a time requirement of $2,550 \times 28,167 = 71,825,850$ seconds (2.28 CPU years), which is almost 100,000 times as long as was required in the algorithm and implementation used here. Unless the function of interest $g(\theta, y_{1:T})$ is invariant to label switching (Geweke, 2007), the MCMC simulator must traverse all the mirror images with nearly equal frequency. As argued in Celeux et al. (2000), for all practical purposes this condition cannot be met with any reliability even with a simulator executed for several CPU centuries.

This example shows that simple "speedup factors" may not be useful in quantifying the reduction in computing time afforded by massively parallel computing environments. For the scope of models set in this work – that is, those satisfying Conditions 1 through 4 – the amenability of sequential posterior simulation to parallelization makes it a highly effective tool. Most important, sequential posterior simulation in a massively parallel computing environment conserves the time, energy, and talent of the investigator for more substantive tasks.

6. CONCLUSION

Recent innovations in parallel computing hardware and associated software provide the opportunity for dramatic increases in the speed and accuracy of posterior simulation. But work toward developing new algorithms that make efficient use of this environment is needed. The posterior simulator presented in this paper represents one promising avenue for future research. It provides accurate approximation of marginal and predictive likelihoods at essentially no additional cost, thereby making formal Bayesian model comparison routine and virtually costless. The simulator also has other attractive features: inherent adaptability to new models; computational efficiency relative to alternatives; reliable and well-grounded measures of numerical accuracy; robustness to irregular posteriors; and a well-developed theoretical foundation. Establishing these properties required a number of contributions to the literature, summarized in Section 1 and then developed in Sections 2, 3, and 4. Section 5 provided an application to a state-of-the-art model illustrating the properties of the simulator.

The methods set forth in the paper reduce computing time dramatically in a parallel computing environment that is well within the means of academic investigators. Relevant comparisons with conventional serial computing environments entail different algorithms, one for each environment. Moreover, the same approximation may be relatively more advantageous in one environment than the other. This precludes generic conclusions about "speed-up factors." In the example in Section 5, for a simple posterior moment, the algorithm detailed in Section 3.3 was nearly 10 times faster than a competing random-walk Metropolis simulator in a conventional serial computing environment. For predictive likelihoods and marginal likelihood it was 100,000 times faster. The parallel computations used a single GPU. Up to eight GPUs can be added to a standard desktop computer at a cost ranging from about $350 (US) for a mid-range card to about $2,000 (US) for a high-performance Tesla card. Computation time is inversely proportional to the number of GPUs.

These contributions address initial items on the research agenda opened up by the prospect of massively parallel computing environments for posterior simulation. In the near term it should be possible to improve on the specific contribution made here in the form of the algorithm detailed in Section 3.3. Looking forward on the research agenda, a major component is extending the class of models for which generic sequential posterior simulation will prove practical and reliable. Large-scale hierarchical models, longitudinal data, and conditional data density evaluations that must be simulated all pose fertile ground for future work.

The research reported here has been guided by two paramount objectives. One is to provide methods that are generic and minimize the demand for knowledge and effort on the part of applied statisticians who, in turn, seek primarily to develop and apply new models. The other is to provide methods that have a firm methodological foundation whose relevance is borne out in subsequent applications. It seems to us important that these objectives should be central in the research agenda.

ACKNOWLEDGMENT

We acknowledge useful comments from Nicolas Chopin, discussions with Ron Gallant, and tutorials in CUDA programming from Rob Richmond. We bear sole responsibility for the content of the paper. An earlier version of this work was posted with the title "Massively Parallel Sequential Monte Carlo for Bayesian Inference." Geweke acknowledges partial financial support from Australian Research Council grants DP110104372 and DP130103356.

REFERENCES

Andrieu, C., Doucet, A., & Holenstein, A. (2010). Particle Markov chain Monte Carlo. *Journal of the Royal Statistical Society, Series B, 72*, 1–33.

Baker, J. E. (1985). Adaptive selection methods for genetic algorithms. In J. Grefenstette (Ed.), *Proceedings of the international conference on genetic algorithms and their applications* (pp. 101–111). Mahwah, NJ: Erlbaum.

Baker, J. E. (1987). Reducing bias and inefficiency in the selection algorithm. In J. Grefenstette (Ed.), *Genetic algorithms and their applications* (pp. 14–21). New York, NY: Wiley.

Berkowitz, J. (2001). Testing density forecasts with applications to risk management. *Journal of Business and Economic Statistics, 19*, 465–474.

Carpenter J., Clifford P., & Fearnhead P. (1999). Improved particle filter for nonlinear problems. *IEEE Proceedings – Radar Sonar and Navigation, 146*, 2–7.

Celeux, G., Hurn, M., & Robert, C. P. (2000). Computational and inferential difficulties with mixture posterior distributions. *Journal of the American Statistical Association, 95*, 957–970.

Chopin, N. (2002). A sequential particle filter method for static models. *Biometrika, 89*, 539–551.

Chopin, N. (2004). Central limit theorem for sequential Monte Carlo methods and its application to Bayesian inference. *Annals of Statistics, 32*, 2385–2411.

Chopin, N., & Jacob, P. (2010). Free energy sequential Monte Carlo, application to mixture modelling. In J. M. Bernardo, M. J. Bayarri, J. O. Berger, A. P. Dawid, D. Heckerman, A. F. M. Smith, & M. West (Eds.), *Bayesian statistics* (p. 9). Oxford: Oxford University Press.

Chopin, N., Jacob, P. I., & Papaspiliopoulis, O. (2013). SMC2: An efficient algorithm for sequential analysis of state space models. *Journal of the Royal Statistical Society, Series B, 75*, 397–426.

Del Moral, P., Doucet, A., & Jasra, A. (2006). Sequential Monte Carlo samplers. *Journal of the Royal Statistical Society, Series B, 68*, 411–436.

Del Moral, P., Doucet, A., & Jasra, A. (2011). On adaptive resampling strategies for sequential Monte Carlo methods. *Bernoulli, 18*, 252–278.

Diebold, F. X., Gunther, T. A., & Tay, A. S. (1998). Evaluating density forecasts with applications to financial risk management. *International Economic Review, 39*, 863–883.

Douc, R., & Moulines, E. (2008). Limit theorems for weighted samples with applications to sequential Monte Carlo methods. *The Annals of Statistics, 36*, 2344–2376.

Durham, G., & Geweke, J. (2011). *Massively parallel sequential Monte Carlo for Bayesian inference*. Working Paper. Retrieved from http://www.censoc.uts.edu.au/pdfs/geweke_papers/gp_working_9.pdf

Durham, G., & Geweke, J. (2013). Improving asset price prediction when all models are false. *Journal of Financial Econometrics, 12*, 278–306.

Fearnhead, P. (1998). *Sequential Monte Carlo methods in filter theory*. Ph.D. thesis, Oxford University.

Flegal, J. M., & Jones, G. L. (2010). Batch means and spectral variance estimates in Markov chain Monte Carlo. *Annals of Statistics, 38*, 1034–1070.

Frühwirth-Schnatter, S. (2001). Markov chain Monte Carlo estimation of classical and dynamic switching and mixture models. *Journal of the American Statistical Association, 96*, 194–209.

Fulop, A., & Li, J. (2011). *Robust and efficient learning: A marginalized resample-move approach.* Working Paper. Retrieved from http://papers.ssrn.com/sol3/papers.cfm?abstract_id=1724203

Gelman, A., Roberts, G. O., & Gilks, W. R. (1996). Efficient metropolis jumping rules. In J. M. Bernardo, J. O. Berger, A. P. Dawid, & A. F. M. Smith (Eds.), *Bayesian statistics* (p. 5). Oxford: Oxford University Press.

Geweke, J. (1989). Bayesian inference in econometric models using Monte Carlo integration. *Econometrica, 57*, 1317–1340.

Geweke, J. (2005). *Contemporary Bayesian econometrics and statistics.* Englewood Cliffs, NJ: Wiley.

Geweke, J. (2007). Interpretation and inference in mixture models: Simple MCMC works. *Computational Statistics and Data Analysis, 51*, 3529–3550.

Geweke, J., & Amisano, G. (2010). Comparing and evaluating Bayesian predictive distributions of asset returns. *International Journal of Forecasting, 26*, 216–230.

Geweke, J., Durham, G., & Xu, H. (2013). *Bayesian inference for logistic regression models using sequential posterior simulation.* Retrieved from http://www.business.uts.edu.au/economics/staff/jgeweke/Geweke_Durham_XU_2013.pdf

Gilks, W. R., & Berzuini, C. (2001). Following a moving target – Monte Carlo inference for dynamic Bayesian models. *Journal of the Royal Statistical Society, Series B, 63*, 127–146.

Gordon, N. G., Salmond, D. G., & Smith, A. F. M. (1993). A novel approach to non-linear and non-Gaussian Bayesian state estimation. *IEEE Proceedings F: Radar and Signal Processing, 140*, 107–113.

Hendeby, G., Karlsson, R., & Gustafsson, F. (2010). Particle filtering: The need for speed. EURASIP Journal on Advances in Signal Processing. doi:10.1155/2010/181403

Herbst, E., Schorfheide, F. (2012). *Sequential Monte Carlo sampling for DSGE models.* Unpublished working paper.

Innovative Computing Laboratory (ICL), University of Tennessee. (2013). *Matrix algebra for GPU and multicore architectures.* Retrieved from http://icl.cs.utk.edu/magma/

Jasra, A., Stephens, D. A., & Holmes, C. C. (2007). On population-based simulation for static inference. *Statistics and Computing, 17*, 263–279.

Kitagawa, G. (1996). Monte Carlo filter and smoother for non-Gaussian nonlinear state space models. *Journal of Computational and Graphical Statistics, 5*, 1–25.

Kloek, T., & van Dijk, H. K. (1978). Bayesian estimates of equation system parameters: An application of integration by Monte Carlo. *Econometrica, 46*, 1–19.

Kong, A., Liu, J. S., & Wong, W. H. (1994). Sequential imputations and Bayesian missing data problems. *Journal of the American Statistical Association, 89*, 278–288.

Lee, L., Yau, C., Giles, M. B., Doucet, A., & Homes, C. C. (2010). On the utility of graphics cards to perform massively parallel simulation of advanced Monte Carlo Methods. *Journal of Computational and Graphical Statistics, 19*, 769–789.

Liu, J. S., & Chen, R. (1995). Blind deconvolution via sequential imputations. *Journal of the American Statistical Association, 90*, 567–576.

Liu, J. S., & Chen, R. (1998). Sequential Monte Carlo methods for dynamic systems. *Journal of the American Statistical Association, 93*, 1032–1044.

Matlab. (2013). *Matlab parallel computing toolbox.* Retrieved from http://www.mathworks.com/products/parallel-computing/description5.html

McGrayne, S. H. B. (2011). *The theory that would not die: How Bayes' rule cracked the enigma code, hunted down Russian submarines, and emerged triumphant from two centuries of controversy.* New Haven, CT: Yale University Press.

Nelson, D. B. (1991). Conditional heteroskedasticity in asset returns: A new approach. *Econometrica*, *59*, 347–370.

Nvidia. (2013). *Nvidia CUDA C Programming Guide, Version 5.0*. Retrived from http://docs.nvidia.com/cuda/pdf/CUDA_C_Programming_Guide.pdf

Roberts, G. O., & Rosenthal, J. S. (2007). Coupling and ergodicity of adaptive Markov chain Monte Carlo algorithms. *Journal of Applied Probability*, *44*, 458–475.

Rosenblatt, M. (1952). Remarks on a multivariate transformation. *Annals of Mathematical Statistics*, *23*, 470–472.

Smith, J. Q. (1985). Diagnostic checks of non-standard time series models. *Journal of Forecasting*, *4*, 283–291.

Souchard, M. A., Wang, Q., Chan, C., Frelinger, J., Cron, A., & West, M. (2010). Understanding GPU programming for statistical computation: Studies in massively parallel massive mixtures. *Journal of Computational Graphics and Statistics*, *19*, 419–438.

Tierney, L. (1994). Markov chains for exploring posterior distributions (with discussion and rejoinder). *Annals of Statistics*, *22*, 1701–1762.

Yang, Y., & Dunson, D. B. (2013). *Sequential Markov chain Monte Carlo*. Unpublished working paper.

MODEL SWITCHING AND MODEL AVERAGING IN TIME-VARYING PARAMETER REGRESSION MODELS

Miguel Belmonte and Gary Koop

Department of Economics, University of Strathclyde, Glasgow, UK

ABSTRACT

This paper investigates the usefulness of switching Gaussian state space models as a tool for implementing dynamic model selection (DMS) or averaging (DMA) in time-varying parameter regression models. DMS methods allow for model switching, where a different model can be chosen at each point in time. Thus, they allow for the explanatory variables in the time-varying parameter regression model to change over time. DMA will carry out model averaging in a time-varying manner. We compare our exact method for implementing DMA/DMS to a popular existing procedure which relies on the use of forgetting factor approximations. In an application, we use DMS to select different predictors in an inflation forecasting application. We find strong evidence of model switching. We also compare different ways of implementing DMA/DMS

Bayesian Model Comparison
Advances in Econometrics, Volume 34, 45–69

ISSN: 0731-9053/doi:10.1108/S0731-905320140000034004

and find forgetting factor approaches and approaches based on the switching Gaussian state space model to lead to similar results.

Keywords: Model switching; forecast combination; switching state space model; inflation forecasting

JEL classifications: C11; C52; E37

1. INTRODUCTION

Bayesian model averaging or model selection (BMA or BMS) methods are commonly used when the researcher is faced with many models. See, for instance, Hoeting, Madigan, Raftery, and Volinsky (1999) and Chipman, George, and McCulloch (2001) for surveys of these methods. Numerous empirical applications use these methods. However, they were developed for regression models or other models where parameters are constant over time. In time series econometrics, motivated by strong empirical evidence of structural breaks or other forms of parameter change in many economic variables, models where parameters change over time have long been used. Models such as the time-varying parameter (TVP) regression model have enjoyed great popularity, particularly in macroeconomics (see, among many others, Cogley, Morozov, & Sargent, 2005; Cogley & Sargent, 2005; D'Agostino, Gambetti, & Giannone, 2011; Koop, Leon-Gonzalez, & Strachan, 2009; Korobilis, 2013; Primiceri, 2005). Just as with constant coefficient models, it is possible that the researcher working with TVP regression models will want to do model averaging and selection. However, it will typically be desirable to do these in a time-varying manner. This leads to an interest in dynamic model averaging (DMA) or dynamic model selection (DMS). With DMA, the weights used in the model averaging procedure can change over time. With DMS, the model selected can change over time. This distinguishes it from conventional model selection methods where one model is selected and assumed to hold at all points in time.

The literature on DMA or DMS is much more limited than that on BMA or BMS. Perhaps the most prominent DMA approach for use with TVP regression models is that of Raftery, Karny, and Ettler (2010). To explain what this algorithm involves, we begin by defining the set of models under consideration. Let y_t be a dependent variable and Z_t be a row vector containing explanatory variables. We have K models which are characterized by having different subsets of Z_t as explanatory variables. Denoting these by $Z_t^{(k)}$ for $k = 1,\ldots,K$, a set of TVP regression models can be written as

$$y_t = Z_t^{(k)}\theta_t^{(k)} + \varepsilon_t^{(k)} \tag{1}$$
$$\theta_{t+1}^{(k)} = \theta_t^{(k)} + \eta_t^{(k)}$$

$\varepsilon_t^{(k)}$ is $N(0,\sigma_\varepsilon^{2(k)})$ and $\eta_t^{(k)}$ is $N(0,\Sigma_\eta^{(k)})$.[1]

DMA and DMS can be done by calculating $\Pr(s_t = k|y^{t-1})$ for $k = 1,\ldots,K$ where $s_t \in \{1,2,\ldots,K\}$ denotes which model applies at each time period and $y^s = (y_1,\ldots,y_s)'$. DMS involves selecting, for forecasting y_t given information available at time $t-1$, the single model with the highest value for $\Pr(s_t = k|y^{t-1})$. DMA involves averaging across models using these probabilities. Different approaches to DMA or DMS arise when different models or methods are used to calculate $\Pr(s_t = k|y^{t-1})$. Raftery et al. (2010), working in an application involving many potential explanatory variables and, hence, a large model space, uses forgetting factor methods to approximate $\Pr(s_t = k|y^{t-1})$. This leads to a computationally simple algorithm which does not require the use of Markov chain Monte Carlo (MCMC) methods. In applications with many potential explanatory variables (e.g., Koop & Korobilis, 2012; Koop & Tole, 2013; Raftery et al., 2010), the algorithm of Raftery et al. (2010) does seem to be the only computationally feasible algorithm currently available. However, as discussed in Section 3 of Raftery et al. (2010), it is an approximate method that does not arise from a particular statistical model of model switching. Furthermore, it is a filtering algorithm as opposed to a smoothing algorithm. That is, it provides the user with $\Pr(s_t = k|y^{t-1})$ for $t = 1,\ldots,T$ as opposed to $\Pr(s_t = k|y^T)$.

The purpose of this paper is to investigate the use of an alternative, model-based, way of allowing for time-varying model switching and compare it to the algorithm of Raftery et al. (2010). This alternative is the family of switching Gaussian state space models described in, among other places, Kim (1994), Kim and Nelson (1999), and Fruhwirth-Schnatter (2001a, 2001b). Switching Gaussian state space models will be described in the following section. Here we note only that they have been occasionally used in econometric applications (see Chapter 13 of Fruhwirth-Schnatter (2006) for a list of applications), but typically for state space models where the system matrices vary across regimes, not for selecting explanatory variables in TVP regression models (an exception being Chan, Koop, Leon-Gonzalez, & Strachan, 2012). An advantage of the use of switching Gaussian state space models is that results are not approximate, being based on a valid Bayesian posterior distribution. A further advantage is that either filtered or smoothed estimates can be obtained using existing algorithms.

A disadvantage of the use of switching Gaussian state space models is that MCMC methods are required. This substantially raises the computational burden and means their usage is limited to model spaces based on relatively few explanatory variables. However, it provides a setting in which we can compare DMA using the algorithm of Raftery et al. (2010) to DMA using switching linear Gaussian state space models. If we find the algorithm of Raftery et al. (2010) to provide results which are quite different from those using switching Gaussian state space models in a setting with a small model space, it will raise concerns about the use of Raftery et al.'s (2010) algorithm in the large model spaces where it is typically used. However, if the two approaches yield similar results, it will increase our confidence in the use of the algorithm of Raftery et al. (2010) in large model spaces.

This paper contains an application involving selecting between or averaging different independently produced forecasts of a dependent variable. That is, Z_t will contain various forecasts of the dependent variable y_t. Methods for combining forecasts provided by different models go back to Bates and Granger (1969) and Granger (2006) provides a recent survey. Recent approaches related to our own include Guidolin and Timmermann (2009), which uses a Markov switching approach to model switching in constant coefficient models and Billio, Casarin, Ravazzolo, and van Dijk (2011) who develop an approach with time-varying forecast weights. Our application is to forecasting US inflation. Papers such as Ang, Bekaert, and Wei (2007) consider various forecasts of inflation (e.g., forecasts produced by professional forecasters, consumer surveys, econometric forecasts, etc.) and investigate which ones forecast best. Ang et al. (2007) find that surveys do. We add to this literature using DMS and DMA methods. Note that, unlike Ang et al. (2007), we can have forecast switching so that, e.g., consumer surveys forecast best at some points in time and econometric models forecast best at other times. We find that there is evidence of model switching which would be missed by conventional approaches. Our empirical application also provides evidence that the algorithm of Raftery et al. (2010) is a reasonable one which yields results which are similar to those provided by the switching Gaussian state space model.

The remainder of this paper is organized as follows. Section 2 describes how switching Gaussian state space models can be used to do DMS or DMA. Section 3 describes our application. It is divided into sections which (i) discuss some general issues in combining inflation forecasts from various sources, (ii) describe the data, (iii) present empirical results using the switching Gaussian state space approach, and (iv) compare the latter approach to DMA and DMS using the methods of Raftery et al. (2010).

2. DMA AND DMS USING SWITCHING LINEAR GAUSSIAN STATE SPACE MODELS

The framework given in Eq. (1) is closely related to the switching linear Gaussian state space model discussed, e.g., in Fruhwirth-Schnatter (2006, pp. 393–394, 406–410) who provides several citations, mostly from the engineering literature, of papers which have used such models. A switching linear Gaussian state space model can be written as

$$y_t = H_t^{[s_t]}\theta_t + \varepsilon_t$$

$$\theta_t = F_t^{[s_t]}\theta_{t-1} + \eta_t$$

where y_t is observed, ε_t is $N(0,\sigma_\varepsilon^{2[s_t]})$, and η_t is $N(0,\Sigma_\eta)$. The errors are independent of each other and at all leads and lags. $s_t \in \{1,\ldots,K\}$ follows a Markov switching specification, i.e., we have a Markov transition matrix with elements $\zeta_{ij} = \Pr(s_t = i|s_{t-1} = j)$ for $i,j = 1,\ldots,K$.

We adapt this specification for use with variable selection in TVP regression models by using particular forms for the system matrices. In particular, we set

$$H_t^{[s_t]} = Z_t G^{[s_t]} \tag{2}$$

$$F_t^{[s_t]} = I$$

In most of our empirical work, we set $Z_t = (z_{1t}, \ldots, z_{Kt})$ to contain K explanatory variables and define $G^{[s_t=k]}$ to be the $K \times K$ matrix which selects the kth explanatory variable. That is, $G^{[s_t=k]}$ is a matrix of zeros except for the $(k;\ k)$th element which is set to one. In the final section of our empirical work, we consider TVP regression models with more than one explanatory variable and $G^{[s_t=k]}$ is defined to pick out the appropriate sets of explanatory variables.

Defined in this way, $\theta_t = (\theta_{1t}, \ldots, \theta_{kt})'$ is a vector of time-varying regression coefficients. The choice $F_t^{[s_t]} = I$ leads to the conventional choice of random walk evolution of these coefficients. We also let Σ_η be a diagonal matrix with kth diagonal element $\sigma^2_{\eta k}$ so that the regression coefficients evolve independently of one another.

In the main part of our empirical work, Z_t will contain different forecasts of inflation. It can be seen that Eq. (2) implies that, when $s_t = k$, the TVP regression model using the kth explanatory variable is used. That is, our model space is composed of K models each containing one explanatory variable. However, we also present some results where we consider TVP regression models with more than one explanatory variable. We allow for all combinations of the K explanatory variables, leading to a model space containing $2^K - 1$ models.

Switching between different TVP regression models is controlled through a Markov switching process with switching probabilities given by ζ_{ij}. Thus, the switching Gaussian state space model, with system matrices defined as in Eq. (2), can be used to do DMS or DMA in the context of single statistical model. And Bayesian methods for posterior inference (filtering and smoothing) in this model are developed in several places, including Fruhwirth-Schnatter (2001a, 2001b). In this paper, we use the algorithm of Fruhwirth-Schnatter (2001a, 2001b) (see the Technical Appendix for details).

It is worth stressing that, although we use the terminology "model switching," the switching Gaussian state space model is a single model and the parameter space retains the same dimension at all points in time. In our empirical work, θ_t remains a $K \times 1$ vector for all t. Thus, the problems associated with switching between model spaces of different dimension (see Green, 1995) do not arise. This is true even if, say, $s_t = 1$ for all t and, hence, a TVP regression model with the first explanatory variable is used in every period. In such a case, the MCMC draws of θ_{1t} will be produced in a data-based manner. But what about $\theta_{2t}, \ldots, \theta_{Kt}$? For periods when $s_t = 1$, they will be drawn from the prior (i.e., the random walk state equation which controls the time variation of the coefficients). We have found such an algorithm to work well (although an informative prior for each $\sigma^2_{\eta k}$ is required). In earlier work, we considered an alternative specification where Σ_η was replaced by $\Sigma_\eta^{[s_t]}$ which was a singular matrix such that $\Sigma_\eta^{[s_t=i]}$ is a matrix of zeros except for the (i, i)th diagonal element. This specification has the property that, if $s_t = i$, then $\theta_{jt} = \theta_{j,t-1}$ for $j \neq i$. Such a specification did not perform as well in our application and, hence, we do not include it here.

3. APPLICATION: SELECTING THE BEST INFLATION FORECASTS

3.1. Introduction

The literature on forecasting inflation is voluminous (see, e.g., Faust & Wright, 2012 for a recent survey). We aim to contribute to the literature on choosing between multiple forecasts of inflation. In an influential paper, Ang et al. (2007) compare various methods for forecasting inflation including surveys (of professional forecasters and of the public at large) and simple time series forecasting methods. Their main conclusion about which methods forecast best is pithily summarized in the first two words of their abstract: "Surveys do." Faust and Wright (2012) come to a similar conclusion using different econometric methods. The purpose of our application is to investigate whether this conclusion holds in the context of a more formal statistical modeling procedure involving DMA and DMS. Most importantly, our framework allows us to investigate whether the best forecasting model changes over time. After all, it is possible that the time series econometrician (whose forecasts are based on past patterns in the data) will forecast well in normal times, but forecast poorly around times of changes such as business cycle turning points. Professional economists, who can use qualitative events observed in real time (e.g., the collapse of Lehman Brothers) to aid in their forecasting, may be better forecasters at turning points. DMS and DMA can directly find patterns such as these where the best forecasting procedure changes over time or over the business cycle. Conventional methods, which just aim to find one best forecast procedure, cannot.

3.2. Data

Care must be taken with variable definitions and timing to make sure the forecasts made by forecasters are matched up with the outcomes they are compared to. Given the influence of the paper by Ang et al. (2007), we follow their choices where possible. The interested reader is referred to Ang et al. (2007) who discuss the relevant issues in detail. As a timing convention, note that all the t subscripts used below are for the times that the forecasts are being made. So, for instance, in 1996Q1 surveys were taken about inflation over the upcoming year through 1997Q1. These are dated as $t = 1996Q1$ in the equations below.

Our dependent variable is CPI inflation. Given that inflation forecasts are typically one-year ahead, we use as our dependent variable an annual inflation rate. To be precise, our dependent variable, π_t^R, is the realized value for inflation over the subsequent year defined as

$$\pi_t^R = \pi_{t+1} + \cdots + \pi_{t+4}$$

where

$$\pi_t = \log\left(\frac{P_t}{P_{t-1}}\right)$$

and P_t is the CPI (Consumer Price Index for All Urban Consumers).

We use four different forecasts of annual inflation rates which can be thought of as coming from four different sets of agents: (i) the professional forecasters, (ii) consumers, (iii) time series econometricians, and (iv) a naive agent.

The professionals' forecasts of inflation are taken from the Survey of Professional Forecasters (SPF) available through the Federal Reserve Bank of Philadelphia website. Detailed explanation about this data source are also available on this website. The inflation forecast we use, π_t^{SPF}, is the median of the one-year ahead inflation forecasts provided by the professionals.

Consumers' forecasts of inflation are taken from the University of Michigan consumer survey. Surveyed individuals are asked by how much they expect prices to change over the next 12 months. The inflation forecast we use, π_t^{CS}, is the median of their forecasts.

There are dozens of different forecasts of inflation produced by time series econometricians. However, it has proved difficult to beat simple forecasting models by much. For instance, Stock and Watson (2010) argue that it is "exceedingly difficult to improve systematically on simple univariate forecasting models." In this spirit, to represent the time series econometrician, we use an autoregressive model. To be specific, π_t^{TS} is the forecast of the time series econometrician using OLS forecasts from an AR(1) model. Forecasts made at time t are made using information available up to and including time $t-1$. Given that π_t^R is an average over four quarters, this means the model used for these forecasts is

$$\pi_t^R = \alpha + \rho\pi_{t-4}^R + \varepsilon_t$$

Finally, we have our naive agent producing simple no-change forecasts, π_t^{NOC}, where the forecaster simply uses the most recently available annual inflation rate as a forecast for next year's inflation. Thus,

$$\pi_t^{NOC} = \pi_{t-1} + \cdots + \pi_{t-4}$$

We stress that π_t^{TS} and π_t^{NOC} will be forecasts made at time t of inflation one year in the future.

All data except π_t^{SPF} is taken from the Federal Reserve Bank of St. Louis' FRED database.[2] Our forecasts runs from 1981Q3 through 2011Q2 (i.e., the last forecast is made in 2011Q2 which can be compared to the actual inflation outcome through 2012Q2). Fig. 1 plots the data.

In terms of the notation used in Section 2, $y_t = \pi_t^R$ and $Z_t = (\pi_t^{SPF}, \pi_t^{CS}, \pi_t^{TS}, \pi_t^{NOC})$.

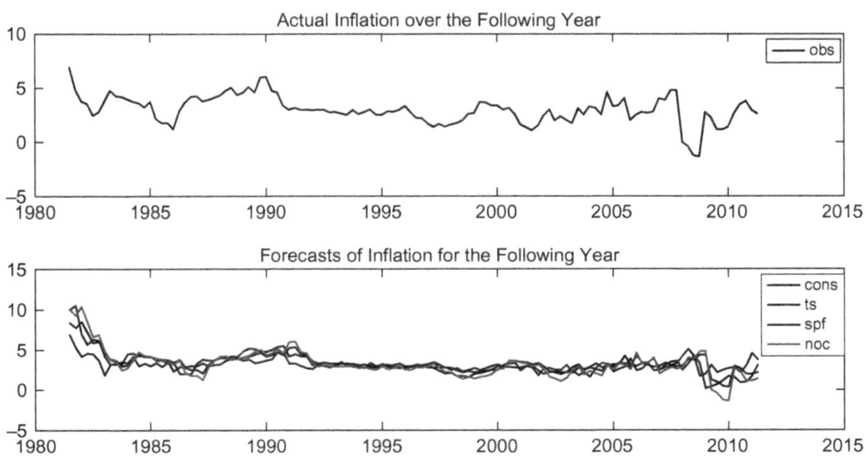

Fig. 1.

3.3. Which Inflation Forecasts Are Best?

Our main interest is in which of the four forecasts has been best at each point in time. To shed light on this, Fig. 2 presents smoothed estimates of the probabilities, $\Pr(s_t = k|y^T)$ for $k = 1,\ldots,4$ and $t = 1,\ldots,T$ using the switching Gaussian state space model.

Our results have some similarities with the general findings of Ang et al. (2007) and Faust and Wright (2012). Consumer surveys, in particular, do tend to be chosen as the best model in many time periods. However, the forecasts of the time series econometrician also do well at many times. The professional forecasts and the no change forecasts are less commonly chosen by DMS, but even they become predominant occasionally. However, our methods allow us to see some interesting time variation in forecast performance. A general pattern that emerges (with some exceptions) is that consumer surveys forecast best in stable times. The times when the consumers are forecasting poorly are the early 1980s, 1990, 2005, and post-2007. With the exception of 2005, these are all recessionary or volatile times.

In the periods when the consumer survey forecasts receive low probability in Fig. 2, it is typically the time series econometrician's forecasts which

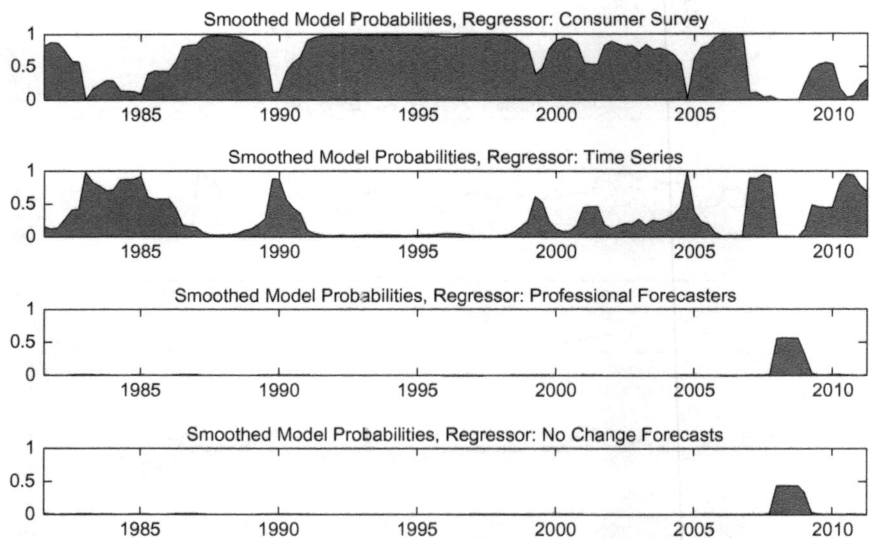

Fig. 2.

are chosen. The one exception to this pattern is the disinflationary period of 2008–2009 associated with the financial crisis and subsequent recession. This is the one period where the professional forecasts and no change forecasts (receiving roughly equal probability) are being selected by the switching state space model. An examination of the original data (see Fig. 1), reveals that both of these forecasts adjusted more rapidly to the disinflation which occurred at this time. However, by the middle of 2009 the professionals and the naive forecasts are again being beaten by the consumer survey and the time series econometrician.

The model probabilities in Fig. 2 are smoothed estimates based on the entire sample. Such a figure is of interest for a retrospective analysis where the researcher looks back on past forecast performance. It is also interesting to present filtered estimates, $\Pr(s_t = k|y^t)$, so as to show which forecasts a researcher in time t (given information available at time t) would have thought were good ones.[3] Fig. 3 presents these filtered probabilities. It can be seen that the main patterns in Fig. 3 are broadly similar to Fig. 2. The consumer survey tends to receive the most probability, followed by the forecasts of the time series econometrician. Periods where the consumer survey is not selected are usually recessionary or unstable times. The main difference is post-2007 where the filtered probabilities are very erratic and indicate the no change forecasts would never have been chosen. The professional forecasters have a brief period of higher probability shortly after the financial crisis, but it is shorter than in Fig. 2.

For the sake of brevity, we do not present parameter estimates but note only that they are reasonable (e.g., the elements of θ_t tend to be around one, although there is some fluctuation over time). Having established that the switching linear state space model does seem to be a sensible way of doing model selection or model averaging in a time-varying manner, we turn to the question of how it compares to the popular DMA algorithm of Raftery et al. (2010).

3.3.1. Comparison to DMA and DMS Using Forgetting Factors

Raftery et al. (2010) introduce an algorithm for doing DMA or DMS which involves the use of forgetting factors. There has been a recent surge in popularity in using forgetting factor methods for model averaging with empirical researchers (see, e.g., Dangl & Halling, 2012; Koop & Korobilis, 2012, 2013; Koop & Tole, 2013; McCormick, Raftery, Madigan, & Burd, 2012; Nicoletti & Passaro, 2012). Raftery et al. (2010, p. 53) stress that their methods are not a special case of a switching state space model such as the one used in the present paper, but are closely related. The switching state

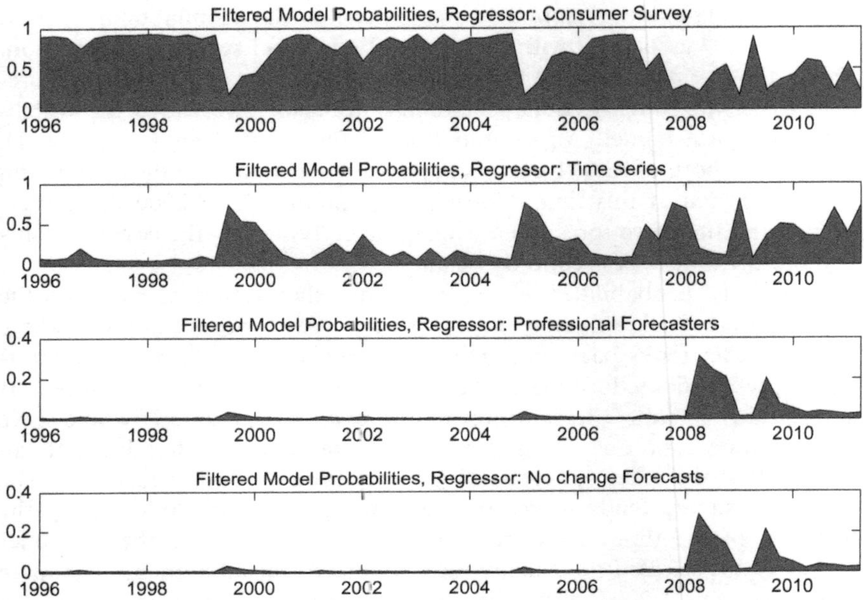

Fig. 3. Using Switching State Space Model.

space model approach involves specification of a matrix of Markov transition probabilities, $\zeta_{ij} = \Pr(s_t = i | s_{t-1} = j)$. The forgetting factor approach does not do so. In cases where the number of models is very large, parsimony considerations mean fully specifying such a matrix is not sensible. The switching state space model involves use of MCMC methods. The forgetting factor approach leads to a filtering algorithm which does not use MCMC. In large model spaces, the computational burden of MCMC methods mean they need to be avoided. Nevertheless, the goal of our switching state space model and Raftery et al.'s (2010) approach is the same: to obtain a method for model selection or model averaging done in a time-varying manner.

The references in the preceding paragraph all contain empirical applications with large model spaces where forgetting factor methods are used. But they do not contain comparisons of the forgetting factor approach with a formal Bayesian model which allows for dynamic model change. With large model spaces this would be computationally infeasible. But with a small model space such as the one used in our empirical application such a comparison can be done. Our aim is to shed light on whether forgetting factor methods lead to similar empirical results as a formal Bayesian approach.

Since forgetting factor methods are established in the literature, we will not provide a description of them here. The reader unfamiliar with DMA using forgetting factors is referred to Raftery et al. (2010) or they can read the brief description provided in Appendix B of this paper. Specification details, such as forgetting factor choices, are discussed in this appendix.

It is important to note that the algorithm of Raftery et al. (2010) is a filtering algorithm. As described in Appendix B, it provides us with $\pi_{t|t,j}$ which is the probability that model j generated at time t, given information through time t. This a similar concept to $\Pr(s_t = k|y^t)$ which can be obtained from our switching Gaussian state space model by running our MCMC algorithm using data through time t. Both are model probabilities conditional on information available at time t. These are plotted in Fig. 4 for the same four models used in the preceding section.

Fig. 4 should be compared to Fig. 3. In a very broad sense, Figs. 3 and 4 tell the same story. Most of the time the consumer surveys are providing

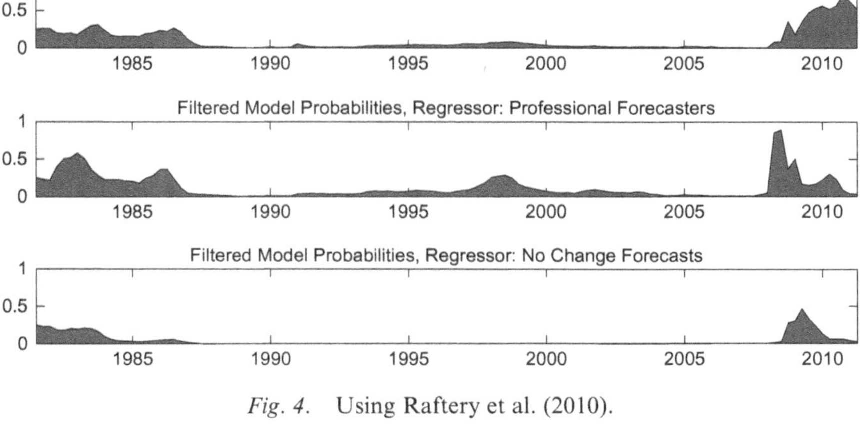

Fig. 4. Using Raftery et al. (2010).

the best forecasts for inflation. The times when the consumer surveys are not forecasting best tend to be unstable or recessionary times. The forecasts of the time series econometrician also receive appreciable weight, especially at the beginning and end of the sample.

However, there are many details in which Figs. 3 and 4 differ. Most prominently, the professional forecasters do better in Fig. 4 than they did in Fig. 3. There are times, most particularly at the beginning and end of the sample, where the forgetting factor approach allocates considerable weight to their forecasts. The second interesting difference is that the switching state space model seems more capable of capturing abrupt switches in model probabilities. For instance, there are several times in Fig. 3 when the probability attached to the consumer surveys switched abruptly from near one to near zero or vice versa (e.g., around 1990, 2000, and 2005). These abrupt switches do not appear using the forgetting factor approach. However, the abrupt switch associated with the financial crisis does appear in both Figs. 3 and 4.

What should the researcher using the forgetting factor approach to DMA make of our results? Insofar as the forgetting factor approach is viewed as a method intended to approximate a switching Gaussian state space model, it suggests the approximation is not bad. However, it is far from providing a close reproduction of exact results provided by a formal Bayesian model. Of course, as we stress above, DMA with forgetting factors is not simply a special case of a switching linear Gaussian state space model and cannot simply be interpreted as an approximation (e.g., where parameters are replaced by estimates). Nevertheless, since the two methods do have the same goal, it is not unreasonable to compare them and to hope that they will tell the same story. In this vein, we find our results (from a small model space) moderately encouraging for the user of DMA with forgetting factors working with a large model space.

3.4. Forecasting Comparison of Different Implementations of DMA/DMS

In this section we compare the one-step ahead forecasting performance of our different implementations of DMA and DMS. For the switching Gaussian state space model, we repeatedly run our MCMC algorithm on an expanding window of data to provide forecasts of y_t using information available at time $t-1$ and use this to calculate the one-step ahead predictive density. The output from this procedure can be thought of as a DMA procedure where we are averaging over $s_t = \{1,\ldots,4\}$. We can also obtain the predictive density for the value of s_t with highest posterior probability, a strategy analogous to DMS.

The forgetting factor approach can be used to do either DMA or DMS (see Appendix B) and involves calculating $\pi_{t|t-1,j}$ for $j=1,\dots,K$ and using these probabilities either to select a single model at each point in time or average across forecasts of all models. The main results in this paper make the same forgetting factor choices as in the preceding section of $\alpha=\lambda=0.99$. However, we also present results for $\alpha=\lambda=1$ and $\alpha=\lambda=0.95$. The first of these Raftery et al. (2010) show to be equivalent to BMA run on an expanding window of data for a constant coefficient model. The second of these allows for a large degree of model switching and time variation in parameters. With regards to the error variance, we use both heteroskedastic and homoskedatic estimates (see Appendix B). The former of these are obtained using a rolling window of 10 observations, the latter an expanding window of data.

To evaluate our forecasts we use mean squared forecast errors (MSFEs) and mean absolute forecast errors (MAFEs) which evaluate the quality of the point forecasts. MSFEs for any approach are presented relative to the MSFE of the no change forecast. To evaluate the quality of the entire predictive density, we also present sums of log predictive likelihoods (logPL). We evaluate our forecasts over the period 1996Q1 to the end of the sample.

Several patterns emerge from Table 1. The sum of log predictive likelihoods would be the preferred Bayesian method of forecast comparison and it indicates that the forgetting factor approach forecasts slightly better than the switching linear Gaussian state space model, provided we allow for heteroskedasticity and do not choose forgetting factor values which allow for too much variation in the coefficients or too much model switching. That is, the forgetting factor method using the benchmark $\alpha=\lambda=0.99$ choices used by Raftery et al. (2010) lead to the best forecast performance (although the $\alpha=\lambda=1$ which leads to conventional BMA in a constant coefficient model on an expanding window of data forecasts only slightly worse). Homoskedastic variants of DMA or DMS forecast quite poorly, emphasizing the importance of allowing for heteroskedasticity.

However, if we look at MSFEs and MAFEs, then a different pattern emerges where the switching linear Gaussian state space model forecasts appreciably better than forgetting factor approaches. Thus, the former methods are better at producing point forecasts. However, forgetting factor methods are clearly doing well in getting higher order moments and the entire shape of the predictive density correct.

Table 1. Comparison of Forecasting Performance.

Approach	MSFE	MAFE	logPL
Switching Linear Gaussian State Space Model			
DMA	0.401	0.514	−91.48
DMS	0.431	0.551	−91.81
Forgetting Factor Approaches (Heteroskedastic)			
DMA, $\alpha=\lambda=1$	0.600	0.656	−90.89
DMA, $\alpha=\lambda=0.99$	0.598	0.655	−90.77
DMA, $\alpha=\lambda=0.95$	0.637	0.680	−93.10
DMS, $\alpha=\lambda=1$	0.600	0.657	−90.89
DMS, $\alpha=\lambda=0.99$	0.598	0.655	−90.75
DMS, $\alpha=\lambda=0.95$	0.653	0.682	−92.46
Forgetting Factor Approaches (Homoskedastic)			
DMA, $\alpha=\lambda=1$	0.637	0.674	−110.60
DMA, $\alpha=\lambda=0.99$	0.647	0.673	−109.53
DMA, $\alpha=\lambda=0.95$	0.657	0.698	−105.63
DMS, $\alpha=\lambda=1$	0.637	0.674	−110.60
DMS, $\alpha=\lambda=0.99$	0.619	0.664	−109.52
DMS, $\alpha=\lambda=0.95$	0.682	0.717	−109.73

With regards to issue of whether model averaging or model selection is better, for the switching linear Gaussian state space model, DMA methods are slightly to be preferred. However, with forgetting factor approaches, the forecast performance of DMA and DMS is very similar to one another (and, indeed, is often identical to three decimal places).

In summary, our forecasting results are somewhat mixed. For the researcher interested in point forecasts, the fully Bayesian estimation procedure for switching state space models is to be preferred since it is leading to substantially lower MSFEs and MAFEs. However, for the researcher interested in the entire predictive density, DMA and DMS methods using forgetting factors are forecasting very well indicating that the approximations and compromises inherent in forgetting factor approaches do not carry a large cost with them.

3.5. Forecasting Comparison of Different Implementations of DMA/DMS in a Larger Model Space

In the preceding sections, we used a small model space containing four models. However, the researcher may often wish to do DMA or DMS in larger model spaces. Accordingly, in this section, we investigate the performance of our two ways of doing DMA/DMS in the larger model space containing 15 models. That is, we augment our preceding model space with all TVP regression models containing 2, 3 or 4 explanatory variables. For the sake of brevity, we only present forecast results. However, we note that smoothed and filtered estimates of model probabilities, comparable to Figs. 2 and 3 indicate that consumer surveys and time series econometrics forecasts still dominate (sometimes individually, sometimes in the model containing the two explanatory variables π_t^{CS} and π_t^{TS}).

Table 2 presents sums of log predictive likelihoods, MSFEs and MAFEs for the same set of approaches as in Table 1. The message of Table 2 is clear. DMA or DMS done using forgetting factor methods is yielding virtually the same results in Tables 1 and 2. However, the forecasting performance of approaches based on the switching linear Gaussian state space model have deteriorated substantially.

In this empirical application, where we usually find models with a single explanatory variable to forecast best, the parsimonious forgetting factor approach is successfully choosing these models and ignoring the less parsimonious models with several explanatory variables. However, the switching linear Gaussian state space approach is not. Remember that the latter

Table 2. Comparison of Forecasting Performance in Larger Model Space.

Approach	MSFE	MAFE	logPL
Switching Linear Gaussian State Space Model			
DMA	0.614	0.982	−113.53
DMS	0.613	0.827	−98.64
Forgetting Factor Approaches (Heteroskedastic)			
DMA, $\alpha=\lambda=1$	0.600	0.656	−90.89
DMA, $\alpha=\lambda=0.99$	0.598	0.655	−90.77
DMA, $\alpha=\lambda=0.95$	0.637	0.680	−93.10
DMS, $\alpha=\lambda=1$	0.600	0.656	−90.88
DMS, $\alpha=\lambda=0.99$	0.598	0.655	−90.75
DMS, $\alpha=\lambda=0.95$	0.652	0.682	−92.46
Forgetting Factor Approaches (Homoskedastic)			
DMA, $\alpha=\lambda=1$	0.637	0.674	−110.60
DMA, $\alpha=\lambda=0.99$	0.647	0.673	−109.53
DMA, $\alpha=\lambda=0.95$	0.657	0.698	−105.63
DMS, $\alpha=\lambda=1$	0.637	0.674	−110.60
DMS, $\alpha=\lambda=0.99$	0.619	0.664	−109.52
DMS, $\alpha=\lambda=0.95$	0.682	0.717	−109.73

approach involves estimating a 15×15 matrix of transition probabilities. With our relatively short data set, the switching Gaussian state space model is over-parameterized and this is leading to poor forecasts.

It is worth noting, too, that the computational time associated with forecasting with this larger version of a switching Gaussian state space model is quite substantial (e.g., several hours), whereas the forgetting factor approach has a trivial computational burden (e.g., several seconds).

4. CONCLUSIONS

The Bayesian empirical researcher often faces a trade-off between the desire to work with a fully specified Bayesian model and the computational burden that use of MCMC methods imposes. In the DMA literature, when the researcher works with large model spaces, it is common to use forgetting factor methods because the computational burden of doing MCMC is simply too great. In this paper, we have worked with relatively small model spaces (where MCMC methods are computationally feasible) to investigate the possible consequences of using approximate forgetting factor methods. We set up a fully specified Bayesian approach, using a switching linear Gaussian state model, which allows for model switching or model averaging in time-varying parameter models. This can be thought of as an alternative to doing DMA using forgetting factor methods.

In a small empirical application involving inflation forecasting using four models, our overall conclusion is that forgetting factor methods and the switching Gaussian state space model are leading to similar empirical results. In terms of model selection, both approaches indicate that the consumer survey provides the best forecasts of inflation most of the time. However, both of them find that time series forecasts and surveys of professionals do tend to forecast better at particular periods (e.g., the recent financial crisis and subsequent recession). In terms of forecasting, the two approaches exhibit a similar performance if we use sums of log predictive likelihoods as a metric. However, MSFEs and MAFEs show a deterioration in forecasts for forgetting factor approaches.

When we move to a larger model space of 15 models, the forecasting performance of the switching linear Gaussian state space model deteriorates substantially. This contrasts with the forgetting factor approach where forecast performance is unaffected by the move to a larger model space. Thus, the switching Gaussian state space model can become over-parameterized even with model spaces of this size.

There are many applications (e.g., Koop & Korobilis, 2012, 2013) where the model space is so large that the use of MCMC methods is impossible. The only feasible way of doing DMA or DMS would involve some sort of approximation such as that involved in the forgetting factor approach. Insofar as the results of the present paper (based on relatively small model spaces) extend to large model spaces, they should provide reassurance that forgetting factor methods are providing reasonable results.

NOTES

1. Note that we have written the error variances, $\sigma_\varepsilon^{2(k)}$ and $\Sigma_\eta^{(k)}$, as being constant. In empirical work it may be desirable to have one or both of them to be time-varying. Details of our treatment of this issue will be given below.
2. Where relevant, monthly data has been made into quarterly data by taking the observation for the last month of the quarter. See Ang et al. (2007, p. 1171)
3. The filtered probabilities are calculated by repeatedly running the MCMC algorithm on an expanding window of data.

ACKNOWLEDGMENTS

This research was supported by the ESRC under grant RES-062-23-2646. Gary Koop is a Fellow at the Rimini Centre for Economic Analysis.

REFERENCES

Ang, A., Bekaert, G., & Wei, M. (2007). Do macro variables, asset markets, or surveys forecast inflation better? *Journal of Monetary Economics*, *54*, 1163–1212.

Bates, J., & Granger, C. (1969). Combination of forecasts. *Operational Research Quarterly*, *20*, 451–468.

Billio, M., Casarin, R., Ravazzolo, F., & van Dijk, H. (2011). Combining predictive densities using nonlinear filtering with applications to US economics data. Tinbergen Institute Discussion Paper No 11-172/4.

Chan, J., & Jeliazkov, I. (2009). Efficient simulation and integrated likelihood estimation in state space models. *International Journal of Mathematical Modelling and Numerical Optimisation*, *1*, 101–120.

Chan, J., Koop, G., Leon-Gonzalez, R., & Strachan, R. (2012). Time varying dimension models. *Journal of Business and Economic Statistics*, *30*, 358–367.

Chipman, H., George, E., & McCulloch, R. (2001). The practical implementation of Bayesian model selection. *IMS Lecture Notes – Monograph Series*, *38*, 66–134.

Cogley, T., Morozov, S., & Sargent, T. (2005). Bayesian fan charts for U.K. inflation: Forecasting and sources of uncertainty in an evolving monetary system. *Journal of Economic Dynamics and Control*, *29*, 1893–1925.

Cogley, T., & Sargent, T. (2005). Drifts and volatilities: Monetary policies and outcomes in the post WWII U.S. *Review of Economic Dynamics*, *8*, 262–302.

D'Agostino, A., Gambetti, L., & Giannone, D. (2011). Macroeconomic forecasting and structural change. *Journal of Applied Econometrics*. doi:10.1002/jae.1257

Dangl, T., & Halling, M. (2012). Predictive regressions with time varying coefficients. *Journal of Financial Economics*, *106*, 157–181.

Faust, J., & Wright, J. (2012). Forecasting inflation. In G. Elliot & A. Timmermann (Eds.), *Handbook of economic forecasting* (Part A, Vol. 2, pp. 2–56). Amsterdam: Elsevier.

Fruhwirth-Schnatter, S. (2001a). Fully Bayesian analysis of switching Gaussian state space models. *Annals of the Institute of Statistical Mathematics*, *53*, 31–49.

Fruhwirth-Schnatter, S. (2001b). Markov chain Monte Carlo estimation of classical and dynamic switching and mixture models. *Journal of the American Statistical Association*, *96*, 194–209.

Fruhwirth-Schnatter, S. (2006). *Finite mixture and Markov switching models*. New York, NY: Springer.

Granger, C. (2006). Invited review: Combining forecasts – Twenty years later. *Journal of Forecasting*, *8*, 167–173.

Green, P. (1995). Reversible jump MCMC computation and Bayesian model determination. *Biometrika*, *82*, 711–732.

Guidolin, M., & Timmermann, A. (2009). Forecasts of US short-term interest rates: A flexible forecast combination approach. *Journal of Econometrics*, *150*, 297–311.

Hoeting, J., Madigan, D., Raftery, A., & Volinsky, C. (1999). Bayesian model averaging: A tutorial. *Statistical Science*, *14*, 382–417.

Kim, C. J. (1994). Dynamic linear models with Markov switching. *Journal of Econometrics*, *60*, 1–22.

Kim, C. J., & Nelson, C. (1999). *State-Space models with regime switching: Classical and Gibbs-sampling approaches with applications*. Cambridge, MA: MIT Press.

Koop, G., & Korobilis, D. (2012). Forecasting inflation using dynamic model averaging. *International Economic Review*, *53*, 867–886.

Koop, G., & Korobilis, D. (2013). Large time-varying parameter VARs. *Journal of Econometrics*, *177*, 185–198.

Koop, G., Leon-Gonzalez, R., & Strachan, R. (2009). On the evolution of the monetary policy transmission mechanism. *Journal of Economic Dynamics and Control*, *33*, 997–1017.

Koop, G., & Tole, L. (2013). Forecasting the European carbon market. *Journal of the Royal Statistical Society, Series A*, *176*, 723–741.

Korobilis, D. (2013). Assessing the transmission of monetary policy shocks using dynamic factor models. *Oxford Bulletin of Economics and Statistics*, *75*, 157–179.

McCormick, T., Raftery, A., Madigan, D., & Burd, R. (2012). Dynamic logistic regression and dynamic model averaging for binary classification. *Biometrics*, *68*, 23–30.

Nicoletti, G., & Passaro, R. (2012). *Sometimes it helps: The evolving predictive power of spreads on GDP*. European Central Bank Working Paper No. 1447.

Primiceri., G. (2005). Time varying structural vector autoregressions and monetary policy. *Review of Economic Studies*, *72*, 821–852.

Raftery, A., Karny, M., & Ettler, P. (2010). Online prediction under model uncertainty via dynamic model averaging: Application to a cold rolling mill. *Technometrics*, *52*, 52–66.

Stock, J., & Watson, M. (2010). Modeling inflation after the crisis. Macroeconomic Policy: Post-crisis and risks ahead. *FRB Kansas City Symposium*, August 26–28, Jackson Hole, Wyoming, KS.

APPENDIX A: BAYESIAN INFERENCE IN THE SWITCHING LINEAR GAUSSIAN STATE SPACE MODEL

The switching linear Gaussian state-space model that we adopt is of the form:

$$p(s_1 = k) = \frac{1}{K}$$
$$\xi_{jk} = p(s_t = k | s_{t-1} = j)$$
$$\theta_1 \sim N(0_K, 2I_K)$$
$$\theta_t = \theta_{t-1} + \eta_t$$
$$y_t = Z_t G^{[s_t = k]} \theta_t + \varepsilon_t$$

for $t = 1,\ldots, T$ and $j, k = 1,\ldots, K$. Error assumptions and definitions of $s_t \in \{1,\ldots, K\}, y_t, Z_t$, and $G^{[s_t = k]}$ are given in Section 2. The remaining parameters of the model are $\psi = (\sigma^2_{\eta 1}, \ldots, \sigma^2_{\eta K}, \sigma^{2[1]}_{\varepsilon}, \ldots, \sigma^{2[K]}_{\varepsilon}, \xi_{11}, \ldots, \xi_{KK})'$. We adopt a notational convention for data and states such that subscripts denote a particular time period and superscripts denote all periods up to that time period. For instance, $s^t = (s_1, \ldots, s_t)'$ denotes all regime indicators up to time t.

We use the Gibbs sampler that sequentially draws from $p(\theta^T | y^T, s^T, \psi)$, $p(s^T | y^T, \theta^T, \psi)$, and $p(\psi | y^T, s^T, \theta^T)$. This technical appendix briefly describes each of these conditional posterior densities. The time-varying parameters are drawn from $p(\theta^T | y^T, s^T, \psi)$ using the algorithm of Chan and Jeliazkov (2009). And $p(s^T | y^T, \theta^T, \psi)$ is drawn as in Fruhwirth-Schnatter (2001a, 2001b). We refer the reader to page 420 of Fruhwirth-Schnatter (2006) for specific details of implementation. Note that this algorithm delivers $p(y_t | s^t, \theta^t, \psi)$ which, when averaged over Gibbs draws, provides us with an estimate of the predictive likelihood.

For $p(\psi | y^T, s^T, \theta^T)$ we use conditionally conjugate priors which lead to the following conditional posteriors. Given inverted Gamma priors for $\sigma^2_{\eta k}$ (for $k = 1, \ldots, K$) with prior hyperparameters c_{0k} and C_{0k} we obtain an inverted Gamma posterior with arguments

$$c_k(S) = c_{0k} + \frac{T}{2}, \quad C_k(S) = C_{0k} + \frac{\sum_{t=1}^{T} (\theta_{k,t+1} - \theta_{k,t})^2}{2}$$

We set the prior hyperparameters to $c_{0k} = 5$ and $C_0 = [0.08, 0.148, 0.45, 0.53]$.

For $\sigma_\varepsilon^{2[k]}$ we also use inverted Gamma priors leading to inverted Gamma conditional posteriors. We set prior hyperparameters $c_{0\varepsilon}^{[k]}$ and $C_{0\varepsilon}^{[k]}$ to $c_{0\varepsilon}^{[k]} = 5$ and $C_{0\varepsilon}^{[k]} = [0.168, 1.480, 7.2, 10.0]$, for $k = 1, .., K$. The resulting posterior has arguments

$$c_\varepsilon^{[k]}(S) = c_{0\varepsilon}^{[k]} + \frac{N_{kk}}{2}, \quad C_\varepsilon^{[k]}(S) = C_{0\varepsilon}^{[k]} + \frac{1}{2}\sum\nolimits_{t:s_t=k}^{T} (y_t - Z_t G^{[s_t=k]}\theta_t)^2$$

where N_{jk} counts the number of transitions from j to k. If $j = k$ it counts the number of periods spent in regime k.

Finally, let ξ be the matrix of Markov transition probabilities ξ_{jk} and let ξ_j be the jth row of this matrix. The conditional conjugate prior for each row is Dirichlet

$$\xi_j \sim \mathscr{D}(e_{j1}, \ldots, e_{jK}), j = 1, \ldots, K$$

Following Fruhwirth-Schnatter (2001b), we adopt a prior reflecting a belief that the probability of staying in a regime is greater than the probability of transition to a new regime. Thus, we set $e_{jj}, j = 1, \ldots, K$ corresponding to the main diagonal to be 4. Hyperparameters off the main diagonal, $e_{ji}, j \neq i$, are set to 0.33. With this choice of prior, the conditional posterior is also Dirichlet with

$$\mathscr{D}(e_{j1} + N_{j1}, \ldots, e_{jK} + N_{jK}), \quad j = 1, \ldots, K$$

APPENDIX B: DYNAMIC MODEL AVERAGING USING FORGETTING FACTORS

This appendix briefly outlines the main features of the DMA algorithm of Raftery et al. (2010) as we implement it in this paper.

Suppose we have $j=1,\ldots,K$ TVP regression models,

$$y_t = Z_t^{(j)}\theta_t^{(j)} + \varepsilon_t^{(j)}$$

$$\theta_{t+1}^{(j)} = \theta_t^{(j)} + \eta_t^{(j)}$$

$\varepsilon_t^{(k)}$ is $N(0,H_t^{(j)})$ and $\eta_t^{(k)}$ is $N(0,Q_t^{(j)})$. We replace $H_t^{(j)}$ by a simple estimate (the sum of squared errors divided by sample size).

If $Q_t^{(j)}$ were known, then an individual model could be estimated in a straightforward manner using the Kalman filter. $Q_t^{(j)}$ appears in the Kalman filtering prediction equation. In particular, if $y^t = (y_1,\ldots,y_t)'$, then

$$\theta_t^{(j)}|y^{t-1} \sim N(\beta_{t|t-1}, V_{t|t-1})$$

where

$$V_{t|t-1} = V_{t-1|t-1} + Q_t^{(j)}$$

This is the only place where Q_t enters the Kalman filtering formulae. If the equation for $V_{t|t-1}$ is replaced by

$$V_{t|t-1} = \frac{1}{\lambda}V_{t-1|t-1}$$

then MCMC methods can be avoided. λ is a forgetting factor. Forgetting factors have long been used in the state space literature to simplify estimation. There are many ways of justifying the use of forgetting factors as leading to sensible approximations. For instance, their use in this context implies that observations j periods in the past have weight λ^j. An alternative way of interpreting λ is to note that it implies an effective window size of $\frac{1}{1-\lambda}$.

The contribution of Raftery et al. (2010) was to develop a filtering algorithm which also used a forgetting factor and allowed for DMA or DMS to be done. These involve calculating $\pi_{t|t-1,j}$ which is the probability that model j should be used for forecasting at time t, given information through time $t-1$. DMA arises if we average forecasts over all models using $\pi_{t|t-1,j}$ as weights. DMS arises if we choose the model with the highest value for $\pi_{t|t-1,j}$. A recursive algorithm involving $\pi_{t|t,j}$ and $\pi_{t|t-1,j}$ can be run, beginning with $\pi_{0|0,j}$ in order to provide the necessary model probabilities. Note that $\pi_{t|t-1,j}$ is a similar concept to the $\Pr(s_t = k|y^{t-1})$ used in the switching state space model and can be used in the same manner.

Raftery et al. (2010) derive a model updating equation of

$$\pi_{t|t,j} = \frac{\pi_{t|t-1,j} p_j(y_t|y^{t-1})}{\sum_{l=1}^{J} \pi_{t|t-1,l} p_l(y_t|y^{t-1})}$$

where $p_j(y_t|y^{t-1})$ is the predictive likelihood for model j produced by the Kalman filter. However, instead of using a Markov transition matrix to model the probability of switching between models, a model prediction equation involving a forgetting factor α is used

$$\pi_{t|t-1,j} = \frac{\pi^{\alpha}_{t-1|t-1,j}}{\sum_{l=1}^{J} \pi^{\alpha}_{t-1|t-1,l}}$$

This algorithm has a large advantage in that no MCMC is required and a complete specification of a Markov transition matrix is not required. It is computationally efficient, involving only the filtering algorithms just described. Its properties are described in more detail in Raftery et al. (2010).

This algorithm requires selection of α, λ, $\pi_{0|0,j}$ and initial conditions for the time-varying parameters. For the last, we use the same values as for the switching state space models. For $\pi_{0|0,j}$ we use the noninformative choice of $\pi_{0|0,j} = \frac{1}{K}$. The main results in the paper set $\alpha = \lambda = 0.99$, although (as noted in the body of the paper) in our forecast comparison exercise we experiment with different values. These forgetting factors are not directly comparable to the parameters in the switching state space models. But, loosely speaking, forgetting factors and priors in switching state space models are both chosen to allow for a moderate degree of parameter and model change.

ASSESSING BAYESIAN MODEL COMPARISON IN SMALL SAMPLES

Enrique Martínez-García and Mark A. Wynne
Research Department, Federal Reserve Bank of Dallas, Dallas, TX, USA

ABSTRACT

We investigate the Bayesian approach to model comparison within a two-country framework with nominal rigidities using the workhorse New Keynesian open-economy model of Martínez-García and Wynne (2010). We discuss the trade-offs that monetary policy – characterized by a Taylor-type rule – faces in an interconnected world, with perfectly flexible exchange rates. We then use posterior model probabilities to evaluate the weight of evidence in support of such a model when estimated against more parsimonious specifications that either abstract from monetary frictions or assume autarky by means of controlled experiments that employ simulated data. We argue that Bayesian model comparison with posterior odds is sensitive to sample size and the choice of observable variables for estimation. We show that posterior model probabilities strongly penalize overfitting, which can lead us to favor a less parameterized model against the true data-generating process when the two become arbitrarily close to each other. We also illustrate that the spillovers from monetary policy across countries have an added confounding effect.

Bayesian Model Comparison
Advances in Econometrics, Volume 34, 71–115

ISSN: 0731-9053/doi:10.1108/S0731-905320140000034006

Keywords: Bayesian methods; posterior model probabilities; Schwarz criterion; new open economy macro; global slack hypothesis

JEL classifications: C11; C13; F41

1. INTRODUCTION

Bayesian methods have become a standard part of the toolkit in quantitative macroeconomics. They are commonly used to estimate the parameters and assess the fit of a given model, but they are also widely employed for comparison across competing models. We can think of a model as a parameterized probability distribution (based on a given theory of how the economy works) that characterizes the data-generating process (DGP) from which the observables that constitute our data are drawn. Hence, by model comparison we mean the evaluation of $k \geq 2$ competing parameterized probability distributions – the models $M_1,..., M_k$ – representing different theories to account for the observed empirical distribution of the data. In other words, model comparison provides guidance on which of the existing theories better accounts for the observed data.

Model selection is a related decision-theory problem that specifies a loss function as a metric to judge the differences across models against the data and pick among competing theories. It is known that under a 0–1 loss function it is optimal to select the model with the highest posterior probability (see, e.g., Kass & Raftery, 1995). Model averaging is another related notion that incorporates model uncertainty by averaging across all possible k models using the weights to reflect how likely each model is given the observed data (see, e.g., Hoeting, Madigan, Raftery, & Volinsky, 1999). Selecting the incorrect model or assigning too large a probability, though, can result in misleading inferences and even in the implementation of suboptimal policies meant to correct for the effect of frictions or economic distortions that may not even be present in the "true" DGP underlying the data. So, this begs the question, when is Bayesian model comparison more prone to fail to detect the true DGP (or its closest match among the available models)?

The Bayesian approach to model comparison consists in placing probabilities on a number of competing models and evaluating the posterior probability of each model (see, e.g., An & Schorfheide, 2007;

Kass & Raftery, 1995). The significance of posterior model probabilities for making comparison across competing models is largely based on the desirable asymptotic properties of these posterior probabilities derived under fairly general regularity conditions. Fernández-Villaverde and Rubio-Ramírez (2004) show that, as the sample size grows arbitrarily large, the Bayesian parameter point estimates converge to their pseudo-true values. They also show that the best model under the Kullback–Leibler distance criterion – the model closest to the "true" DGP in the Kullback–Leibler sense – is the one with the highest posterior model probability. Moreover, these asymptotic properties hold even if the models being compared are non-nested, non-linear, and do not even include a model for the "true" DGP.

In this paper, we illustrate the *less-desirable* small sample properties of Bayesian posterior model probabilities. We work with simulated data in controlled experiments and make our case using a standard log-linearized two-country New Open-Economy Macro (NOEM) model with nominal rigidities as the "true" DGP. We compare the NOEM model against three alternative (log-linear) specifications that either assume flexible prices (instead of nominal rigidities), posit a closed-economy setting for each country (instead of an open-economy one), or both. All three competing models are nested in the NOEM model and the dimensionality of their parameter space is lower. We consider these three alternative specifications because they evoke important concerns for policymaking – such as the role of globalization (openness to trade) and monetary policy in the presence/absence of nominal rigidities.

We design a number of experiments to illustrate how model comparison depends not only on the length of the time series used for estimation but also on the selection of the observable macro variables on which the compared models are estimated. We show that in small samples the Bayesian posterior model probabilities are more likely to favor a more parsimonious specification over the NOEM model (our "true" DGP) when the simulated data are generated under a parameterization that brings the probability distribution of the DGP *close* to that of some of the alternative model specifications (theories) under consideration.

In our particular illustrations, that means posterior model probabilities can favor a closed-economy model whenever the degree of trade openness is *low enough* or can favor a model that abstracts from nominal rigidities whenever monetary policy is *near-optimal* and/or the degree of price stickiness is *low*. More generally, our work suggests that model comparison, model selection, and model averaging can be distorted in economically relevant ways whenever model comparison strongly penalizes the more richly parameterized models. Furthermore, what the preferred model ends up being is not straightforward because their implied probability distributions tend to be nonlinear in the parameters – and, accordingly, there may be more than one model that appears empirically *close* to the "true" one.

The remainder of the paper proceeds as follows: Section 2 outlines the workhorse model of Martínez-García and Wynne (2010), and describes its building blocks. Several alternative nested specifications are proposed for model comparison whereby monetary policy effectiveness changes by removing features such as household's preference for imported varieties or rigidities in firm price-setting behavior. In Section 3 we illustrate our findings showing that in small samples posterior model probabilities may fail to pick the more-heavily parameterized NOEM model against the alternative nested specifications, even though the NOEM model is the "true" DGP for the data. These confounding results also appear when we try an alternative selection of observables. In Section 4 we discuss our findings, make recommendations for applied work with these techniques, and draw policy implications for the class of open-economy models that we investigate. Section 5 provides a brief summary of the technical insights gained from our exercise and its policy implications, and concludes. We also provide a companion online appendix for the interested reader where further detail on the model and the implementation strategy is given (see Martínez-García & Wynne, 2014).

2. ECONOMIC MODEL

We adopt the model of Martínez-García and Wynne (2010). This is a two-country, symmetric NOEM model with complete asset markets and nominal rigidities in the spirit of Clarida, Galí, and Gertler (2002), subject to country-specific productivity and monetary shocks. The stylized model abstracts from capital accumulation, and assumes a cashless economy and perfectly flexible exchange rates. Labor is immobile across countries, but all varieties of goods produced in each country can be traded. The model provides a tractable economic environment that departs from monetary neutrality and allows international spillovers through trade.

The model features two standard distortions in the goods markets – monopolistic competition in production and constrained price-setting behavior subject to Calvo (1983) contracts and producer currency pricing (as in Clarida et al., 2002):

- The introduction of an optimal labor subsidy for firms funded with lump-sum (non-distortionary) taxes eliminates the mark-up distortion caused by monopolistic competition. It also ensures that the deterministic steady state of the model is the same under either flexible prices or nominal rigidities. Hence, the key assumption on which the non-neutrality of monetary policy hinges is price stickiness modelled à la Calvo (1983).
- The law of one price holds at the variety level because all prices are set in the producer's own currency. Deviations from purchasing-power parity (PPP) arise solely due to differences in preferences that result in the composition of the consumption basket varying across countries, with local households consuming a larger share of the locally produced varieties than of the imported ones (home bias). The degree of openness to trade that allows for the endogenous propagation of country-specific shocks internationally is also directly tied to the appetite of households for imported goods.[1]

Because the setup of the model we use is otherwise extensively discussed in Martínez-García and Wynne (2010), here we shall put the emphasis on the key equations of its log-linearized representation and their economic interpretation. The companion online appendix (i.e., Martínez-García & Wynne, 2014) provides further details on the building blocks of the model as well as on our approach to data simulation, Bayesian estimation, and Bayesian model comparison.

The workhorse (log-linearized) model.[2] The basic structure of the New Keynesian model is given by a log-linearized system of three-equations – which includes a Phillips curve, an IS curve, and an interest rate-based monetary policy rule – that characterize the dynamics of output, inflation, and the short-term nominal interest rate. Goodfriend and King (1997), Clarida, Galí, and Gertler (1999), and Woodford (2003) among others contributed to the derivation of those equations from explicit optimizing behavior on the part of firms (price-setters) and households in the presence of nominal rigidities.

Clarida et al. (2002) extends the three-equation workhorse New Keynesian model to a two-country setting. Building on that contribution, Martínez-García and Wynne (2010) show that the same basic structure of three log-linearized equations can be generalized to describe the dynamics of output, inflation, and the short-term interest rate when a country is open to trade. The monetary policy rule remains focused on domestic objectives even in the open-economy model presented by Martínez-García and Wynne (2010) – but both the Phillips curve and the IS curve differ from their closed-economy counterparts due to the interactions across countries that take place through trade and the resulting spillovers into inflation and aggregate demand. The model of Martínez-García and Wynne (2010) showcases the interconnectedness that arises through trade in goods, while keeping most of the simplicity and tractability of the workhorse (closed-economy) New Keynesian model.

In the framework of Martínez-García and Wynne (2010), the open-economy Phillips curve can be written for each country as follows:

$$\hat{\pi}_t \approx \beta \mathbb{E}_t(\hat{\pi}_{t+1}) + \ldots$$

$$\Phi\left[(1-\xi)\left(\varphi+\left(\frac{\sigma-(\sigma-1)(1-2\xi)}{\sigma-(\sigma-1)(1-2\xi)^2}\right)\right)\hat{x}_t+\xi\left(\varphi+\left(\frac{\sigma+(\sigma-1)(1-2\xi)}{\sigma-(\sigma-1)(1-2\xi)^2}\right)\right)\hat{x}_t^*\right] \tag{1}$$

$$\hat{\pi}_t^* \approx \beta \mathbb{E}_t(\hat{\pi}_{t+1}^*) + \ldots$$

$$\Phi\left[\xi\left(\varphi+\left(\frac{\sigma+(\sigma-1)(1-2\xi)}{\sigma-(\sigma-1)(1-2\xi)^2}\right)\right)\hat{x}_t+(1-\xi)\left(\varphi+\left(\frac{\sigma-(\sigma-1)(1-2\xi)}{\sigma-(\sigma-1)(1-2\xi)^2}\right)\right)\hat{x}_t^*\right] \tag{2}$$

where $\hat{\pi}_t$ and $\hat{\pi}_t^*$ denote Home and Foreign inflation (i.e., quarter-over-quarter changes in the consumption price index), and $\hat{x}_t$ and $\hat{x}_t^*$ define the Home and Foreign output gaps or slack (i.e., the deviations of output from its potential under flexible prices). The composite coefficient $\Phi \equiv \left(\frac{(1-\alpha)(1-\beta\alpha)}{\alpha}\right)$ is the common term on the slope of the open-economy Phillips curve, $0<\beta<1$ is the subjective intertemporal discount factor, and $0<\alpha<1$ is the Calvo price stickiness parameter. The differences in slope coefficients for domestic and foreign slack that arise in Eqs. (1)–(2) are related to the inverse of the Frisch elasticity of labor supply $\varphi>0$, the elasticity of intratemporal substitution between Home and Foreign goods $\sigma > 0$, and the share of imported goods in the consumption basket $0 \leq \xi \leq \frac{1}{2}$.[3]

Price stickiness breaks monetary policy neutrality in the short run, establishing a Phillips curve relationship between nominal (inflation) and real variables (slack). The assumption that household preferences for consumption goods are defined over imported as well as domestic varieties is what gives rise to the *global slack hypothesis* in this framework – i.e., to the idea that in a world open to trade the relevant trade-off for monetary policy captured by the Phillips curve is between a country's inflation and global (rather than local) slack. Not surprisingly, the structural parameters α and ξ feature prominently among the structural parameters that determine the slope of the open-economy Phillips curve in Eqs. (1)–(2). These parameters characterize respectively the fraction of firms that cannot update their prices in any given period (price stickiness) and the import shares (openness), although the role each plays in the dynamics of the model is different.

- The parameter α enters through the common term for the slope Φ. This structural parameter captures the degree of price stickiness, and price stickiness is the key distortion that introduces monetary non-neutrality. Under flexible prices (absent nominal rigidities), monetary policy has no real effects. Therefore, the real effects of monetary policy in the model tend to be negligible as α becomes arbitrarily close to zero – since a larger fraction of firms becomes unconstrained to change prices every period.
- The parameter ξ appears in the composite terms that differentiate the slope for domestic and foreign slack. This structural parameter determines the import share (the extent of trade openness) and explains deviations from in the model. In a closed-economy setting there is no endogenous mechanism for the international transmission of shocks. Even if trade was permitted in this model, an analogous situation would arise with no endogenous international propagation of shocks if all imports were excluded from the consumption basket – i.e., when $\xi = 0$.[4] Therefore, international propagation tends to be attenuated as the import share ξ becomes arbitrarily close to zero.

The open-economy IS Eqs. (3)–(4) illustrate how the output gaps, $\hat{x}_t$ and $\hat{x}_t^*$, are tied to shifts in consumption demand over time and across countries,

$$(1-2\xi)\mathbb{E}_t[\hat{x}_{t+1}-\hat{x}_t] \approx (1-\xi)(\sigma-(\sigma-1)(1-2\xi))[\hat{r}_t-\hat{\bar{r}}_t] - \ldots$$
$$\xi(\sigma+(\sigma-1)(1-2\xi))[\hat{r}_t^*-\hat{\bar{r}}_t^*] \tag{3}$$

$$(1-2\xi)\mathbb{E}_t[\hat{x}_{t+1}^*-\hat{x}_t^*] \approx -\xi(\sigma+(\sigma-1)(1-2\xi))[\hat{r}_t-\hat{\bar{r}}_t] + \ldots$$
$$(1-\xi)(\sigma-(\sigma-1)(1-2\xi))[\hat{r}_t^*-\hat{\bar{r}}_t^*] \tag{4}$$

where the real interest rates in the Home and Foreign country are defined by the Fisher equation as $\hat{r}_t \equiv \hat{i}_t - \mathbb{E}_t[\hat{\pi}_{t+1}]$ and $\hat{r}_t^* \equiv \hat{i}_t^* - \mathbb{E}_t[\hat{\pi}_{t+1}^*]$ respectively, and $\hat{i}_t$ and $\hat{i}_t^*$ are the Home and Foreign short-term nominal interest rates. The natural real rates that would prevail under flexible prices are denoted $\hat{\bar{r}}_t$ for the Home country and $\hat{\bar{r}}_t^*$ for the Foreign country. Price stickiness introduces in the IS equations a wedge between the real interest rate (the opportunity cost of consumption today vs. consumption tomorrow) and the natural real rate of interest that captures its distortionary effects on aggregate demand as shown in Eqs. (3)–(4). However, the Calvo parameter α, which determines the degree of nominal rigidities present, does not appear explicitly in the equations. In turn, the demand for imported goods ξ plays a prominent role in the open-economy IS equations because it affects the contributions of the demand distortions arising in the local and export markets to the output gap of each given country.

The Home and Foreign Taylor (1993)-type monetary policy rules complete the specification of the NOEM model. Monetary policy pursues the goal of domestic stabilization (even in a fully integrated world) and, hence, solely responds to changes in the local economic conditions as determined by each country's inflation and output gap. As is commonly done in the literature, we assume *intrinsic* or endogenous inertia in the policy rules described in Eqs. (5)–(6) resulting from policymakers intentionally smoothing out their policy response to changing economic conditions,

$$\hat{i}_t \approx \rho_i \hat{i}_{t-1} + (1-\rho_i)[(1+\psi_\pi)\hat{\pi}_t + \psi_x \hat{x}_t] + \hat{\varepsilon}_t^m \tag{5}$$

$$\hat{i}_t^* \approx \rho_i \hat{i}_{t-1}^* + (1-\rho_i)[(1+\psi_\pi)\hat{\pi}_t^* + \psi_x \hat{x}_t^*] + \hat{\varepsilon}_t^{m*} \tag{6}$$

where $\hat{\varepsilon}_t^m$ and $\hat{\varepsilon}_t^{m*}$ are the Home and Foreign monetary policy shocks modeled with a bivariate normal distribution with zero mean and positive covariance across countries. The policy parameters $\psi_\pi > 0$ and $\psi_x > 0$ represent the sensitivity of the monetary policy rule to movements in inflation and the output gap respectively, while $0 \leq \rho_i < 1$ represents the policy smoothing parameter.

The natural rates $\hat{\bar{r}}_t$ and $\hat{\bar{r}}_t^*$ can be expressed as functions of expected changes in Home and Foreign potential output, i.e.,

$$\hat{\bar{r}}_t \approx (1-\xi)\left(\frac{\sigma-(\sigma-1)(1-2\xi)}{\sigma-(\sigma-1)(1-2\xi)^2}\right)\left(\mathbb{E}_t\left[\hat{\bar{y}}_{t+1}\right]-\hat{\bar{y}}_t\right)+\ldots$$

$$\xi\left(\frac{\sigma+(\sigma-1)(1-2\xi)}{\sigma-(\sigma-1)(1-2\xi)^2}\right)\left(\mathbb{E}_t\left[\hat{\bar{y}}_{t+1}^*\right]-\hat{\bar{y}}_t^*\right) \tag{7}$$

$$\hat{\bar{r}}_t^* \approx \xi\left(\frac{\sigma+(\sigma-1)(1-2\xi)}{\sigma-(\sigma-1)(1-2\xi)^2}\right)\left(\mathbb{E}_t\left[\hat{\bar{y}}_{t+1}\right]-\hat{\bar{y}}_t\right)+\ldots$$

$$(1-\xi)\left(\frac{\sigma-(\sigma-1)(1-2\xi)}{\sigma-(\sigma-1)(1-2\xi)^2}\right)\left(\mathbb{E}_t\left[\hat{\bar{y}}_{t+1}^*\right]-\hat{\bar{y}}_t^*\right) \tag{8}$$

reflecting the fact that natural rates respond to expected changes in – rather than the level of – real economic activity as measured by potential output. Potential output refers to the output that would have been produced under flexible prices, and accordingly $\hat{\bar{y}}_t$ and $\hat{\bar{y}}_t^*$ denote the corresponding Home and Foreign potential output in the model. Home and Foreign potential output can be expressed solely in terms of real shocks since monetary shocks have no real effects absent nominal rigidities, i.e.,

$$\hat{\bar{y}}_t \approx \left(1+(\sigma-1)\left(\frac{2\xi(1-\xi)}{\varphi(\sigma-(\sigma-1)(1-2\xi)^2)+1}\right)\right)\hat{a}_t-\ldots$$

$$(\sigma-1)\left(\frac{2\xi(1-\xi)}{\varphi(\sigma-(\sigma-1)(1-2\xi)^2)+1}\right)\hat{a}_t^* \tag{9}$$

$$\hat{\bar{y}}_t^* \approx -(\sigma-1)\left(\frac{2\xi(1-\xi)}{\varphi(\sigma-(\sigma-1)(1-2\xi)^2)+1}\right)\hat{a}_t + \ldots$$

$$\left(1+(\sigma-1)\left(\frac{2\xi(1-\xi)}{\varphi(\sigma-(\sigma-1)(1-2\xi)^2)+1}\right)\right)\hat{a}_t^* \quad (10)$$

where $\hat{a}_t$ and $\hat{a}_t^*$ denote the corresponding Home and Foreign productivity shocks in the model.

The natural rates of interest and potential output are invariant to monetary policy or monetary policy shocks. In turn, the natural rates only depend on productivity shocks that are modeled as a VAR(1) without spillovers but with positive covariance in their innovations across countries. Natural rates and potential output summarize the dynamics of a competing, nested model that abstracts from nominal rigidities – in effect, a stylized International Real Business Cycle (IRBC) model without capital accumulation. The model presented here also nests another competing class of models which effectively assume a closed economy whenever we set the import share ξ equal to zero. We include all those nested variants in our Bayesian model comparison exercise.

Moreover, $\hat{y}_t = \hat{\bar{y}}_t + \hat{x}_t$ and $\hat{y}_t^* = \hat{\bar{y}}_t^* + \hat{x}_t^*$ are respectively the actual Home and Foreign output variables. Domestic terms of trade (defined as the price of imports relative to the price of exports) is proportional to the output differential across countries, $\widehat{tot}_t \approx \left(\frac{1}{\sigma-(\sigma-1)(1-2\xi)^2}\right)(\hat{y}_t - \hat{y}_t^*)$, capturing the relative scarcity of Home- versus Foreign-produced goods. The domestic trade balance $\widehat{tb}_t \equiv \hat{y}_t - \hat{c}_t \approx \xi\left(\frac{\sigma+(\sigma-1)(1-2\xi)}{\sigma-(\sigma-1)(1-2\xi)^2}\right)(\hat{y}_t - \hat{y}_t^*)$ is proportional to the output differential across countries illustrating the net movement in goods that takes place across borders whenever relative scarcity of Home- versus Foreign-produced goods arises in order to intratemporally smooth consumption.[5] For further details on the trade features of this class of open-economy New Keynesian models the interested reader is referred to Martínez-García and Søndergaard (2009).

Model Solution. We can replace Eqs. (7)–(10) into Eqs. (1)–(6) to express the system of equations that characterizes the model as follows:

$$M\hat{Z}_t = N\mathbb{E}_t[\hat{Z}_{t+1}] + Q\hat{\varepsilon}_t \quad (11)$$

where

$$\hat{Z}_t = (\hat{\pi}_t, \hat{\pi}_t^*, \hat{y}_t, \hat{y}_t^*, \hat{i}_{t-1}, \hat{i}_{t-1}^*, \hat{a}_{t-1}, \hat{a}_{t-1}^*)'$$

$$\hat{\varepsilon}_t = (\hat{\varepsilon}_t^a, \hat{\varepsilon}_t^{a*}, \hat{\varepsilon}_t^m, \hat{\varepsilon}_t^{m*})'$$

and M, N, and Q are conforming matrices. For reasonable parameter values the matrix M is invertible, and (11) can be rewritten as

$$\hat{Z}_t = \Gamma \mathbb{E}_t[\hat{Z}_{t+1}] + \Psi \hat{\varepsilon}_t \tag{12}$$

where $\Gamma = M^{-1}N$ and $\Psi = M^{-1}Q$. Blanchard and Kahn (1980) provide conditions under which a unique stable solution exists for the system of equations in (12). Although it is not easy to derive analytically the parameter restrictions that guarantee existence and uniqueness, numerical experiments show that the policy parameter ψ_π is key and that the lower bound on ψ_π above which the model attains determinacy depends on the policy parameter ψ_x. In an open-economy model with interest rate smoothing in the monetary policy rule, the Taylor principle (i.e., $\psi_\pi > 1$) remains broadly consistent with satisfying the Blanchard–Kahn conditions for determinacy for a wide range of plausible values of the structural parameters of the model. We parameterize the model for simulation to ensure existence and uniqueness of the solution, and we accordingly set the range of priors for estimation to avoid as best as possible the regions of the parameter space that result in indeterminacy or no-solution.

We partition $\hat{Z}_t$ into two blocks with $\hat{Z}_{1t} = (\hat{\pi}_t, \hat{\pi}_t^*, \hat{y}_t, \hat{y}_t^*)'$ and $\hat{Z}_{2t} = (\hat{i}_{t-1}, \hat{i}_{t-1}^*, \hat{a}_{t-1}, \hat{a}_{t-1}^*)'$. Assuming the Blanchard–Kahn conditions are indeed satisfied and imposing $\lim_{J \to +\infty} \Gamma^J \mathbb{E}_t[\hat{Z}_{1t+J}] = 0$, we solve the system of equations in (12) to characterize the solution of the NOEM model in state-space form as follows:

$$\hat{Z}_{2t} = A_1(\theta)\hat{Z}_{2t-1} + B_1(\theta)\hat{\varepsilon}_t \tag{13}$$

$$\hat{Z}_{1t} = C_1(\theta)\hat{Z}_{2t} + D_1(\theta)\hat{\varepsilon}_t \tag{14}$$

where $A_1(\theta)$, $B_1(\theta)$, $C_1(\theta)$, and $D_1(\theta)$ are conforming matrices, and θ is the vector of structural parameters of the model that enter into those matrices. Fernández-Villaverde, Rubio-Ramírez, Sargent, and Watson (2007) explore the link between Dynamic Stochastic General Equilibrium (DSGE) models and state-space representations like this one. The solution in Eqs. (13)–(14) shows that inflation and output in both countries, $\hat{Z}_{1t}$, can be characterized as linear functions of a vector of state variables, $\hat{Z}_{2t}$, and structural shock innovations, $\hat{\varepsilon}_t$. Since the vector of structural shock innovations, $\hat{\varepsilon}_t$, is normally distributed, the Gaussian state-space representation of the solution in Eqs. (13)–(14) implies that inflation and output are also normally distributed processes (see Hamilton, 1994 for further discussion on the Gaussian state-space model).

Model Simulation. We use the same benchmark parameterization of the model described in Martínez-García and Wynne (2010) with only a small modification: in our exercise, we assume log-utility on consumption and accordingly set the elasticity of intertemporal substitution to one. We explore the sensitivity of standard Bayesian model comparison with respect to the value of the parameters ξ and ψ_π replacing, in each case, the parameterization used in Martínez-García and Wynne (2010) with points along an interval that spans the region of interest of the parameter space. We provide further details on the choice of parameter values and intervals in the companion online appendix (i.e , Martínez-García & Wynne, 2014).

These two parameters – one structural ξ, one policy ψ_π – are crucial in the model for different reasons. The structural parameter ξ defines how *close* the countries are to autarky; and, as we have indicated before, it plays a significant role in the specification of the open-economy Phillips curves and IS equations. The structural parameter α indicates the degree of nominal rigidity and this friction is the reason why monetary policy has real effects in the model. The parameter α directly affects the overall slope of the open-economy Phillips curve in a similar way as in the closed-economy case. However, we recognize that the distortion that arises is conditional not only on the structure of the economy (for instance, the degree of integration through trade ξ or the size of the nominal rigidity α) but – most importantly – on monetary policy. Since the policy parameter ψ_π determines the tolerance for inflation of the domestic policymakers in this environment, it has a direct influence on how much slack accumulates – measured by the output gap. It also affects how *close* monetary policy is to attaining the optimal allocation under flexible prices. We choose to focus primarily on this policy parameter here rather than directly on α.

We use the log-linear approximation of the workhorse model of Martínez-García and Wynne (2010) – henceforth, the NOEM model – as our DGP and simulate data at each point of the relevant interval of the parameter space for each of the two parameters under consideration. We keep in all cases the realization of the shocks invariant and all other structural parameters unchanged at their benchmark values. We simulate the full model over 11,000 periods and drop the first 1,000 observations of each series to exclude any effect of the initial conditions on the simulation. We also select three subsamples of 160 observations each, which correspond to 40 years of quarterly observations – a plausible upper-bound length for many time series of international macro data which often can be much shorter than that. The simulation is implemented with code written for Dynare (see, e.g., Adjemian et al., 2011). Working with simulated, rather than actual, data allows for a more precise assessment of the Bayesian posterior model probabilities and their sensitivity to implementation since we always know the true DGP.

Model Estimation. The 10,000-period simulated sample allows us to illustrate the asymptotic behavior of the posterior model probability, while the simulated subsamples of 160 observations illustrate the small sample inference problems that could arise in the data. Bayesian estimation and model comparison is implemented with the Dynare software too. We assume a uniform prior over all competing models: the NOEM model (the true DGP, M_1), a variant with flexible prices and openness to trade (M_2), a variant with nominal rigidities and autarky derived under the assumption $\xi = 0$ (M_3), and a variant with flexible prices and $\xi = 0$ (M_4).

The system of equations that characterizes M_1 and the variants M_2, M_3, and M_4 as special cases of the specification for the NOEM model (M_1) can be found in the companion online appendix (i.e., Martínez-García & Wynne, 2014). The solution for each model variant $k = 1, 2, 3, 4$ fits into the Gaussian state-space representation form given in Eqs. (13)–(14) where $A_k(\theta)$, $B_k(\theta)$, $C_k(\theta)$, and $D_k(\theta)$ are the corresponding conforming matrices for each. The set of structural parameters θ is common to all models, but not all of the parameters affect the dynamics in each of the k specifications and this is reflected in the matrices $A_k(\theta)$, $B_k(\theta)$, $C_k(\theta)$, and $D_k(\theta)$ accordingly. We compute the marginal density of each model with a Laplace approximation after estimating these four nested variants of the model including the true DGP (M_1). The Laplace approximation works rather well in practice, in particular for highly peaked, unimodal posterior densities; and, we view this approximation as reasonable for our illustrations too.

As is conventionally done in the Bayesian literature, rather than imposing "non-informative" priors on the structural parameters, we choose fairly "informative" priors to incorporate other sources of information and to reflect current views on the structural parameters themselves. The prior mean is set to match the true parameter value of the DGP used to simulate the data (which corresponds with the parameterization indicated above). For the parameters of interest ξ and ψ_π (and also for α), the mean of the prior is set to vary along the interval that we evaluate. The shape and the dispersion of the prior distributions are fixed in all our experiments.[6] The same priors for the parameters are used in the estimation of

Table 1. Prior Distributions.

Structural Parameters	Prior Density	Domain	Prior Mean	Prior Std. Def
Non-policy parameters				
β	Fixed	–	0.99	–
φ	Gamma	$\mathbb{R}^+$	2	2
σ	Gamma	$\mathbb{R}^+$	1.5	1
ξ	Beta	(0, 0.5)	0.06, range: (0, 0.5)	0.01
α	Beta	(0, 1)	0.75, range: (0, 1)	0.07
Policy parameters				
ρ_i	Beta	(0, 1)	0.78	0.1
ψ_π	InvGamma	$\mathbb{R}^+$	0.33, range: (0, 6)	2
ψ_x	InvGamma	$\mathbb{R}^+$	1.29	2
Shock parameters				
δ_a	Beta	(0,1)	0.95	0.05
σ_a	InvGamma	$\mathbb{R}^+$	0.7	2
ρ_{a,a^*}	Beta	(0, 1)	0.25	0.18
σ_m	InvGamma	$\mathbb{R}^+$	0.38	2
ρ_{m,m^*}	Beta	(0, 1)	0.5	0.22

Note: The table reports only the prior mean and prior standard deviation for each structural parameter. For any plausible choice of these two moments of the prior, there is a mapping onto the prior distribution parameters v and s that matches both of them and fully characterizes the prior distribution itself. For the Normal distribution, the mean is $\mu = v$ and the variance is $\sigma^2 = s^2$. For the Beta distribution, the mean is $\mu = v/(v+s)$ and the variance is $\sigma^2 = vs/((v+s)^2(v+s+1))$. For the Gamma distribution, the mean is $\mu = vs$ and the variance is $\sigma^2 = vs^2$. For the Uniform distribution, the upper and lower bound of the support are v and s respectively, while the mean is $\mu = (v+s)/2$ and the variance is $\sigma^2 = (v-s)^2/12$. For the Inverse Gamma distribution, the mean is $\mu = s/(v-1)$ and the variance is $\sigma^2 = s^2/((v-1)^2(v-2))$.

the four competing models that we compare. All our choices on the prior distributions are summarized in Table 1. Further discussion on the rationale behind the selection of prior distributions can be found in the companion online appendix (i.e., Martínez-García & Wynne, 2014).

3. FINDINGS

3.1. Sample Size of Observables for Estimation

The key features of the NOEM model, the DGP for the simulated data (model M_1), that distinguish it from the competing models M_2, M_3, and M_4 are openness to trade and monetary non-neutrality due to the presence of nominal rigidities. Conventional practice would be to include a selection of nominal and real variables for both the Home and Foreign country in the estimation in order to facilitate the empirical assessment of these four models. We also require that the observables be measured variables in all competing models. In order to avoid stochastic singularity in Bayesian estimation we must have the same number of observable variables as structural shocks. Since we have monetary and productivity shocks that are country-specific in all models considered, we choose to estimate all competing theories with four observable variables: Home and Foreign output as well as Home and Foreign inflation. However, we subsequently argue that a standard choice of variables such as the one postulated here – while reasonable ex ante – has implications for Bayesian model comparison that are worth considering further.

Monetary policy under flexible prices and a zero import share $\xi = 0$ (model M_4) or with flexible prices and open to trade (model M_2) has no real effects, therefore influencing nominal variables only. Home and foreign inflation offers insights only on the differences in the implementation of monetary policy across countries. In other words, nominal variables may not suffice to distinguish between autarky (M_4) and openness to trade (M_2) under flexible prices. The economies represented by models M_2 and M_4 are already at their respective potential and the output gap is naturally zero – but they still differ in the allocations they attain. As can be inferred from Eqs. (5)–(6), productivity shocks from both countries endogenously influence the potential attained by each in M_2 but not in M_4 where only local productivity shocks matter. Still, potential output comoves across countries in model M_4 if solely because of the exogenous covariance of the productivity innovations – so comovement by itself does not rule out an autarky solution either. Naturally, a lower import share ξ in model M_2 tends to result in output allocations that are increasingly *more similar* between models M_2 and M_4, making it harder to tell them apart based on the selected macro observables.

The optimal monetary policy for the workhorse closed-economy New Keynesian model with nominal rigidities is to set inflation to zero.[7] This policy prescription seemingly carries over to the open-economy model under autarky. The optimal monetary policy for the case with nominal rigidities and a zero import share, $\xi = 0$ (model M_3) is to set inflation to zero in both countries ensuring that the economy attains the same allocation as under flexible prices and autarky (model M_4). Therefore, a more aggressive policy response to inflation – a higher value of ψ_π to keep inflation at bay – should result in allocations that are increasingly *more similar* between models M_3 and M_4, making it harder to tell them apart based on the observed nominal and real macro data.

Setting the inflation rate on domestic consumption to zero in both countries for the NOEM model with trade (model M_1) is not necessarily going to attain the same allocation as under flexible prices and trade (model M_2). One way to illustrate this point is through closer inspection of the inflation rate equation. We can express the Home and Foreign inflation rate of consumption goods as $\hat{\pi}_t \approx \hat{\pi}_t^H + \xi \Delta \widehat{tot}_t$ and $\hat{\pi}_t^* \approx \hat{\pi}_t^{F*} - \xi \Delta \widehat{tot}_t$ respectively, where $\hat{\pi}_t^H$ and $\hat{\pi}_t^{F*}$ denote Home and Foreign inflation of the locally produced goods (i.e., quarter-over-quarter changes in the output price index) and $\Delta \widehat{tot}_t$ represents changes in the terms of trade. Setting monetary policy to bring down inflation to zero in both countries (i.e., $\hat{\pi}_t \approx \hat{\pi}_t^* \approx 0$) does not ensure that the rate of inflation on the locally produced goods $\hat{\pi}_t^H$ and $\hat{\pi}_t^{F*}$ becomes zero as well, except in the case where $\xi = 0$ (model M_3 with nominal rigidities or model M_4 under flexible prices).

As noted before, the terms of trade capture the relative scarcity of Home- versus Foreign-produced goods, so the inflation rate on consumption goods can be further rewritten as follows:

$$\hat{\pi}_t \approx \hat{\pi}_t^H + \xi \left(\frac{1}{\sigma - (\sigma - 1)(1 - 2\xi)^2} \right) (\Delta \hat{y}_t - \Delta \hat{y}_t^*) \tag{15}$$

$$\hat{\pi}_t^* \approx \hat{\pi}_t^{F*} - \xi \left(\frac{1}{\sigma - (\sigma - 1)(1 - 2\xi)^2} \right) (\Delta \hat{y}_t - \Delta \hat{y}_t^*) \tag{16}$$

which implies that inflation and the growth differential across countries must be related in equilibrium.[8] We cannot assume that optimal monetary policy implies that $(\Delta \hat{y}_t - \Delta \hat{y}_t^*) \approx 0$ because output is driven by country-specific shocks, which generally do not produce identical growth rates for both countries in every period. As a result, in the NOEM model with trade (model M_1), a monetary policy set to bring down consumption inflation toward zero would generally not attain a zero inflation rate on the locally produced goods $\hat{\pi}_t^H$ and $\hat{\pi}_t^{F*}$ – so long as local firms are subject to price stickiness and cannot adjust their prices, there will be some loss relative to the flexible price case. However, given a monetary policy that strongly reacts to consumption inflation, the efficiency loss caused by the distortion tends to be lower the smaller the value of the import shares ξ is.

Based on that, setting inflation to zero in both countries under the NOEM model specification (model M_1) should result in an allocation that is increasingly *close* to the allocation attained under flexible prices and trade (model M_2) as the import share ξ becomes arbitrarily close to zero (as it would also occur for models M_3 under price stickiness and M_4 under flexible prices). Therefore, a more aggressive policy response to inflation – a higher value of ψ_π to keep inflation at bay – should result in allocations that are increasingly *more similar* between all models – M_1, M_2, M_3, and M_4 – as the import share ξ becomes arbitrarily close to zero, making it harder to tell them apart based on any of the observed macro data that we have.[9]

3.1.1. Experiment with the Policy Parameter ψ_π

To investigate Bayesian model comparison, we evaluate the implications of increasing the similarity between all four competing models with the selection of observables indicated before. We simulate data and compare all four models on an interval that spans ψ_π between 0 and 6 under the benchmark parameterization that sets the import share in the consumption basket ξ at a low value of 0.06 in order to increase the similarity between all four competing models as tolerance for inflation declines – i.e., as ψ_π increases while keeping the Calvo-parameter α unchanged at 0.75. All the posterior model probabilities from this experiment are summarized in Fig. 1.

As expected, posterior model probabilities favor the true DGP (the NOEM model, M_1) as the sample size gets asymptotically large – which is what we find with a long sample of 10,000 simulated observations. Interestingly, the international transmission mechanism is weak enough that under reasonable parameterizations of the monetary policy rule the closed-economy model M_3 can still appear as the preferred one. Moreover, we show that the more parsimonious model M_3 may get the upper hand in short samples of 160 simulated observations based on the computed posterior model probabilities – see, in particular, short sample 3 in Fig. 1 – whether monetary policy is more or less aggressive in fighting inflation.

The crucial difference between these two models is that M_1 (the true model) features nominal rigidities that result in monetary policy non-neutrality and is open to trade, while monetary policy remains non-neutral in M_3 but there is no endogenous transmission of shocks across countries because households in each country do not demand imported varieties of goods from each other. Therefore, if we were to wrongly conclude on the basis of the evidence available that M_3 is preferred by the data, we may also wrongly conclude that a loosening of monetary policy has no real spillover effects on the economic activity of the other country (when it actually does!).

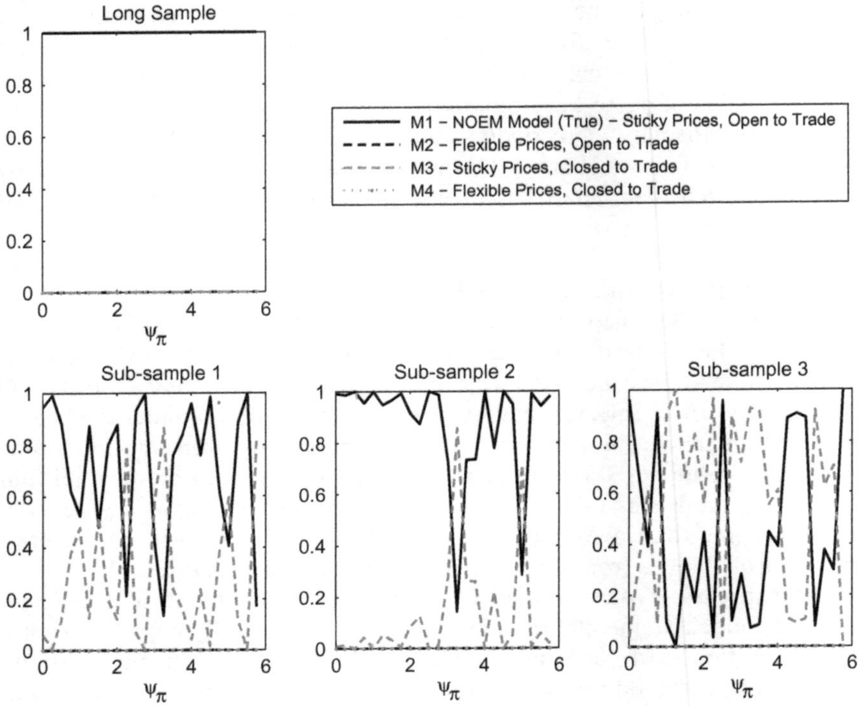

Fig. 1. Posterior Model Probabilities with Respect to the Monetary Policy Response to Inflation. *Note*: The model is simulated over 10,000 periods with code written for Dynare version 4.2.4 and Matlab version 7.13.0.564. The long sample refers to the 10,000 observations while the three short samples are selected to cover three sub-periods including 160 observations each. The set of observables include Home and Foreign inflation, Home and Foreign output. The figure plots the computed Bayesian posterior model probabilities for an interval over the parameter ψ_π. The code for the simulation is available upon request from the authors.

The implications of that policy *mistake*, of course, would only become obvious if the policy change were to be implemented and take effect. This may result in an incorrect identification of the source of business cycle fluctuations as endogenous international spillovers would be attributed in the M_3 model to the country-specific shocks and, in particular, to the exogenous covariance of the innovations. It would be too late to find out ex post that the expected dynamic implications of this policy shock were predicated on a misspecified model that did not take into account households' true preferences for imported varieties from other countries and the potential impact that intratemporal smoothing consumption through trade can have.

3.1.2. Experiment with the Structural Parameter ξ

A smaller share of imported goods in the consumption basket, under the NOEM model (model M_1) shuts down the key channel for the endogenous international transmission of shocks, resulting in an allocation closer to autarky (as in model M_3). Therefore, a lower value of ξ should result in allocations that make it increasingly more difficult to distinguish between models M_1 and M_3. Similar to what we before did for the policy parameter, we simulate data and compare all four alternative models on an interval that spans ξ between 0 and $\frac{1}{2}$. All the posterior model probabilities from this experiment are summarized in Fig. 2.

While the asymptotic results validate the true DGP (the NOEM model, M_1) when we look at a long sample of 10,000 observations, we see that again it is possible to argue in favor of the more parsimonious closed-economy model M_3 in short samples based on the computed posterior model probabilities – see Fig. 2 – whenever the import shares are small enough. As before, we would be missing out the endogenous transmission mechanism that comes from trade by mistakenly selecting model M_3. The crucial difference from the policymakers point of view between these two models is that M_1 (the true model) defines the relevant trade-off for monetary policy to be between domestic inflation and global slack (the *global slack hypothesis*), while model M_3 represents the standard closed-economy view which postulates that the monetary policy trade-off that arises from nominal rigidities is between domestic inflation and domestic slack. Selecting the wrong model in this case would result in an incorrect identification of the sources of business cycle fluctuations and how they are transmitted across countries, as we argued before. However, it can also lead policymakers to ignore the role and consequences of foreign factors in the dynamics of inflation when setting monetary policy or in evaluating a policy change.

One of the major concerns for us would be that model comparison in small samples may contribute to such policy *mistakes*. However, we also recognize that this is a selection error that could have easily been avoided just by looking at trade itself. Since model M_1 implies nonzero imports while model M_3 imposes zero imports, both predictions are incompatible,

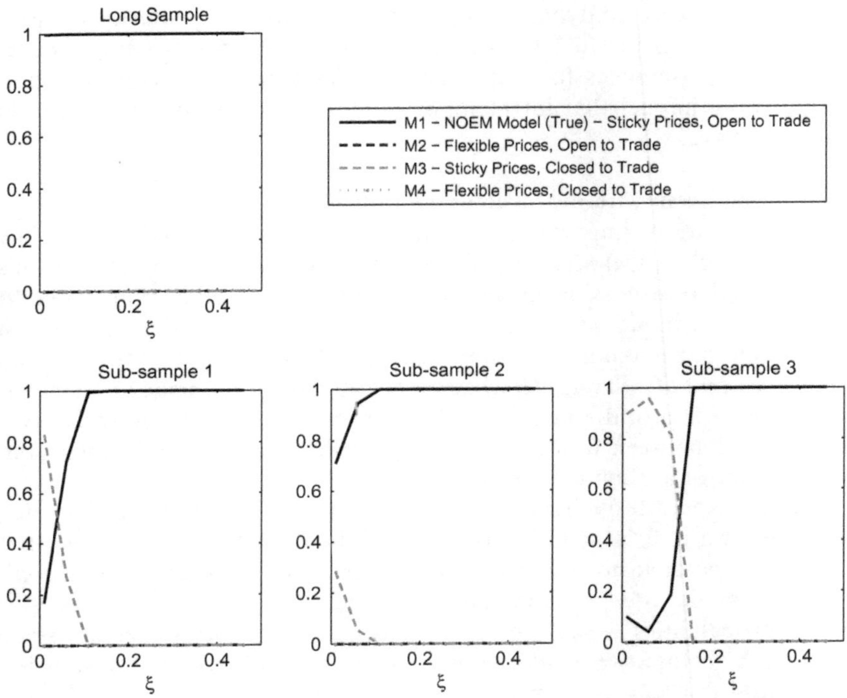

Fig. 2. Posterior Model Probabilities with Respect to the Degree of Openness. *Note*: The model is simulated over 10,000 periods with code written for Dynare version 4.2.4 and Matlab version 7.13.0.564. The long sample refers to the 10,000 observations while the three short samples are selected to cover three sub-periods including 160 observations each. The set of observables include Home and Foreign inflation and Home and Foreign output. The figure plots the computed Bayesian posterior model probabilities for an interval over the parameter ξ. The code for the simulation is available upon request from the authors.

and so one of the two models can be easily refuted in the data. Therefore, more generally, we expect that the selection of observables for estimation can help us avoid some of these *mistakes*.

3.2. Selection of Observables for Estimation

In the benchmark implementation described so far, we make model comparisons based on Home and Foreign output as well as Home and Foreign inflation. Now we experiment with the selection of an alternative set of four observables replacing Foreign output with the terms of trade for Bayesian estimation. Guerrón-Quintana (2010) shows that Bayesian estimation and structural identification can be sensitive to the selection of observables, and not too surprisingly we find that posterior model probabilities are also sensitive to our selection of observables in small samples.

Using terms of trade data as an observable is meant to reveal further information about the trade channel for the international transmission of shocks. Figs. 3 and 4 replicate the experiments behind Figs. 1 and 2 respectively – everything remains the same in our implementation and estimation except for the fact that we are using now a different set of observables to estimate each model. The evidence confirms that the posterior model probabilities are unperturbed by the alternative combinations of observables used for estimation when arbitrarily large samples are available. However, we see that the information content of the terms of trade can work to either revert or worsen the erroneous preference documented earlier toward the more parsimonious, closed-economy models that may arise in small samples.

Using terms of trade data as an observable, we also investigate in Fig. 5 a range of values for the parameter α that determines the degree of price stickiness present in the economy. In this case, posterior model probabilities favor the true DGP (the NOEM model, M_1) as the sample size gets arbitrarily large – but, interestingly, we find that 10,000 quarterly observations (2,500 years of quarterly data) may not be *large enough* to pick the true DGP if α is *too low*. We also show that the international transmission mechanism is weak enough that under reasonable parameterizations of the stickiness parameter α, the closed-economy model M_3 can become the preferred one in small samples (see short sample 2 in Fig. 5). Moreover, we also show that the flexible price specifications M_2 and M_4 may get the upper hand as well in sub-samples of 160 observations when α is low – see short samples 1 and 3 in Fig. 5.

The crucial difference between these two alternative models (M_2 and M_4) and model M_1 (the true model) is that the true DGP features nominal rigidities that result in monetary policy non-neutrality, while monetary policy is neutral whenever prices are flexible. Therefore, if we were to wrongly conclude on the basis of the evidence available that either M_2 or M_4 are preferred by the data over M_1, we may also wrongly conclude that a loosening of monetary policy has no real effects on economic activity (when it actually does!). The implications of that policy *mistake*, of course, would only become obvious if a policy change were to be implemented and take effect

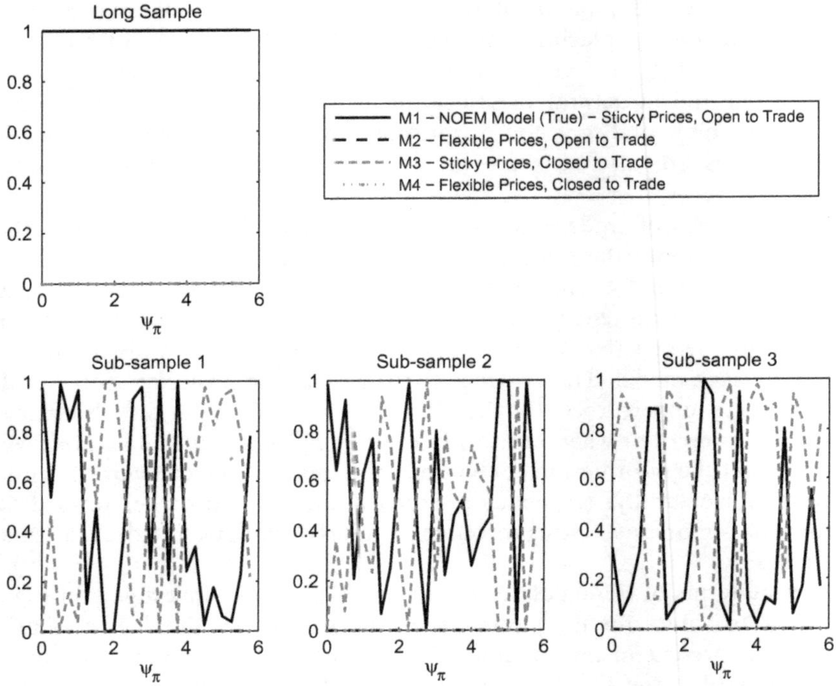

Fig. 3. Posterior Model Probabilities with Respect to the Monetary Policy Response to Inflation (with ToT Data). *Note*: The model is simulated over 10,000 periods with code written for Dynare version 4.2.4 and Matlab version 7.13.0.564. The long sample refers to the 10,000 observations while the three short samples are selected to cover three sub-periods including 160 observations each. The set of observables include Home and Foreign inflation, Home output, and terms of trade. The figure plots the computed Bayesian posterior model probabilities for an interval over the parameter ψ_π. The code for the simulation is available upon request from the authors.

by which time it would already be too late to find out that such a policy shift was predicated on a misspecified model.

Common sense suggests that one may want to experiment with a number of possible combinations of observable variables as a robustness check. In practice, though, the selection may be already significantly limited due to data problems (quality) and due to availability limitations. However, exploring alternative sets of observables whenever feasible is only a

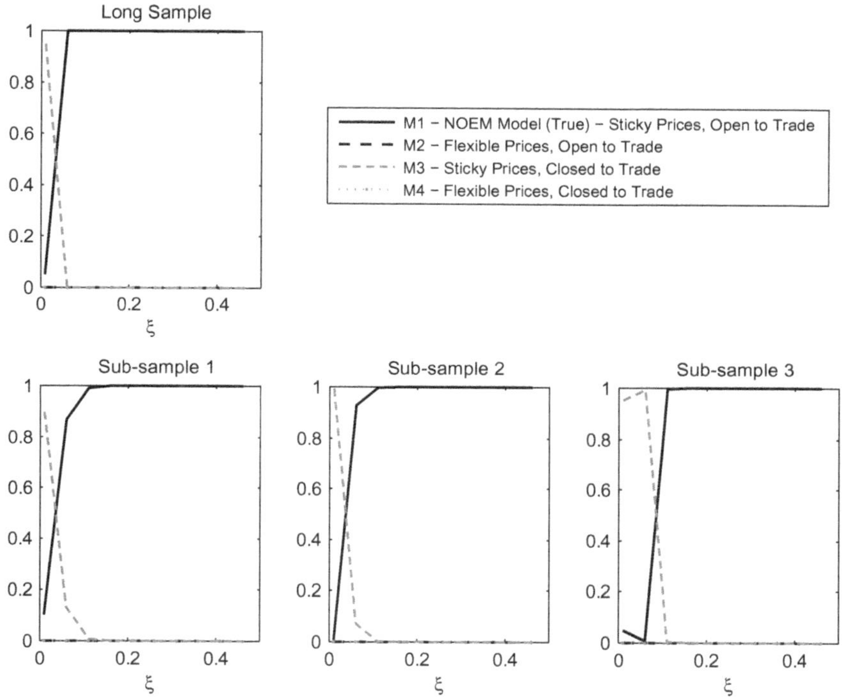

Fig. 4. Posterior Model Probabilities with Respect to the Degree of Openness (with ToT Data). *Note*: The model is simulated over 10,000 periods with code written for Dynare version 4.2.4 and Matlab version 7.13.0.564. The long sample refers to the 10,000 observations while the three short samples are selected to cover three sub-periods including 160 observations each. The set of observables include Home and Foreign inflation, Home output, and terms of trade. The figure plots the computed Bayesian posterior model probabilities for an interval over the parameter ξ. The code for the simulation is available upon request from the authors.

practical recommendation that can help us determine how robust the support for a particular model is – it does not say anything about the deeper question of how we should choose the model most favored by the data whenever alternative combinations of observables produce contradictory evidence (i.e., when they produce significantly different posterior model probabilities). It does not offer us further guidance on how to select the *appropriate* set of observables for a given model either.

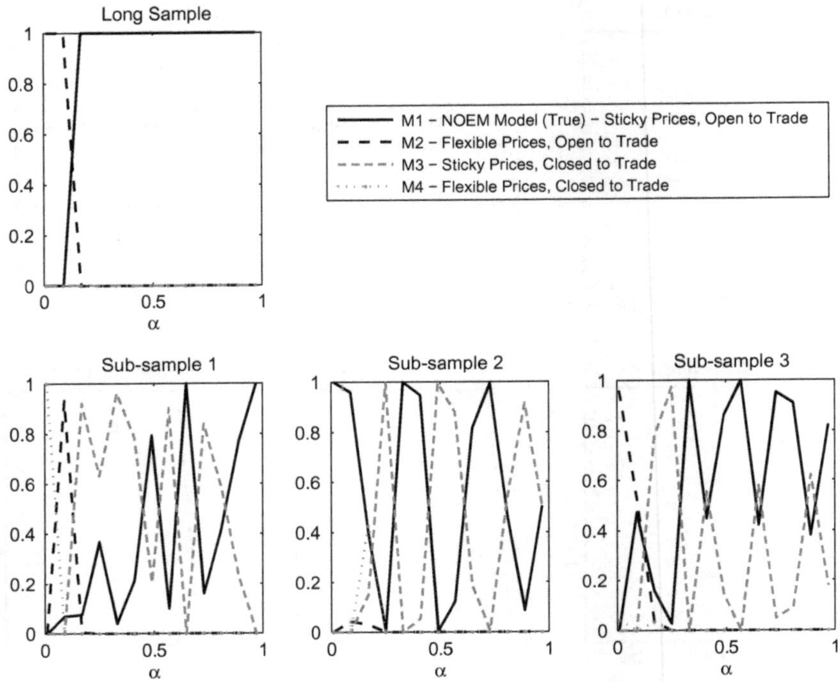

Fig. 5. Posterior Model Probabilities with Respect to the Degree of Price Stickiness (with ToT Data). *Note*: The model is simulated over 10,000 periods with code written for Dynare version 4.2.4 and Matlab version 7.13.0.564. The long sample refers to the 10,000 observations while the three short samples are selected to cover three sub-periods including 160 observations each. The set of observables include Home and Foreign inflation, Home output, and terms of trade. The figure plots the computed Bayesian posterior model probabilities for an interval over the parameter α. The code for the simulation is available upon request from the authors.

In our exercise, however, we have the advantage – uncommon in applied macroeconometrics work – to know the true DGP underlying the data, and so we can dig a little deeper into these results based on simulated data. The macro observables that are common to all four models are all ultimately related to two core variables per country that characterize the dynamics of the NOEM model – inflation and the output gap (which given a specification for potential output can be related to the observable measure of output), whose dynamic path is characterized by a solution of the form presented in Eqs. (13)–(14). Not surprisingly, as different models become more similar in the path they imply for output and inflation, they also appear closer when we use an alternative set of observables that are in effect linear combinations of output and inflation themselves together with the structural shock innovations. Naturally, if Bayesian model comparison methods fail to select the correct specification in short samples with the standard selection of observables that includes the core variables, they may tend to produce false signals in small samples with other alternative selection of observable variables – as we have seen here.

Variable selection, in this case, could contribute to attenuate the problem or simply help us detect whether a selection problem exists (when the same model selection technique gives different predictions with alternative observables), but it cannot, in general, avoid the problem entirely for us. In other words, when model specifications become arbitrarily *close*, the selection of observables for estimation cannot help us consistently avoid the preference toward more parsimonious specifications that we have found in our small sample experiments.

4. DISCUSSION

We have a collection of $k \geq 2$ models each of which is fully described with a parameterized joint probability density over the vector of observable (endogenous) variables Z, i.e.,

$$M_i = \{f_i(z \mid \theta_i) : \theta_i \in \Theta_i\}, \forall i = 1, \ldots, k \tag{17}$$

where θ_i is the vector of unknown parameters of model M_i, Θ_i is the parameter space and $d_i = \dim(\Theta_i)$ its dimension, $f_i(z \mid \theta_i)$ is the parameterized probability density, and z is a given realization of the vector of observable variables Z.

The likelihood function for model M_i, given n observations of the observable variables $z^n = (z_1, \ldots, z_n)$, is the probability of z^n occurring under the probability density that describes model M_i given the vector of parameters θ_i, i.e., $L_i(\theta_i) \equiv f_i(z^n \mid \theta_i)$. We refer to the log-likelihood function for model M_i as $l_i(\theta_i)$ and represent it as follows:

$$l_i(\theta_i) \equiv \ln f_i(z^n \mid \theta_i) = \sum\nolimits_{j=1}^{n} \ln f_i(z_j|\theta_i), \forall i = 1, \ldots, k \tag{18}$$

We assign prior probabilities, $\Pr(M_i)$, to all model specifications $i=1,\ldots,k$, and also prior probabilities to the parameters θ_i that characterize each model, $f_i(\theta_i)$.

The marginal likelihood $m_i \equiv f_i(z^n \mid M_i)$ of any model M_i, $i=1,\ldots,k$, is referred to as the model evidence, and it is defined by the expectation taken over the likelihood function $L_i(\theta_i) \equiv f_i(z^n \mid \theta_i)$ with respect to the prior distribution of the parameters $f_i(\theta_i)$, i.e.,

$$m_i \equiv f_i(z^n \mid M_i) = \int_{\Theta_i} f_i(z^n \mid \theta_i) f_i(\theta_i) d\theta_i, \forall i = 1,\ldots,k \tag{19}$$

The posterior probability for model M_i can be calculated using Bayes' Theorem as,

$$\Pr(M_i \mid Z^n = z^n) = \frac{f_i(z^n|M_i)\Pr(M_i)}{\sum_{p=1}^{k} f_p(z^n|M_p)\Pr(M_p)} = \frac{m_i \Pr(M_i)}{\sum_{p=1}^{k} m_p \Pr(M_p)}, \forall i = 1,\ldots,k \tag{20}$$

where the marginal likelihood m_i providing evidence for model M_i multiplied by the prior assigned to that particular model is normalized with respect to the sum of the product between the marginal likelihood of the model and the model prior of all $k(\geq 2)$ models under consideration.

The Bayesian posterior odds for model M_1 versus the alternative model M_i, $i=2,\ldots,k$, summarize the relative support that the data provides for one specification over the other with the ratio of their posterior probabilities, i.e.,

$$\frac{\Pr(M_1 \mid Z^n = z^n)}{\Pr(M_i \mid Z^n = z^n)} = \frac{m_1}{m_i} \frac{\Pr(M_1)}{\Pr(M_i)}, \forall i = 2,\ldots,k \tag{21}$$

The posterior odds in favor of model M_1 against an alternative specification M_i, $i=2,\ldots,k$, can be expressed as the product of the prior odds $\frac{\Pr(M_1)}{\Pr(M_i)}$ in favor of M_1 times the corresponding Bayes Factor defined by the ratio $B_{1i} = \frac{m_1}{m_i}$, as can be seen in Eq. (21). The marginal likelihood is key to calculate the Bayes Factor – which is the quotient of the marginal likelihoods of two alternative models. Therefore, marginal likelihoods are also crucial to derive the Bayesian posterior odds in Eq. (21) conventionally used for Bayesian model selection.

Though there are alternative approaches to compute the Bayes Factor and the Bayesian posterior odds – such as the (generalized) Savage–Dickey density ratio discussed by Verdinelli and Wasserman (1995) – the method based on the marginal likelihood remains the most common in applied macro-econometrics work with Bayesian techniques. We rely on the computation of the marginal likelihood in all our experiments as well. Here we review some aspects of the estimation and the computation of the marginal likelihood that can directly affect the assessment of competing models based on Bayesian posterior odds, with special emphasis on understanding what contributes to explain the false signals in model selection that we have encountered in our experiments with small samples.

4.1. Interpreting Our Findings: The Role of Sample Size

After specifying the priors over the models and over the model parameters, the practical difficulty in calculating posterior probabilities or posterior odds is computing the marginal likelihood defined in Eq. (19). Only in very special cases we can calculate the marginal likelihood analytically – most notably for the exponential likelihood family with conjugate priors, as in the case of Gaussian linear models (see, e.g., Zellner, 1971). In practice, analytical solutions are often intractable and computational methods are needed.

Among the different methods available to approximate the marginal likelihood, we can list: (a) asymptotic approximations (Laplace's method, Schwarz criterion, BIC); (b) numerical integration (e.g., Gaussian quadrature), importance sampling, and annealed importance sampling (see, e.g., Geweke, 1989; Neal, 2001); (c) posterior distribution simulations (e.g., Markov Chain Monte Carlo (MCMC) methods like the Metropolis–Hastings algorithm and the Gibbs sampler); and (d) variational inference (see, e.g., Corduneanu & Bishop, 2001), expectation propagation (see, e.g., Minka, 2001).

We use the Laplace approximation to compute the marginal likelihood of a given specification and derive the Bayesian posterior odds for model comparison. Asymptotic approximation methods, such as Laplace's method, rely on normal asymptotic approximations of the marginal likelihood. These methods work well in most familiar problems, are accurate, easy to compute, and fast. They provide adequate approximations especially for well-behaved posterior densities that are highly peaked and unimodal, since asymptotic approximations rely on a normal density to approximate the posterior density.

The Gaussian state-space representation in Eqs. (13)–(14) implies that the likelihood of the model, $L_i(\theta_i) = f_i(z^n \mid \theta_i)$, is characterized by a normal distribution under the DGP as well as under any of the alternative specifications we propose for model comparison. Hence, in our case the likelihood of the models we investigate is known to be Gaussian. The specification of the prior distributions for the model parameters then plays an important role in retaining the highly peaked and unimodal shape on the posterior density and, therefore, to ensure that the Laplace approximation is reasonably accurate.

In our illustrations of Bayesian model comparison the choice of the Laplace approximation method appears reasonable on grounds of computational accuracy. For other models, however, approximation methods such as Laplace's method may not attain accurate estimates of the marginal likelihood. In that case, alternative ways to compute the marginal likelihood should be pursued in order to avoid model selection errors due to inaccurate estimates of the marginal likelihood. An evaluation of the advantages and disadvantages of alternative methods to compute the marginal likelihood – especially when models are less well-behaved than the ones considered here – goes beyond the scope of this paper. We leave it for future research.

Apart from the reasonable accuracy attained in our exercise, we also discuss this asymptotic approximation method in greater detail here to gain further insight on the role of sample size and the penalization of overfitting that is inherent in Bayesian posterior odds calculations.

4.1.1. Laplace's Approximation Method: Accuracy and Sample Size

The Laplace's (or Gaussian) method, which we apply in our experiments with Bayesian model comparison, is based on the idea that asymptotically the posterior distribution can be approximated with a multivariate Gaussian distribution (see, e.g., Kass, Tierney, & Kadane, 1988). Let $\hat{\theta}_i$ be the posterior mode which is defined as the vector of parameters θ_i that maximizes the posterior probability $f_i(\theta_i|z^n)$ which characterizes model M_i. The posterior probability is proportional to the likelihood function times the model parameters' priors, i.e., $f_i(\theta_i|z^n) \propto f_i(z^n|\theta_i)f_i(\theta_i)$, so the optimization problem to derive the posterior mode can be defined as

$$\hat{\theta}_i = \arg\max_{\theta_i} \{\ln(f_i(z^n|\theta_i)f_i(\theta_i))\} \tag{22}$$

where $h_i(\theta_i) \equiv \ln(f_i(z^n|\theta_i)f_i(\theta_i))$ is a log-transformation that also maximizes the posterior probability. The first-order conditions of the maximization problem in Eq. (22) imply that $\nabla h_i(\hat{\theta}_i) = 0$.

Expanding $h_i(\theta_i)$ as a quadratic function around $\hat{\theta}_i$ we obtain that

$$h_i(\theta_i) \approx h_i(\hat{\theta}_i) + \nabla h_i(\hat{\theta}_i)(\theta_i - \hat{\theta}_i) - \frac{1}{2}(\theta_i - \hat{\theta}_i)' H(\hat{\theta}_i)(\theta_i - \hat{\theta}_i) \tag{23}$$

where $H(\hat{\theta}_i) = -D^2 h_i(\theta_i)$ is the negative Hessian of second derivatives of $h_i(\theta_i)$ evaluated at $\hat{\theta}_i$. Replacing the first-order conditions derived from (22) and exponentiating Eq. (23) yields an approximation of $f_i(z^n|\theta_i)f_i(\theta_i)$ that has the form of a normal density with mean $\hat{\theta}_i$ and covariance matrix $H(\hat{\theta}_i)^{-1}$. Integrating that expression we obtain the Laplace approximation of the marginal likelihood, i.e.,

$$\ln(m_i) \approx \ln(f_i(z^n|\hat{\theta}_i)) + \ln(f_i(\hat{\theta}_i)) + \frac{d_i}{2}\ln(2\pi) - \frac{1}{2}\ln\left|H(\hat{\theta}_i)\right| \equiv \ln(m_i)\big|_{Laplace} \tag{24}$$

where d_i is the dimension of the parameter space Θ_i of model M_i for any $i = 1, \ldots, k$.

Kass et al. (1988) and Kass, Tierney, and Kadane (1990) show that, under certain regularity conditions, errors in this approximation are bounded by $O_P(n^{-1})$ where n is the number of observations used in the estimation. We can also obtain an $O_P(n^{-1})$ approximation of the marginal likelihood with $\hat{\theta}_i^{MLE}$ being the maximum likelihood estimator (MLE) and $H(\hat{\theta}_i^{MLE})$ being the observed information matrix (i.e., the negative of the Hessian matrix evaluated at the MLE estimator, $\hat{\theta}_i^{MLE}$) in Eq. (24)). The inverse of the Fisher information matrix (i.e., the inverse of the expected information matrix which converges as n grows to the inverse of the asymptotic covariance matrix) can also be used in Eq. (24) but at the expense of incurring a greater approximation error in the computation of the marginal likelihood of order $O_P\left(n^{-\frac{1}{2}}\right)$.

Thus, when Laplace's method is applied to both the numerator and denominator of the Bayes Factors $B_{1i} = \frac{m_1}{m_i}$ in Eq. (21) to compare M_1 against any other alternative specification M_i, $i = 2, \ldots, k$, the resulting approximation of the Bayes Factors retains an approximation error of order $O_P(n^{-1})$ (or of order $O_P\left(n^{-\frac{1}{2}}\right)$ if the Fisher information matrix is used).[10] For many problems for which the sample size n is moderate and the likelihood is reasonably approximated by that of a normal distribution, the Laplace method produces accurate and easy to compute approximations of the marginal likelihood and the Bayes Factors.[11]

Hence, the Laplace approximation to computing marginal likelihoods seems reasonable in our illustrations in part because the Gaussian state-space representation of the solution ensures the normality of the likelihood and the posterior densities are expected to be well-behaved. It is reasonable also because the sample sizes more relevant to us are sufficiently large so that the approximation error is negligible and the computed Bayes Factors are adequately accurate.

In providing guidance on the sample size required to attain an adequate approximation with the Laplace method, we follow the recommendations of Kass and Raftery (1995). Kass and Raftery (1995) warn us that sample sizes of less than $5d_i$ observations may be insufficient to attain an accurate approximation of the marginal likelihood with the Laplace method, where d_i is the dimension of the parameter space of model M_i. In turn, sample sizes greater than $20d_i$ should be *large enough* to ensure the method works well in most cases in which the likelihood function itself is not too different from that of a normal distribution. However, we must recognize that a sample size of $20d_i$ observations appears increasingly out of reach in practice for most heavily parameterized medium- and large-scale DSGE models.

In the experiments reported in this paper, the most parameterized specification is the DGP (model M_1) which includes 12 parameters (not counting the calibrated intertemporal discount factor, β). All other specifications have fewer than 12 parameters. We set the small sample size in our experiments to $n = 160$ quarterly observations (40 years of quarterly data). This implies that all models under consideration are above the minimum threshold of $5d_i$ observations suggested by Kass and Raftery (1995) and, in fact, come close to the $20d_i$ threshold in our case.

We are neither interested in very long sample sizes that should lead to the correct outcome in model selection but are generally not available for applied work, nor in the very short samples where the posterior densities are still largely dominated by the priors we place on the model parameters. In turn, we examine in our experiments a sample range in between those, which is more realistic and relevant for applied work (given the length of data that is generally available). Our notion of a small sample size in practice satisfies the following broad criteria:

(a) The sample size is *large enough* so that the Laplace approximation works well given an expected approximation error of order $O_P(n^{-1})$ (or of order $O_P(n^{-\frac{1}{2}})$) with the Fisher information matrix and surpasses the lower threshold recommended by Kass and Raftery (1995).
(b) The sample size is *large enough* so that there is enough data to overwhelm the priors.
(c) The sample size is *not too large* so that the penalization for overfitting that we highlight in this paper still has bite to tilt the posterior odds in favor of the most parsimonious specification (and at the expense of selecting the wrong model).

Under this notion of small samples, the Laplace method suffices for our purpose of providing an accurate assessment of the problem of *false signals* in Bayesian model selection – a problem that arises, as can be seen in our illustrations, whenever very large sample sizes of observations are not available for the estimation and a problem that would be otherwise masked by the priors for very short sample sizes.

4.1.2. Laplace's Approximation Method: Overfitting Penalization and Sample Size

Apart from the appropriateness of the Laplace method given the notion of a small sample that we investigate here, this asymptotic approximation also helps us shed some light on the role that sample size n and the dimensionality of the parameter space d_i of a given model M_i, $i = 1,\ldots, k$, play on the calculations of the marginal likelihood, the Bayes Factors, and the Bayesian posterior odds for model comparison.

As the sample size n grows, the different terms of the Laplace approximation to the marginal likelihood grow at different rates. The log-likelihood function should grow proportionally to n; the size of the penalization for overfitting that arises from the Hessian term $\ln|H(\tilde{\theta}_i)|$ increases at the rate of $d_i \ln(n)$, which also depends on the dimensionality of the parameter space; while the remaining approximation terms are invariant with sample size but depend on the choice of priors and the dimensionality d_i.

More generally, the different terms of the Laplace approximation of the marginal likelihood in Eq. (24) grow with sample size n as indicated here,

$$\ln(m_i)\big|_{Laplace} \approx \underbrace{\ln(f_i(z^n|\tilde{\theta}_i))}_{O(n)} + \underbrace{\ln(f_i(\tilde{\theta}_i))}_{O(1)} + \underbrace{\frac{d_i}{2}\ln(2\pi)}_{O(1)} - \frac{1}{2}\underbrace{\ln|H(\tilde{\theta}_i)|}_{O(d_i \ln(n))} \tag{25}$$

For any given sample size for which this approximation holds, there is a penalty for the dimensionality of the model d_i that comes from the last two terms in the right-hand side of Eq. (25) and varies with n. When Laplace's method is applied to both the numerator and denominator of the Bayes Factors $B_{1i} = \frac{m_1}{m_i}$, $i = 2,\ldots,k$ using Eq. (25), the resulting approximation to compare the model evidence of M_1 against that of any other alternative specification M_i, $i = 2,\ldots,k$ can be expressed as follows:

$$\ln(B_{1i})\big|_{Laplace} \approx \underbrace{(l_1(\tilde{\theta}_1) - l_i(\tilde{\theta}_i))}_{O(n)} + \underbrace{(\ln(f_1(\tilde{\theta}_1)) - \ln(f_i(\tilde{\theta}_i)))}_{O(1)} + \ldots$$

$$\underbrace{\left(\frac{(d_1 - d_i)}{2}\ln(2\pi)\right)}_{O(1)} - \frac{1}{2}\underbrace{(\ln|H(\tilde{\theta}_1)| - \ln|H(\tilde{\theta}_i)|)}_{O((d_1 - d_i)\ln(n))} \tag{26}$$

Similarly, this can be extended to approximate the Bayesian posterior odds defined in Eq. (21).

At moderate sample sizes for which the Laplace approximation seems appropriate, the penalty for overfitting can become the deciding factor to understand why Bayesian model comparison may favor parsimony even at the expense of selecting the wrong model. As the sample size n keeps growing, the differences in the log-likelihood function $(l_1(\tilde{\theta}_1) - l_i(\tilde{\theta}_i))$ should grow proportionally to n while the size of the penalty increases at the rate of $(d_1 - d_i)\ln n$. Hence, the overfitting penalty embedded here is a relatively harder threshold to meet in samples of moderate length such as the ones we explore in all our illustrations whenever the probability densities that characterize each competing model are arbitrarily *close* to each other.

In other words, for moderate sample sizes it might occur that the Bayesian posterior odds favor the less parameterized model if the log-likelihood differences between the models under comparison are too small to outweigh the overfitting penalty found in Eq. (26). Otherwise, researchers would require unrealistically large sample sizes to be able to consistently identify the correct model when the true specification is more heavily parameterized than the alternative. That explains mechanically why in our experiments we validate the asymptotics in Fernández-Villaverde and Rubio-Ramírez (2004) but still find that the more parsimonious model could be the one picked up against the more complex true specification even when using 40 years of quarterly data for that!

4.1.3. BIC's Approximation Method: An Alternative Trade-off Between Accuracy at a Given Sample Size and the Role of Priors for Model Selection

A more efficient, but (in general) less accurate asymptotic approximation is obtained by: (a) using a consistent, likelihood-based estimator $\tilde{\theta}_i$ to evaluate the approximation (naturally, the MLE estimator, $\tilde{\theta}_i = \hat{\theta}_i^{MLE}$ can used for this); (b) retaining only those terms in Eq. (24) that increase with the sample size n, i.e., dropping $\ln(f_i(\tilde{\theta}_i)) + \frac{d_i}{2}\ln(2\pi)$ which do not increase with n; and, (c) using the fact that for large n, the determinant $|H(\tilde{\theta}_i)|$ is proportional to n^{d_i}. This approximation is called the Schwarz criterion and takes the form,

$$\hat{m}_i \approx l_i\left(\hat{\theta}_i^{MLE}\right) - \left(\frac{d_i}{2}\right)\ln(n) \tag{27}$$

where $l_i(\tilde{\theta}_i) \equiv \ln(f_i(z^n|\tilde{\theta}_i))$ is the log-likelihood function evaluated at the value of the estimator $\tilde{\theta}_i$. The right-hand side in Eq. (27) is equal to the Schwarz criterion for model selection where d_i is the dimension of the parameter space Θ_i of model M_i for any $i = 1,\ldots,k$. This approximation was first derived by Schwarz (1978) (see also Akaike, 1978).

Kass and Wasserman (1995) show that under regularity conditions similar to those for the Laplace approximation, the Schwarz criterion satisfies that

$$m_i = \hat{m}_i + O_P(1) \tag{28}$$

where $\tilde{\theta}_i$ is a consistent, likelihood-based estimator (or simply the MLE estimator $\hat{\theta}_i^{MLE}$ as indicated before). Moreover, the relative error of the approximation tends to zero in probability, i.e., $\frac{|\hat{m}_i - m_i|}{|m_i|} \rightarrow_P 0$. Notice that minus twice the Schwarz criterion is the Bayesian Information Criterion (or BIC). Hence, the BIC provides an $O_P(1)$ approximation for the marginal likelihood as well. The Schwarz criterion, and by extension the BIC, are in effect $O_P(1)$ approximations to the marginal likelihood.

The BIC approximation is appealing for model comparison in a number of respects that we highlight here:

First, it does not depend on the prior assigned to the vector of parameters. So, this procedure can be applied to compute the marginal likelihood even when the priors $f_i(\theta_i)$ are difficult to set precisely or are debated in the literature. This is an important consideration in applied work where we often do not have strong reasons to favor one particular prior distribution over others.

Second, the BIC is related to the Minimum Description Length (MDL) stochastic complexity measure proposed by Rissanen (1987). In recent years, MDL has received much attention in the literature on statistical model selection as it allows for a unified treatment of model selection and statistical inference. The MDL measure provides a quantification of the goodness of fit that can be attained with a given probability distribution to account for the statistical regularities observed in the data. From the work of Rissanen (1996) and Qian-Künsch (1998) it follows that the MDL-proposed measure of stochastic complexity of the observed data relative to a given parameterized model can be expressed as minus the maximum log-likelihood plus a model complexity term that is determined by the Fisher information matrix and the MLE estimator of the model parameters. In this sense, the BIC approximation we consider here is minus the MDL measure of stochastic complexity. Hence, our findings using the BIC can be interpreted in light of what the MDL principle stands for as well.

Third, the Laplace and the BIC approximations should be asymptotically equivalent for large sample sizes, i.e.,

$$\ln(m_i)\big|_{Laplace} \overset{n\to\infty}{\to} \underbrace{l_i(\tilde{\theta}_i) - \left(\frac{d_i}{2}\right)\ln(n)}_{=Schwarz\ criterion = -\frac{1}{2}BIC} \tag{29}$$

under some regularity conditions. The BIC approximation may be viewed as a rough approximation of the log of the marginal likelihood. We say that the BIC and the Laplace method are asymptotically correct, though, because they both select a model whose posterior probability is a maximum whenever n becomes sufficiently large. Moreover, as indicated by Eq. (29), the BIC and Laplace methods must agree on the selected model as sample size n becomes arbitrarily large.

Fourth, the BIC approximation to the Bayes Factor B_{1i} that compares model M_1 against the alternative model M_i for any $i = 2,\ldots,k$, i.e.,

$$BIC_{1i} = -2\left[l_1(\tilde{\theta}_1) - l_i(\tilde{\theta}_i) - \left(\frac{d_1 - d_i}{2}\right)\ln(n)\right] \tag{30}$$

satisfies, as shown by Kass and Raftery (1995), that as $n \to \infty$,

$$\frac{-\frac{1}{2}BIC_{1i} - \ln(B_{1i})}{(B_{1i})} \underset{P}{\to} 0, \forall i = 2,\ldots,k \tag{31}$$

In contrast to the Laplace approximation, the relative error of $\exp\left(-\frac{1}{2}BIC_{1i}\right)$ in approximating the Bayes Factor B_{1i} is generally of order $O_P(1)$.[12] For the moderate and large sample sizes n for which this result holds, the error bounds of the approximation would not increase with the sample size itself. This is a rough approximation, but one that should give us a reasonable indication of the evidence for the sample sizes that we use in our illustrations of Bayesian model comparison in this paper.

Under some conditions applying to nested models, such as the ones considered in this paper, the BIC approximation under unit information priors is accurate to order $O_P\left(n^{-\frac{1}{2}}\right)$ (see Kass & Raftery, 1995; Kass & Wasserman, 1995).[13] Thus, if one is willing to consider these priors as suitable, then the BIC (and the Schwarz criterion) can be thought as providing a reasonably good approximation to the log of the Bayes Factors that is comparable in terms of the accuracy attained for moderate and large sample sizes to that of the Laplace method using the Fisher information matrix.

Fifth, the BIC approximation is quite intuitive and easy to interpret retaining the penalization for overfitting indicated before with the Laplace approximation in Eq. (26). The BIC approximation contains a term evaluating how much better (or worse) one model with parameters set to their consistent, likelihood-based estimates fits the data relative to an alternative model also evaluated with parameters at their consistent, likelihood-based estimates (i.e., $l_1(\tilde{\theta}_1) - l_i(\tilde{\theta}_i)$) and another term that punishes the added complexity of one model over the other (i.e., $\left(\frac{d_1 - d_i}{2}\right)\ln(n)$).[14]

This confirms the simple interpretation given before of one of the plausible explanations of the false signals problem that we have illustrated in the experiments described in the previous section. It suggests posterior model probabilities can favor the wrong model specification in part because of the penalization of complexity that comes with it, as can be inferred from Eq. (30).

4.2. Other Considerations in Evaluating Bayesian Posterior Odds for Model Comparison

Our discussion thus far provides a qualitative interpretation of the reported findings, but one that is ultimately dependent on the accuracy of the approximation of the marginal likelihood used. We have argued that using the Laplace approximation is reasonable given the characteristics of the solution to the models we are investigating and the fact that we explore moderate and large sample sizes for which the approximation should be valid. We conclude that unless we have an arbitrarily large sample size, standard Bayesian posterior odds may still favor parsimony even when the true model specification is more complex. This is well-understood on the basis of the Laplace approximation. What we do not have from this is a quantitative rule to determine the sample size that would be needed to accurately and consistently select the true model overcoming the penalization for overfitting. We are unable to be much more specific than this since assessing the sample sizes required to avoid the problem of *false signals* is likely to be model-dependent and to vary for different families of probability distributions.

To wrap up, we discuss a number of related points regarding the implementation of the Bayesian estimation of a model (such as parameter identification, the choice of priors, the selection of observables, etc.) that can affect the fit of the competing specifications under comparison and consequently also lead to erroneous model selection for small sample sizes.

4.2.1. Parameter Identification

Identification can refer to the mapping from the deep parameters of the model to the reduced-form parameters that characterize a unique solution as in Eqs. (13)–(14). As indicated before, Blanchard and Kahn (1980) provide conditions under which such a unique stable solution exists. In this regard, the conventional practice is to set the range of the prior distributions to avoid or minimize the draws that come from regions of the parameter space for which no solution exists or where indeterminacy arises. Notice that although the unique solution is linear in the variables, the reduced-form parameters are generally nonlinear functions of the deep parameters – reflecting the cross-equation restrictions implied by the model.

Identification also refers in our context to the mapping from the solution to the observable data and the conditions under which a unique likelihood function $L_i(\theta_i) \equiv f_i(z^n|\theta_i)$ exists. Identification problems in this latter sense arise if distinct parameter values do not result in different probability distributions of the data, i.e., θ_i is identified if $f_i(z^n \mid \theta_i^{(1)}) = f_i(z^n \mid \theta_i^{(2)})$ implies that $\theta_i^{(1)} = \theta_i^{(2)}$ for all z^n (see, e.g., Hsiao, 1983, pp. 226–227). If identification fails to hold, no estimation procedure can pin down uniquely the vector of parameters θ_i irrespective of the sample size. Bayesian estimation only circumvents the problem by using priors and, as Canova and Sala (2009) points out, may end up concealing the problems of identification that way.

It is recognized that lack of identification leads to wrong inferences and can significantly affect our estimates of a model (see, e.g., Martínez-García, Vilán, & Wynne, 2012; Ríos-Rull, Schorfheide, Fuentes-Albero, Kryshko, & Santaeulàlia-Llopis, 2012). The lack of identification can also be a problem for Bayesian model selection because we need to compute the marginal likelihood of models to derive their posterior odds with a badly shaped likelihood function due to lack of identification. The issue is rarely addressed in applied work where identification is not usually explicitly verified before estimation. We argue that checking identification of the model should be standard practice given the potential problems derived from lack of identification.

Several methods already exist to check identification in linearized models using: (i) the autocovariogram (Andrle, 2010; Iskrev, 2010);[15] (ii) the spectral density (Komunjer & Ng, 2011; Qu & Tkachenko, 2012); and (iii) Bayesian indicators (Koop, Pesaran, & Smith, 2013). For a review and methodological comparison of these techniques, the interested reader is referred to Mutschler (2014).

4.2.2. Variable Selection

Guerrón-Quintana (2010) illustrates how the set of observables chosen for estimation affects the way in which the structural parameters enter into the log-likelihood function and, therefore, conditions the model estimation via likelihood-based methods. Our experiments show that the dangers that Guerrón-Quintana (2010) warned us about in regards to estimation also play a role in Bayesian model comparison as differences in the set of observables can affect the differences in the log-likelihood functions across models that we can tease out from the data. Our simulations indicate that the selection of observables might help with model comparison in small samples, but it does not necessarily resolve the problem that arises when more parsimonious specifications are preferred over the more heavily parameterized ones that characterize the true DGP of the observed data.

All our experiments were conducted after estimating the competing model specifications on the same set of observables to maintain comparability. We suggest, however, that data not included in the set of observables can be used for cross-validation. For instance, we use output and inflation to estimate the four models under consideration in the experiments plotted in Figs. 1 and 2. Trade data – while not directly used in the estimation – can serve for cross-validation of the model selection implied by Bayesian posterior odds given that preference for a closed-economy specification would be inconsistent with a non-zero trade series.

One could argue that model selection can be refined in that same way – e.g., Bayesian estimation and model comparison is not warranted with closed-economy models when there are open-economy alternatives if the data suggests nonzero trade (irrespective of whether we actually end up using the trade data for the estimation or not). Unlike what happens in our experiments, in practice none of the models available for comparison describes the exact DGP underlying the observed data. We suggest that even in those circumstances performing Bayesian model comparison with different sets of observable variables can offer additional insights about the robustness of the evidence in favor of a given model against the alternatives.

4.2.3. Prior Selection

In this paper we assume "non-informative" prior probabilities on the models, i.e., $\Pr(M_i) = \frac{1}{k}$ for all $i = 1,\ldots,k$, and we keep the distribution of priors on parameters $f_1(\theta_i)$ across all model specifications invariant. The set of structural parameters that characterizes each competing model is a subset of the set of parameters for the true model (the DGP), i.e., $\Theta_i \subseteq \Theta_1$ for all $i = 2, 3, 4$. In fact, the set of parameters for each competing model M_i can be described simply as $\Theta_i = \{\theta_1 \in \Theta_1 : \theta_{1l} = 0 \text{ for some } l = 1,\ldots,\bar{l} \text{ where } 1 \leq \bar{l} < d_1\}$. Our experiments make the illustration simpler because the distributions of all competing models $f_i(z|\theta_i)$ for all $i = 2, 3, 4$ are in effect limiting cases of the distribution of the true model $f_1(z|\theta_i)$. We then merely choose points on an interval that parameterize the true DGP (model M_1) *closer and closer* to the probability distribution of at least one of the alternative (more parsimonious) models to highlight the importance of the penalization for overfitting that arises even with moderate sample sizes.[16]

We ignore very short samples where the posterior distribution may still be dominated by the priors and base our investigation on moderate sample sizes for which the Laplace approximation works reasonably well. We view this as most relevant for applied work and do not explore the role of priors (and prior selection) further in our current analysis. We leave that for future research.

4.2.4. Nested versus Non-nested Models in Bayesian Model Comparison
When the distributions of the true model and a competing one become arbitrarily close to each other, for a given sample size, the differences in the log-likelihood function ought to be smaller between the two models. Then, the penalty for overfitting ends up dominating our results and favoring the more parsimonious one over the more heavily parameterized one (even if that is the true model). Bayesian model comparison through posterior model probabilities embodies a strong preference toward the lowest dimensional model (Occam's razor). Our experiments show that as a consequence we may fail to find support for the true (more complex) model in small samples in spite of the good asymptotic properties demonstrated in Fernández-Villaverde and Rubio-Ramírez (2004).

Our illustrations, however, are largely based on comparisons between nested model specifications. When competing models are non-nested and can be represented by probability distributions that do not overlap, then the posterior probability of the true model converges more quickly to the true one. This fact follows from standard asymptotic theory, as noted in Kass and Wasserman (1995). In those instances, we expect the severity of the *false signals* problem highlighted in this paper to be lessened. The simple logic behind this is that the more dimensions along which two models differ, the easier it becomes to find a way to tell them apart.

5. CONCLUDING REMARKS

In this paper we compare models with Bayesian posterior model probabilities working with a stylized specification of an open economy model that generates a short-run relationship between global slack and domestic inflation – the open-economy New Keynesian model of Martínez-García and Wynne (2010). Using a standard parameterization of the model, we generate artificial data which we then use to estimate four competing models (including the true model from which the data is simulated and three nested, simpler variants) with standard Bayesian techniques. We find that Bayesian model comparison based on posterior model probabilities is sensitive to the choice of observables and to sample size. While asymptotically the posterior probability of the true model converges to one, we show that in small samples (of moderate length) the posterior model probabilities penalization for overfitting may lead us to favor a more parsimonious model instead.

It has been argued in the literature that when the evidence favors the more parsimonious model, the costs in terms of fit cannot be too large because the probability distribution of the preferred model and the true model must be close. We believe, though, that this has consequences that go beyond our ability to fit the data. Selecting the wrong model (model selection) or accounting for model uncertainty (through model averaging) on the basis of posterior model probabilities that seemingly support the wrong specification affects our ability to use these models for things that we care about such as policy analysis or forecasting.

That is particularly important, e.g., when we think that Bayesian model comparison may have trouble to find support in the data for frictions in the goods market – nominal rigidities – if monetary policy is near optimal, even when those frictions are a feature of the economy. This can affect how we evaluate the costs of alternative monetary policies or how we forecast the future path of standard aggregate macro variables as the trade-offs that policymakers would face hinge on whether those frictions are present or not.

In our view, a strong preference for parsimonious models is not always and everywhere a desirable feature – even if they fit the data well. We see the primary contribution of our paper as illustrating how these "wrong choices" can occur in small samples and why it matters. We caution that variable selection may not help us eliminate the problem of *false signals* in model selection with small samples. We leave it for future research to investigate the small sample properties of other criteria for model comparison.

NOTES

1. We distinguish here between the endogenous international propagation that comes from trade and the purely exogenous international propagation that arises – even in the absence of trade–from the specification of correlated exogenous shock processes in both countries. By endogenous international propagation we refer more precisely to the effect that a shock impacting the foreign country has on the domestic macro aggregates as a result of the economic agents' response to that shock.

2. All variables are defined in logs as deviations from steady-state.

3. The inverse of the intertemporal elasticity of substitution is equal to 1 under the assumption of log-utility on consumption.

4. In that case, there would be no reason for these countries to trade with each other and in equilibrium there would be no exchange of goods anyway because the households of one country would not demand imports from the other country.

5. The terms of trade and the trade balance are related within the model as follows:

$$\widehat{tb}_t \approx (\sigma + (\sigma - 1)(1 - 2\xi))\xi\widehat{tot}_t$$

Hence, the so-called Harberger–Laursen–Metzler (HLM) effect arises naturally within this model: given a marginal propensity to consume less than unity, an improvement in a country's terms of trade raises current income but current consumption increases less than current income causing private savings to increase and improving the trade balance.

6. In keeping the prior invariant, we tie our hands to facilitate model comparison and preclude the priors themselves from becoming a source of additional degrees of freedom to *fine-tune* the estimation and the computation of posterior model probabilities.

7. That is the case assuming, as Martínez-García and Wynne (2010) do, that an optimal labor subsidy for firms funded with lump-sum (non-distortionary) taxes is set to eliminate the mark-up distortion.

8. In the model, the inflation differential is the same whether measured in terms of consumption or the output of the locally produced goods–i.e., $\hat{\pi}_t - \hat{\pi}_t^* \approx \hat{\pi}_t^H - \hat{\pi}_t^{F*}$. It is worth noting as well that the model implies growth differentials across countries should be reflected in the differentials between the inflation rates calculated on consumption and output (akin to the CPI and the GDP deflator respectively), i.e., $\hat{\pi}_t - \hat{\pi}_t^H \approx \xi\left(\frac{1}{\sigma - (\sigma - 1)(1 - 2\xi)^2}\right)(\Delta\hat{y}_t - \Delta\hat{y}_t^*)$ and $\hat{\pi}_t^* - \hat{\pi}_t^{F*} \approx -\xi\left(\frac{1}{\sigma - (\sigma - 1)(1 - 2\xi)^2}\right)(\Delta\hat{y}_t - \Delta\hat{y}_t^*)$.

9. The fact that we include inflation among our observables, however, may help distinguish the models with price stickiness from those under flexible prices if inflation is set to zero under flexible prices. However, more generally, flexible prices only imply that the output gap ought to be zero, but do not determine the inflation path *per se*. If we adopt the same monetary policy specification as for the NOEM model, this pins down a non-zero inflation rate that becomes increasingly less informative about the presence of nominal rigidities in the model as the solution of all four models becomes more similar.

10. See, e.g., the discussion on page 778 of Kass and Raftery (1995) of the approximation error of the Bayes Factors of nested models under the Laplace method.

11. The Gaussian state-space representation of the solution implies the normality of the likelihood for the models investigated in this paper. Sample size is a determinant factor on the appropriateness of the normal approximation for the posterior distribution. Slate (1994) provides guidance on the sample size requirements needed to obtain posterior normality and guarantee the accuracy of the Laplace's method for the exponential distribution family. The normal, gamma, and beta among other well-known distributions belong to the exponential family.

12. We can re-write the posterior model probability in Eq. (20) corresponding to model M_i for any $i = 2, \ldots, k$ in terms of Bayes Factors with respect to model M_1, B_{i1}, as follows:

$$\Pr(M_i \mid Z^n = z^n) = \frac{B_{i1} m_1 \Pr(M_i)}{\sum_{p=1}^{k} B_{p1} m_1 \Pr(M_p)} = \frac{e^{-\ln(B_{1i})}\Pr(M_i)}{\sum_{p=1}^{k} e^{-\ln(B_{1p})}\Pr(M_p)}, \quad \forall i = 2, \ldots, k$$

where in the second equality we use the fact that $B_{i1} = \frac{1}{B_{1i}}$ for all $i = 2, \ldots, k$. Then, it is possible to use the approximation result in Eq. (31) to express the posterior model probability in terms of the BIC as defined in Eq. (30), i.e.,

$$\Pr(M_i \mid Z^n = z^n) \approx \frac{e^{-\frac{1}{2}BIC_{1i}} \Pr(M_i)}{\sum_{p=1}^{k} e^{-\frac{1}{2}BIC_{1p}} \Pr(M_p)} \propto e^{-\frac{1}{2}BIC_{1i}} = e^{\frac{1}{2}BIC_{i1}}$$

Posterior model probabilities and the BIC are related up to an approximation of order $O_P(1)$ as well, and should be asymptotically equivalent (under weak conditions).

13. The unit information prior is a data-dependent prior (typically multivariate Normal) with mean at the MLE estimator, and precision equal to the information provided by one observation.

14. The BIC is part of a family of competing penalized likelihood functions that also includes the Akaike Information Criterion, the Deviance Information Criterion (DIC), or the Takeuchi Information Criterion (TIC). These functions differ mostly on the penalty they impose for overfitting. The AIC has a fixed penalty that does not grow with $\ln(n)$, i.e., $l_i(\tilde{\theta}_i) - d_i$ where d_i is the dimensionality of the parameter space. Although it can be shown that AIC is optimal in the sense of minimizing the Kullback–Leibler (KL) distance criterion, when it comes to model selection it is not consistent asymptotically, unlike the BIC. For sample sizes of eight or more observations, BIC has a higher penalty for overfitting than the AIC. Hence, since a sample size of less than eight observations is unrealistic, we can say that the BIC penalizes complex models more than other well-known model selection criteria such as the AIC.

15. For the implementation of the local identification procedure of Iskrev (2010) adopted by the software package Dynare and their implementation of an optional Monte Carlo exploration of the space of model parameters, see Ratto and Iskrev (2011).

16. We use two different ways of accomplishing this. When we compare closed-economy versus open-economy models we do so by bringing the import share closer and closer to zero. In the case where we are comparing the NOEM model against the International Real Business Cycle model we do not alter the degree of price stickiness directly but in turn bring the two distributions closer together by choosing to implement an increasingly more aggressive monetary policy that is closer to the optimal policy.

ACKNOWLEDGMENTS

We would like to thank Nathan Balke, María Teresa Martínez-García, and Valentín Martínez Mira for helpful suggestions. Diego Vilán was a co-author in a related project and contributed to the early stages of the development of this paper. We gratefully acknowledge the outstanding research assistance provided by Valerie Grossman, the help of Bradley Graves and Kuhu

Parasrampuria, and the Federal Reserve Bank of Dallas' support. All remaining errors are ours alone. The views expressed in this paper are those of the authors and do not necessarily reflect the views of the Federal Reserve Bank of Dallas, or the Federal Reserve System.

REFERENCES

Adjemian, S., Bastani, H., Juillard, M., Mihoubi, F., Perendia, G., Ratto, M., & Villemot, S. (2011). *Dynare: Reference manual, version 4*. Dynare Working Papers 1, CEPREMAP. Retrieved from http://www.dynare.org/wp-repo/dynarewp001.pdf

Akaike, H. (1978). A new look at the Bayes procedure. *Biometrika*, *65*, 53–59.

An, S., & Schorfheide, F. (2007). Bayesian analysis of DSGE models. *Econometric Reviews*, *26*(2-4), 113–172.

Andrle, M. (2010). *A Note on identication patterns in DSGE models*. European Central Bank Working Paper No. 1235. Retrieved from http://www.ecb.europa.eu/pub/pdf/scpwps/ecbwp1235.pdf

Blanchard, O. J., & Kahn, C. M. (1980). The solution of linear difference models under rational expectations. *Econometrica*, *48*(5), 1305–13011.

Calvo, G. A. (1983). Staggered prices in a utility-maximizing ramework. *Journal of Monetary Economics*, *12*(3), 383–398.

Canova, F., & Sala, L. (2009). Back to square one: Identication issues in DSGE models. *Journal of Monetary Economics*, *56*(4), 431–449.

Clarida, R., Galí, J., & Gertler, M. (1999). The science of monetary policy: A new Keynesian perspective. *Journal of Economic Literature*, *37*(4), 1661–1707.

Clarida, R., Galí, J., & Gertler, M. (2002). A simple framework for international monetary policy analysis. *Journal of Monetary Economics*, *49*(5), 879–904.

Corduneanu, A., & Bishop, C. (2001). Variational Bayesian model selection for mixture distributions. In T. Richardson & T. Jaakkola (Eds.), *Proceedings of the eighth international conference on artificial intelligence and statistics* (pp. 27–34). San Francisco, CA: Morgan Kaufmann Publishers Inc. Retrieved from http://research.microsoft.com/pubs/67239/bishop-aistats01.pdf

Fernández-Villaverde, J., & Rubio-Ramírez, J. F. (2004). Comparing dynamic equilibrium models to data: A Bayesian approach. *Journal of Econometrics*, *123*(1), 153–187.

Fernández-Villaverde, J., Rubio-Ramírez, J. F., Sargent, T., & Watson, M. (2007). A, B, C, (and D)'s for understanding VARs. *American Economic Review*, *97*, 1021–1026.

Geweke, J. (1989). Bayesian inference in econometric models using Monte Carlo integration. *Econometrica*, *57*, 1317–1340.

Goodfriend, M., & King, R. (1997). The new neoclassical synthesis and the role of monetary policy, NBER chapters. In *NBER Macroeconomics Annual 1997* (Vol. 12, pp. 231–296). Cambridge, MA: MIT Press, National Bureau of Economic Research, Inc. Retrieved from http://www.nber.org/chapters/c11040.pdf

Guerrón-Quintana, P. A. (2010). What you match does matter: The effects of data on DSGE estimation. *Journal of Applied Econometrics*, *25*(5), 774–804.

Hamilton, J. D. (1994). State-space models. In R. Engle & D. McFadden (Eds.), *Handbook of econometrics* (Vol. 4, Chap. 50, pp. 3039–3080). Amsterdam: Elsevier Science B.V.

Hoeting, J. A., Madigan, D., Raftery, A. E., & Volinsky, C. T. (1999). Bayesian model averaging: A tutorial. *Statistical Science*, *14*(4), 382–417.

Hsiao, C. (1983). Identification. In Z. Griliches & M. D. Intriligator (Eds.), *Handbook of econometrics* (Vol. 1, Chap. 4). Amsterdam: North-Holland.

Iskrev, N. I. (2010). Local identification in DSGE models. *Journal of Monetary Economics*, *57*(2), 189–202.

Kass, R. E., & Raftery, A. E. (1995). Bayes factors. *Journal of the American Statistical Association*, *90*(430), 773–795.

Kass, R. E., Tierney, L., & Kadane, J. B. (1988). Asymptotics in Bayesian computation. In J. Bernardo, M. DeGroot, D. Lindley, & A. Smith (Eds.), *Bayesian statistics* (Vol. 3). Oxford: Oxford University Press.

Kass, R. E., Tierney, L., & Kadane, J. B. (1990). The validity of posterior asymptotic expansions based on Laplace's method. In S. Geisser, J. S. Hodges, S. J. Press, & A. Zellner (Eds.), *Bayesian and likelihood methods in statistics and econometrics*. New York, NY: North-Holland.

Kass, R. E., & Wasserman, L. (1995). A reference Bayesian Test for nested hypotheses and its relationship to the Schwarz criterion. *Journal of the American Statistical Association*, *90*(431), 928–934.

Komunjer, I., & Ng, S. (2011). Dynamic identification of dynamic stochastic general equilibrium models. *Econometrica*, *79*(6), 1995–2032.

Koop, G., Pesaran, M. H., & Smith, R. P. (2013). On Identification of Bayesian DSGE models. *Journal of Business and Economic Statistics*, *31*(3), 300–314.

Martínez-García, E., & Søndergaard, J. (2009). Investment and trade patterns in a sticky-price, open-economy model. In G. Calcagnini & E. Saltari (Eds.), *The economics of imperfect markets. The effect of market imperfections on economic decision-making. Contributions to economics*. Heidelberg: Springer (Physica-Verlag).

Martínez-García, E., Vilán, D., & Wynne, M. A. (2012). Bayesian estimation of NOEM models: Identication and inference in small samples. *Advances in Econometrics*, *28*, 137–199.

Martínez-García, E., & Wynne, M. A. (2010, September). *The global slack hypothesis*. Federal Reserve Bank of Dallas Staff Papers No. 10. Retrieved from http://www.dallasfed.org/assets/documents/research/staff/staff1002.pdf

Martínez-García, E., & Wynne, M. A. (2014, August). *Technical note on 'assessing Bayesian model comparison in small samples'*. Federal Reserve Bank of Dallas, Globalization and Monetary Policy Institute Working Paper No. 190. Retrieved from http://www.dallasfed.org/assets/documents/institute/wpapers/2014/0190.pdf

Minka, T. P. (2001). Expectation propagation for approximate Bayesian inference. In *Proceedings of the seventeenth conference on uncertainty in artificial intelligence* (pp. 362–369). San Francisco, CA: Morgan Kaufmann Publishers Inc. Retrieved from http://arxiv.org/ftp/arxiv/papers/1301/1301.2294.pdf

Mutschler, W. (2014). *Identification of DSGE models – A comparison of methods and the effect of second order approximation*. Mimeo, University of Münster.

Neal, R. M. (2001). Annealed importance sampling. *Statistics and Computing*, *11*(2), 125–139.

Qian, G., & Künsch, H. (1998). Some notes on Rissanen's stochastic complexity. *IEEE Transactions on Information Theory*, *42*(2), 782–786.

Qu, Z., & Tkachenko, D. (2012). Identification and frequency domain quasi-maximum likelihood estimation of linearized dynamic stochastic general equilibrium models. *Quantitative Economics*, *3*(1), 95–132.

Ratto, M., & Iskrev, N. I. (2011). *Identification analysis of DSGE models with Dynare*. European commission and Banco de Portugal. Retrieved from https://www.ifk-cfs.de/. leadmin/downloads/events/conferences/mon.spol2011/RATTO_IdentifFinal.pdf

Ríos-Rull, J.-V., Schorfheide, F., Fuentes-Albero, C., Kryshko, M., & Santaeulàlia-Llopis, R. (2012). Methods versus substance: Measuring the effects of technology shocks. *Journal of Monetary Economics*, *59*(8), 826–846.

Rissanen, J. (1987). Stochastic complexity (with discussion). *Journal of the Royal Statistical Society, Series B*, *49*(3), 223–265.

Rissanen, J. (1996). Fisher information and stochastic complexity. *IEEE Transactions on Information Theory*, *42*, 40–47.

Schwarz, G. (1978). Estimating the dimension of a model. *The Annals of Statistics*, *6*(2), 461–464.

Slate, E. (1994). Parameterizations for natural exponential families with quadratic variance functions. *Journal of the American Statistical Association*, *89*, 1471–1482.

Taylor, J. B. (1993). Discretion versus policy rules in practice. *Carnegie-Rochester Conference Series on Public Policy*, *39*, 195–214.

Verdinelli, I., & Wasserman, L. (1995). Computing Bayes factors using a generalization of the Savage-Dickey density ratio. *Journal of the American Statistical Association*, *90*(430), 614–618.

Woodford, M. (2003). *Interest and prices, Foundations of a theory of monetary policy*. Princeton, NJ: Princeton University Press.

Zellner, A. (1971). *An introduction to Bayesian inference in econometrics*. Hoboken, NJ: Wiley.

BAYESIAN SELECTION OF SYSTEMIC RISK NETWORKS

Daniel Felix Ahelegbey[a] and Paolo Giudici[b]

[a]*Graduate School of Economics and Management, University of Venice, Venice, Italy*

[b]*Department of Economics and Management, University of Pavia, Pavia, Italy*

ABSTRACT

The latest financial crisis has stressed the need of understanding the world financial system as a network of interconnected institutions, where financial linkages play a fundamental role in the spread of systemic risks. In this paper we propose to enrich the topological perspective of network models with a more structured statistical framework, that of Bayesian Gaussian graphical models. From a statistical viewpoint, we propose a new class of hierarchical Bayesian graphical models that can split correlations between institutions into country specific and idiosyncratic ones, in a way that parallels the decomposition of returns in the well-known Capital Asset Pricing Model. From a financial economics viewpoint, we suggest a way to model systemic risk that can explicitly take into account frictions between different financial markets, particularly suited to study the ongoing banking union process in Europe. From a

Bayesian Model Comparison
Advances in Econometrics, Volume 34, 117–153

ISSN: 0731-9053/doi:10.1108/S0731-905320140000034007

computational viewpoint, we develop a novel Markov chain Monte Carlo algorithm based on Bayes factor thresholding.

Keywords: Bayesian analysis; financial econometrics; financial risk management; graphical Gaussian models; model construction and estimation

JEL classifications: C58; C51; C11

1. MOTIVATION

In the latest financial crisis, started in 2007, the core capital of banks has proved to be insufficient to cover impairment losses arising from loans and security portfolios. Consequently, several banks have strengthened their capital base or reduced their asset exposure. Other banks have been bailed out by state aids or have defaulted. To reduce the risk of similar crises in the future and to enhance the resilience of the banking sector, a new regulatory framework, the so-called Basel III package, has been proposed, implying more stringent capital requirements for financial institutions (Basel Committee on Banking Supervision, 2011). The effectiveness of the new regulatory framework to prevent banking default and financial crisis is an open problem, particularly as regulations themselves are still in progress, and may thus benefit from the results of research findings in the field.

The study of bank failures is important for two reasons. First, an understanding of the factors related to bank failure enables regulatory authorities to supervise banks more efficiently. In other words, if supervisors can detect problems early enough, regulatory actions can be taken to prevent a bank from failing and, therefore, to reduce the costs of its bail-in, faced by shareholders, bondholders, and depositors; as well as those of its bail-out, faced by the governments and, therefore, by the taxpayers. Second, the failure of a bank very likely induces failures of other banks or of parts of the financial system as a whole. Understanding the determinants of a single bank failure may thus help to understand the determinants of financial systemic risks, which could be due to microeconomic, idiosyncratic factors, or macroeconomic imbalances. When problems are detected, their causes can be removed or isolated, to limit "contagion effects."

The literature on predictive models for single bank failures is relatively recent: until the 1990s, most authors emphasize the absence of default risk of a bank (see, e.g., Gup, 1998; Roth, 1994), in the presence of a generalized expectation of state interventions. However, in the last years we have witnessed the emergence of financial crisis in different areas of the world, and a correlated emphasis on systemic financial risks. In addition, government themselves are less willing than before to save banks, partly for their financial shortages and partly for a growing negative sentiment from the public opinion. As a consequence of all of these, recent research trends have seen a growing body of literature on bank failures, and systemic risks originating from the above mentioned sentiments and actions.

The studies on bank failures can be classified into three main streams: financial market models, scoring models, and macroeconomic models.

Financial market models originate from the seminal paper of Merton (1974), in which the market value of a bank's asset, typically modeled as a diffusion process, is insufficient to meet its liabilities. Due to its practical limitations, Merton's model has been evolved into a reduced form (see, e.g., Vasicek, 1984), leading to a widespread diffusion of the resulting model, and the related implementation in Basel II credit portfolio model. For a review of this evolution see, for example, the book by Resti and Sironi (2007). In order to implement market models, diffusion process parameters and, therefore, bank default probabilities can be obtained on the basis of share price data that can be collected almost in real time from financial markets. Market data are public, relatively easy to collect, and are quite objective. On the other hand, they may not reflect the true fundamentals of the underlying financial institutions, and may lead to a biased estimation of the probability of failure. Indeed, the recent paper by Idier, Lamé, and Mésonnier (2013) and Fantazzini and Maggi (2012) show that market models may be good in very short-term predictions, but not in medium and long-term ones, where the importance of fundamental financial data emerge.

Scoring models are based only on financial fundamentals, taken from the publicly available balance sheet information. Their diffusion followed the seminal paper by Altman (1968), which has induced the production of scoring models for banks themselves: notable examples are Sinkey (1975), Tam and Kiang (1992), Rose and Kolari (1985), and Cole and Gunther (1998). The development of the Basel regulation and the recent financial crisis have further boosted the literature on scoring models for banking failure predictions. Recent examples include Arena (2005) and Davis and Karim (2008) who use logit models; Vazquez and Federico (2012) who use a probit model and Klomp and Haan (2012) who use a principal component factor approach. Scoring models have been extended in different ways: interesting developments include the incorporation of macroeconomic components (see, e.g., Kanno, 2013; Kenny, Kostka, & Masera, 2013; Koopman, Lucas, & Schwaab, 2011; Mare, 2012); and the explicit consideration of the credit portfolio, as in the symbol model of De Lisa, Zedda, Vallascas, Campolongo, and Marchesi (2011) – that allows stress tests of banking asset quality and capital, as emphasized in the recent paper by Halaj (2013). The problem with scoring models is that they are mostly based on balance sheet data, which are different from the market and has a low frequency of update (annual or, at best, quarterly) and depends on subjective strategic choices. They may thus be good to predict defaults (especially in the medium term) but not in the assessment of systemic risks, which occur very dynamically and with short notice.

Our focus here will not be on the prediction of single defaults but rather, on how such prediction are correlated with each other, in a systemic perspective. The research literature on systemic risk is very recent, and follows closely the developments of the recent financial crisis. A comprehensive review is discussed in Brunnermeier and Oehmke (2012) who also provide a historical comparison of different crisis. Specific measures of systemic risk have been proposed, in particular, by Adrian and Brunnermeier (2010), Acharya, Pedersen, Philippon, and Richardson (2010), Brownlees and Engle (2011), Huang, Zhou, and Zhu (2011), Billio, Getmansky, Lo, and Pelizzon (2012) and, from a different perspective, Segoviano and Goodhart (2009). All of these approaches are built on financial market price information, on the basis of which they lead to the estimation of appropriate quantiles of the estimated loss probability distribution of a financial institution, conditional on a crash event on the financial market. These literature developments have led to and are still contributing to the identification of the Systemically Important Financial Institutions (SIFIs), at the global and regional level. They however do not address the issue of how risks are transmitted between different institutions.

Systemic risk can indeed also be defined as the risk that the failure of one significant financial institution can cause or significantly contribute to the failure of other significant financial institutions, as a result of their linkages to each other (see, e.g., Billio et al., 2012). Trying to address this aspect of systemic risks, researchers have recently proposed network models that can help model the systemic risk in financial systems which display complex degrees of connectedness. In particular, Billio et al. (2012) propose several econometric measures of connectedness based on principal component analysis and Granger-causality networks. They find that hedge funds, banks, and insurance companies have become highly interrelated over the past decade, likely increasing the level of systemic risk through a complex and time-varying network of relationships. Chen, Cummins, Viswanathan, and Weiss (2013) and Barigozzi and Brownlees (2013) follow similar approaches.

The adoption of a robust financial network approach in the above context is recommended not only because of its proper emphasis on financial interdependencies, but also for its possibility to describe how the structure of these interdependencies evolves in time. If this is achieved, we would be able to address systemic financial risks in two directions: on one hand, to understand the role that a network structure plays in the spread of financial shocks; on the other hand, to understand the impact of simulated stress events on a network of interdependencies (see, e.g., Castrén & Kavonius, 2009; Cont, Moussa, & Santos, 2011; Gai & Kapadia, 2009; Georg & Poschmann, 2010; Markose, Giansante, Gatkowski, & Shaghaghi, 2010; Nier, Yang, Yorulmazer, & Alentorn, 2007; Peltonen, Scheicher, & Vuillemey, 2013). Simulated networks, as those described in the previous paper, can help in studying the impact of crisis scenarios on the system, in a stress-testing framework. However, they typically assume a fixed network structure which may not be the one supported by the data.

To learn financial networks from the data we propose empowering network models with multivariate graphical models. Graphical models embed the idea that interactions among random variables in a system can be represented in the form of graphs, whose nodes represents the variables and whose edges shows their interactions. For an introduction to graphical models see, for example, Edwards (1990), Lauritzen (1996), Lauritzen and Wermuth (1989), Pearl (1988), Wermuth and Lauritzen (1990), and Whittaker (1990).

Graphical models can be employed to accurately estimate the adjacency matrix, aimed at measuring interconnectedness between different financial institutions and, in particular, to assess central ones that may be the most contagious or the strongest source of contagion (as in Billio et al., 2012). Network models use the correlation matrix estimated from the data to derive the adjacency matrix. Although useful, this approach takes into account only the marginal (indirect) effect of a variable on another, without looking at the (direct) effect of other variables. In our context, it does not distinguish between the direct and the indirect effect of a country on other ones. Graphical models, instead, focus on the partial correlation matrix that is obtained by measuring only the direct correlation between two variables. A partial correlation coefficient can express the change in the expected value of a dependent variable, caused by a unitary change of the independent variables, when the remaining variables are held constant. In so doing, the effect of a bank on another is split into a direct effect (estimated by the partial correlation) and an indirect effect (what is left in the marginal correlation). Here we follow this approach and derive the adjacency matrix, the main input of a financial network model, not from the correlation matrix but, rather, from the partial correlation matrix obtained from the application of graphical models to the available data.

To achieve this aim we consider multivariate Gaussian graphical models, defined in terms of Markov properties, that is, conditional independencies associated with the underlying graph (Lauritzen, 1996; Whittaker, 1990). While traditional network models assume fixed graphical structures (such as fully connected graphs), the structure of a graphical model is typically random, and can be learned from the data, as a good fitting structure. Such a model selection can be performed by testing, in a stepwise procedure, the statistical significance of conditional independencies, which are equivalent to specified zeroes among certain partial correlation coefficients which, in turn, are equivalent to missing edges in the network representation.

The use of graphical models can thus help to have a deeper understanding of the relationships between variables by distinguishing direct from indirect relationships. From their appearance in the 1990s, several methodological advances have been made for graphical models. Less so in terms of applications, especially in financial economics. In our opinion, this requires to solve two main problems.

First, the assumption of a random graph may be an important added value in situations where little a priori knowledge is present, as is the case for systemic risk. In addition, results should not be conditioned on single models, but, rather, should be model averaged, to avoid suboptimal inferences.

Second, graphical models do not allow to decompose asset return correlations into market specific and idiosyncratic effects, as in the classical CAPM models (Sharpe, 1964). This assumption seems to be restrictive in finance.

The first problem can be solved with the use of more advanced, Bayesian, graphical models, as shown in Madigan and York (1995), Giudici and Green (1999) and, more recently, Ahelegbey, Billio, and Casarin (2014). In particular, Madigan and York (1995) and Giudici and Green (1999) propose a Bayesian model able to consider all possible graphical structures, choose the best fitting ones and, if necessary, average inferential results over the set of all models.

The methodological contribution of this paper can solve the second problem as well. We propose Bayesian hierarchical graphical models that allow correlations to be decomposed, as in a CAPM-like model, into a country (market) effect plus a bank-specific (idiosyncratic) effect.

On the other hand, the applied contribution of this paper is in the understanding of whether and how a distress probability is transmitted between different banks, that belong to different countries, with different regulatory systems. The world financial market is not perfect: many frictions exists between different countries, mainly due to different regulations, given the fundamental relevance of banks for the economies to which they belong. A very interesting case study, in this respect, is the Eurozone, where the European Central Bank (ECB) has recently taken over the supervision of the largest banks (with total assets greater than 30bn euro) in each of the member states. Thus, eventually, the euro banking market will evolve into a single market but, at the time being, it is still fragmented. It thus becomes timely and rather interesting to study the degree of convergence toward a European banking union, looking at the comovements between stock returns of the banks in the area, that may give important insights.

Finally, we remark that the paper also contributes to computational statistics as, in order to apply the proposed hierarchical model to the large p/small n database at hand, a novel Markov chain Monte Carlo algorithm, based on Bayes factor thresholding, has been developed and implemented in a Matlab routine, available upon request.

The paper is organized as follows. In Section 2, we introduce Bayesian graphical models, and show their theoretical implications: in particular, we show how a hierarchical approach can decompose the variance–covariance matrix of the model into a variance–covariance matrix between countries and, conditionally on it, on a variance–covariance matrix between banks. In Section 3 we introduce a simulated data application that is helpful for the set-up of our computational methodology. Section 4 describes the empirical results obtained with the application of the Bayesian hierarchical graphical models to the data. Finally, Section 5 contains some concluding remarks and future research directions.

2. METHODOLOGY

In this section we review the basics of Bayesian graphical models and, then, describe the methodology and the computational approach we propose.

2.1. Bayesian Graphical Models

From a statistical viewpoint, while correlations can be estimated, on the basis of the N observed time series of data, assuming that, at each time point, observations follow a multivariate Gaussian model, with unknown variance–covariance matrix Σ, partial correlations can be estimated assuming that the same observations follow a graphical Gaussian model, in which Σ is constrained by the conditional independence described by a graph.

Let $X = (X_1, \ldots, X_N) \in R^N$ be a random vector distributed according to a multivariate normal distribution $\mathcal{N}_N(\mu, \Sigma)$. In this paper, without loss of generality, we will assume that the data are generated by a stationary process, and, therefore, $\mu = 0$. In addition, we will assume throughout that the covariance matrix Σ is nonsingular.

Let $G = (V, E)$ be an undirected graph, with vertex set $V = \{1, \ldots, N\}$, and edge set $E = V \times V$, a binary matrix, with elements e_{ij}, that describes whether pairs of vertices are (symmetrically) linked between each other ($e_{ij} = 1$), or not ($e_{ij} = 0$). If the vertices V of this graph are put in correspondence with the random variables $X_1, \ldots, X_N$, the edge set E induces conditional independence on X via the so-called Markov properties (see, e.g., Lauritzen, 1996; Whittaker, 1990) or, from an econometric viewpoint, Carvalho and West (2007) and Corander and Villani (2006)).

More precisely, the pairwise Markov property determined by G states that, for all $1 \le i < j \le N$,

$$e_{ij} = 0 \Leftrightarrow X_i \perp X_j \mid X_{V \setminus \{i,j\}}$$

that is, the absence of an edge between vertices i and j is equivalent to independence between the random variables X_i and X_j, conditionally on all other variables $X_{V \setminus \{i,j\}}$.

In our context, all random variables are continuous and it is assumed that $X \sim \mathcal{N}_N(0, \Sigma)$. Let the elements of Σ^{-1}, the inverse of the variance–covariance matrix, be indicated as $\{\sigma^{ij}\}$. It can be shown (see, e.g., Whittaker, 1990) that the following equivalence also holds:

$$X_i \perp X_j \mid X_{V \setminus \{i,j\}} \Leftrightarrow \rho_{ijV} = 0$$

where

$$\rho_{ijV} = \frac{-\sigma^{ij}}{\sqrt{\sigma^{ii}\sigma^{jj}}}$$

denotes the *ij*th partial correlation, that is, the correlation between X_i and X_j conditionally on the remaining variables $X_{V\backslash\{i,j\}}$.

Therefore, by means of the pairwise Markov property, given an undirected graph $G=(V,E)$, a graphical Gaussian model can be defined as the family of all N-variate normal distributions $\mathcal{N}_N(0,\Sigma)$ that satisfy the constraints induced by the graph on the partial correlations, as follows:

$$e_{ij}=0 \Leftrightarrow \rho_{ijV}=0$$

for all $1 \leq i < j \leq N$.

Graphical model uncertainty can be taken into account, along with parameter uncertainty, within a Bayesian approach, whose main practical advantage is that inferences on quantities of interest can be averaged over different models, each of which has a weight that corresponds to its Bayesian posterior probability. (see, e.g., Giudici & Castelo, 2003; Giudici & Green, 1999; Madigan & York, 1995).

To achieve this aim, the first task is to recall the expression of the marginal likelihood of a graphical Gaussian model, and specify prior distributions over the parameter Σ as well as on the graphical structures G.

For a given graph G, consider a sample x of size n from $P=N_N(0,\Sigma)$, and let S_n be the corresponding observed variance–covariance matrix. For a subset of vertices $A \subset N$, let Σ_A denote the variance–covariance matrix of the variables in X_A, and define Σ_A the corresponding observed matrix.

When the graph G is decomposable the likelihood of a graphical Gausssian model specified by P nicely decomposes as follows (see, e.g., Dawid & Lauritzen, 1993):

$$p(x \mid \Sigma, G)=\frac{\prod_{C \in C} p(x_C \mid \Sigma_C)}{\prod_{S \in S} p(x_S \mid \Sigma_S)}$$

where C and S respectively denote the set of cliques and separators of the graph G, and

$$P(x_C \mid \Sigma_C)=(2\pi)^{-\frac{n|C|}{2}} \mid \Sigma_C \mid^{-\frac{n}{2}} \exp\left[-\frac{1}{2} tr\{S_C \Sigma_C^{-1}\}\right]$$

and similarly for $P(x_S \mid \Sigma_S)$.

Dawid and Lauritzen (1993) propose a convenient prior for the parameters of the above likelihood, which is named hyper Wishart. It can be obtained from a collection of clique specific marginal Wishart as follows:

$$\frac{\prod_{C \in C} l(\Sigma_C)}{\prod_{S \in S} l(\Sigma_S)}$$

where $l(\Sigma_C)$ is the density of a Wishart distribution, with hyper-parameters T_C and α, and similarly for $l(\Sigma_S)$. For the definition of the hyper-parameters, we follow Giudici and Green (1999) and let T_{0C} and T_{0S} be the sub-matrices of a large matrix T_0 of dimension $N \times N$, and choose $\alpha > N$. To complete prior specification, for $P(G)$ we assume a uniform prior over all possible graphical structures.

Dawid and Lauritzen (1993) show that, under the previous assumptions, the posterior distribution of the variance–covariance matrix Σ, is a hyper-Wishart distribution with $\alpha + N$ degrees of freedom and a scale matrix given by

$$T_n = T_0 + S_n$$

where S_n is the sample variance–covariance matrix.

The proposed prior distributions can also be used to integrate the likelihood with respect to the unknown random parameters, obtaining the so-called marginal likelihood of a graph, which will be the main metric for model selection. This follows from

$$P(X \mid G) = P(X \mid \Sigma, G)P(\Sigma)$$

Giudici and Green (1999) show that such marginal likelihood is equal to

$$p(x \mid G) = \frac{\prod_{C \in C} p(x_C)}{\prod_{S \in S} p(x_S)}$$

where

$$P(x_C) = (\pi)^{-\frac{n|C|}{2}} \frac{k(|C|, \alpha + n)}{k(|C|, \alpha)} \frac{det(T_{0C})^{\alpha/2}}{\det(T_{nC})^{(\alpha+n)/2}}$$

where $k(\cdot)$ is a normalization constant, given by $k(x,y) = \prod_{j=1}^{x} \Gamma\left(\frac{y+1-j}{2}\right)$, T_{0C} and T_{nC} are sub-matrices corresponding to x_C of a large matrix T_0 and T_n.

The metric expressed by the above marginal likelihood is the basic ingredient for Gaussian graphical model selection and averaging, as will now be shown.

According to the conventional Bayesian paradigm, being the model space discrete, the best graphical model will be that with the highest a posteriori probability. By Bayes rule, the posterior probability of a graph is given by

$$P(G \mid X) \propto P(X \mid G)P(G)$$

and, therefore, since we assumed a uniform prior over the graph structures, maximizing the posterior probability is equivalent to maximizing the marginal likelihood metric. For graphical model selection purposes we shall thus search in the space of all possible graphs for the structure such that

$$G^* = \arg\max_G P(G \mid X) \propto P(X \mid G)$$

The Bayesian paradigm does not force conditioning inferences on the (best) model chosen. The assumption of G being random, with a prior distribution on it, allows any inference on quantitative parameters to be model averaged with respect to all possible graphical structures, with weights that correspond to the posterior probabilities of each graph. This because, by Bayes' theorem

$$P(\Sigma \mid X) = P(\Sigma \mid X, G)P(G \mid X)$$

The above allows to overcome the main drawback of non-Bayesian graphical models, namely, the fact that, once a model is chosen, all inferences will be conditional on that model, even if it has a little support from the data (although maximal).

However, in real situations, the number of possible graphical structures may be very large and we may need to restrict the number of models to be averaged. This can be done efficiently, for example, following a simulation-based procedure for model search, such as Markov chain Monte Carlo (MCMC) sampling, described in Madigan and York (1995). One of the standard MCMC methods is the Metropolis–Hastings (MH) algorithm, which is based on an acceptance–rejection scheme. In our context, given an initial graph, the algorithm samples a new graph using a proposal distribution.

We remark that, to guarantee irreducibility of the Markov chain, we follow Giudici and Castelo (2003) and test whether the proposal is a decomposable graph. Following Giudici and Green (1999) and Frydenberg and Lauritzen (1989) we apply the concept of junction tree (see Lauritzen, 1996 or Cowell, Dawid, Lauritzen, & Spiegelhalter, 1999). More specifically, the addition of an edge between two nodes is allowed if the two nodes belong to two different connected components or if the cliques they belong to are connected by a separator. On the other hand, the removal of an edge is allowed if such edge belongs to a single clique. After each iteration the junction forest is updated according to the new organization of the cliques.

The newly sampled graph is then compared with the old graph, with a decision rule to either reject or accept the proposed sample. The proposed graph G_{new} is either accepted, becoming (G_{old}) in the next step, or rejected in which case the previous graph is maintained as G_{old}. The decision to accept or reject a proposed graph depends on an acceptance probability.

By assuming a uniform graph prior, the log acceptance probability is given by

$$\log(Ac) = \max\{\log(P(X \mid G_{\text{new}})) - \log(P(X \mid G_{\text{old}})), 0\}$$

2.2. Hierarchical Graphical Models

One of the appealing features of graphical models for multivariate time series analysis is to represent graphically the logical implications as well as the conditional independence relationships among the variables. In high dimensional settings, it is extremely hard to extract meaningful information from the complex interaction among the variables. In addition, learning such complex interactions is computationally intensive when using standard structure learning schemes. There is a high chance that the learning algorithm gets trapped, spending much time to learn local optimum structures which might not be representative. A possible solution to the trap problem is to add more structure to the graphical model. For example, in the Bayesian setting, the prior distribution over the graph space can be enriched and used to penalize complex structures. Another approach, particularly suited for the applied problem considered in this paper, is to employ hierarchical graphical models: this will allow most of structural learning to be localized to groups of variables which however are not independent and, thus, can borrow inferential strength from each other.

Several researchers have discussed and applied hierarchies in graphical models (see, e.g., Gyftodimos & Flach, 2002; Hensman, Lawrence, & Rattray, 2013; Zhang, Wells, Turnbull, Sparrow, & Cassano, 2005). The approach proposed in this paper is in the spirit of Guo, Levina, Michailidis, and Zhu (2011, 2013) but from a Bayesian perspective.

Suppose we have a large data set of n random variables, $X=(X_1,\dots,X_n)$ that can be grouped into k categories, $(Z_1,\dots,Z_k)$, such that the kth category contains nk variables $Z_k=(Y_1,\dots,Y_{nk})$. Without loss of generality, we assume that each variable in X belongs to exactly one of the categories, that is, $\sum_{i=1}^{k} n_i = n$, and that the observations in the same category are centered along each variable. To exemplify, the random variables in X can be banks that can be grouped into k different countries (represented as Z) with each country consisting of a number of banks (represented as Y), so that $X=(Y,Z)$.

In this context, we consider Z as a compressed representation of the banking system in different countries. Based on the classification, we build a two-level hierarchical model composed of a country-specific component that explains relationships between banking systems of different countries, and an idiosyncratic component that models relationships between banks in the same country. More formally

$$P(X)=P(Y,Z)=P(Z)P(Y \mid Z)$$

To achieve this aim, we introduce a one-way decomposition of the covariance between any two banks, X_i, $X_j \in X$, that belong, respectively, to countries Z_a and Z_b. Let G_z be a $k \times k$, 0\1 matrix, representing the structure of between country relationships. In our context, all random variables are continuous and it is assumed that $X \sim \mathcal{N}_n(0,\Sigma)$, $Z \sim \mathcal{N}_k(0,\Phi)$, and $Y \sim \mathcal{N}_n(0,\Psi)$. We assume that:

$$\sigma_{i,j} = g^z_{a,b} \circ (\phi_{a,b} \circ \psi_{i,j}) + (1-g^z_{a,b}) \circ \psi_{i,j}, \quad X_i \in Z_a,\ X_j \in Z_b$$
$$\sigma_{i,j} = (\phi_{a,a} \circ \psi_{i,j}), \quad X_i, X_j \in Z_a, X_i \neq X_j$$

where $(\circ)$ is the Hadamard product, $g^z_{a,b}=G_z(a,b)$ captures the link between country Z_a and Z_b, $\phi_{a,b}$ is the covariance between Z_a and Z_b, and

$\psi_{i,j}$ is the covariance between banks Y_i and Y_j. This decomposition treats $\phi_{a,b}$ as a common factor controlling any linkage between two banks Y_i and Y_j from different countries. On the other hand, $\psi_{i,j}$ measures the covariance between banks Y_i and Y_j, independent of a country-specific effect.

A nice aspect of the above model is that it can be further decomposed, to accommodate for further grouping effects. For instance, in our specific application, the term $\phi_{a,b}$ may be further decomposed into a macroeconomic country effect plus a banking sector specific ones, calculating the difference between macroeconomic log returns (e.g., of the Gross Domestic Product of a country) and capitalization weighted log returns of the banking sector of a country. In addition, the term ψ_{y_i,y_j} can also be similarly decomposed in groups, according to bank-specific balance sheet variables (such as asset size, leverage, etc.).

2.3. Efficient Structural Inference Scheme

We now focus on sampling the multivariate instantaneous relationships among random variables by allowing for simultaneous interactions. We will do so by extending the work by Ahelegbey et al. (2014). In large dimensional settings, a common drawback of the classical M−H sampling scheme is the likelihood of spending much time to learn local optimum structures which might not be representative of the global optimum structure. Second, sampling structures with simultaneous interactions (undirected links) leads to difficulties in diagnosing convergence of the chain. According to Madigan and York (1995), irreducibility is guaranteed on condition that the structures satisfy acyclic constraints. This makes the standard scheme not feasible in sampling undirected structures. In this paper, we propose a Markov chain Monte Carlo scheme that employs a Bayes factor thresholding.

A sketch of the idea is as follows: We initialize the algorithm by assuming a fully connected structure (G_{old}). At each iteration, we delete the edge between two variables X_i and X_j to produce a new structure (G_{new}). We then compute the posterior of the two structures and compute the Bayes factor. By assuming a uniform graph prior, it can be shown that the log Bayes factor is given by

$$\log(\text{BF}) = \log(P(\mathcal{X} \mid G_{\text{new}})) - \log(P(\mathcal{X} \mid G_{\text{old}}))$$

If log(BF) $> \tau$, where $\tau \geq 0$, then the model of the new structure is much preferred which means that the edge between X_i and X_j must be deleted. However, if log(BF) $\leq \tau$, then the edge between X_i and X_j can be retained. Thus the mechanism automatically accepts edges between variables leading to improvements in structural learning.

To sample our two-level hierarchical model, we marginalize out analytically the parameters of the structural model to obtain an efficient Gibbs sampling algorithm (see, e.g., Casella & Robert, 2004). That is, we sample the parameters and the relationships in blocks (see, e.g., Roberts & Sahu, 1997). Let G_z and G_y be the structure of the between country and between bank relationships. Thus, our approach involves sampling from the posterior distribution of the between country structural relationships (G_z) given Ψ, then updating Ψ given (G_z). The next step involves sampling from the posterior distribution of the between banks structural relationships (G_y) given Ψ, Φ, and G_z. Let $\mathscr{X} = (\mathscr{Z}, \mathscr{Y})$ where $\mathscr{X}$ is the observed data set of the variables arranged into $\mathscr{Z}$ and $\mathscr{Y}$ representing the country-specific and bank-specific data set respectively. The resulting collapsed Gibbs sampler (Liu, 1994) consists of the following steps:

1. Sample between country structural relationships: $(G_z \mid \mathscr{Z})$,
2. Update between country parameters: $(\Phi \mid \mathscr{Z}, G_z)$,
3. Sample between banks structural relationships: $(G_y \mid \mathscr{Y}, \mathscr{Z}, \Phi, G_z)$,
4. Update between banks parameters: $(\Psi \mid \mathscr{Y}, \mathscr{Z}, \Phi, G_z, G_y)$.

Following the idea by Wang and Li (2012), we implement the algorithm such that steps 1 and 2 can be combined and steps 3 and 4 as well. A pseudo presentation of the algorithm is shown in Algorithm (1)

Algorithm 1. Sampling Network

Initialize T_0 to be identity matrix
Compute $T_n = T_0 + S_n$
Initial with a fully connected graph (upper triangle) G
for $t = 1$ to Total iterations **do**
  for each variable x_i **do**
    extract the clique x_{iC} centered around x_i
    for each variable $x_j \in x_{iC} \backslash x_i$ **do**
      extract the clique $x_{iC\backslash j} = x_{iC} \backslash x_j$
      Compute log Bayes factor, $\log(\text{BF}) = \log(P(x_{iC\backslash j})) - \log(P(x_{iC}))$
      if $\log(\text{BF}) > \tau$ **then**
        $x_j \notin x_{iC}$, set $G_{ij} = 0$ and update $T_{0,ii}$
      else
        $x_j \in x_{iC}$, set $G_{ij} = 1$ and update $T_{0,ii}$ and $T_{0,ij}$
      Update T_n following Carvalho, Massam, and West (2007)

2.4. Centrality Measures

Let A be $n \times n$, 0\1 matrix, which we will call the adjacency matrix. Network statistics can be derived using A: in particular, meaningful summary measures can be obtained using an appropriate singular value decomposition of such matrix. The simplest measure of node centrality is the degree which measures the number of connection of a given node in a network. In the framework of undirected network, the degree (d_i) of node i is simply

$$d_i = \sum_{j=1}^{n} a_{i,j}$$

where j represents all other nodes, n is the total number of nodes, and a_{ij} is the (i,j) element of the adjacency matrix of the network, which is defined as 1 if node i is connected to node j, and 0 otherwise.

The centrality measure that has been proposed in financial network modeling to explain the capacity of an agent to cause systemic risk, that is, a large contagion loss on other agents, is the eigenvector centrality (see, e.g., Billio et al., 2012; Furfine, 2003). The eigenvector centrality measure is a measure of the importance of a node in a network. It assigns relative scores to all nodes in the network, based on the principle that connections to few high scoring nodes contribute more to the score of the node in question than equal connections to low scoring nodes.

More formally, for the ith node, the centrality score is proportional to the sum of the scores of all nodes which are connected to it, as in the following equation:

$$x_i = \frac{1}{\lambda} \sum_{j=1}^{n} a_{i,j} x_j$$

where x_j is the score of a node j, $a_{i,j}$ is the (i,j) element of the adjacency matrix of the network, λ is a constant, and n is the number of nodes of the network. The previous equation can be rewritten for all nodes, more compactly, as

$$Ax = \lambda x$$

where A is the adjacency matrix, λ is the eigenvalue of the matrix A, with associated eigenvector x, an n-vector of scores (one for each node).

Note that, in general, there will be many different eigenvalues λ for which a solution to the previous equation exists. However, the additional requirement that all the elements of the eigenvector be positive (a natural request in our context) implies (by the Perron–Frobenius theorem) that only the eigenvector corresponding to the largest eigenvalue provides the desired centrality measures. Therefore, once an estimate of A is provided, network centrality scores can be obtained from the previous equation, as elements of the eigenvector associated to the largest eigenvalue.

In our two-level hierarchical model, it follows that the centrality measure decomposes into a country-specific component, and a between banks component.

In addition to centrality measures we will consider, as a summary measure of systemic risk of a bank, the number of its links (present edges) with other banks in the systems.

3. SIMULATION

To illustrate how our methodology and computational algorithm work, in this section we illustrate it with a 6-node simulation experiment of Wang and Li (2012). The experiment considers a model of $n=6$ variables, with $T=18$ observations, and an estimated sum of squares $S=T\hat{\Sigma}=TA^{-1}$, where the upper triangular matrix of A is defined as

$$A=\begin{pmatrix} 1 & 0.5 & 0 & 0 & 0 & 0.4 \\ & 1 & 0.5 & 0 & 0 & 0 \\ & & 1 & 0.5 & 0 & 0 \\ & & & 1 & 0.5 & 0 \\ & & & & 1 & 0.5 \\ & & & & & 1 \end{pmatrix}$$

Using our proposed algorithm, we run 2,000 Gibbs iterations with 200 burn-in ones and considered three thresholds $\tau=0,1,2$. The result of the posterior probabilities of edge presence in the graph are

$$(\hat{p}_{ij} \mid \tau = 0) = \begin{pmatrix} 1 & 0.94 & 0.02 & 0.01 & 0.02 & 0.83 \\ & 1 & 0.94 & 0.02 & 0.01 & 0.02 \\ & & 1 & 0.93 & 0.02 & 0.01 \\ & & & 1 & 0.93 & 0.01 \\ & & & & 1 & 0.94 \\ & & & & & 1 \end{pmatrix}$$

$$(\hat{p}_{ij} \mid \tau = 1) = \begin{pmatrix} 1 & 0.96 & 0.05 & 0.03 & 0.04 & 0.92 \\ & 1 & 0.96 & 0.04 & 0.02 & 0.04 \\ & & 1 & 0.96 & 0.04 & 0.03 \\ & & & 1 & 0.96 & 0.04 \\ & & & & 1 & 0.96 \\ & & & & & 1 \end{pmatrix}$$

$$(\hat{p}_{ij} \mid \tau = 2) = \begin{pmatrix} 1 & 0.95 & 0.76 & 0.76 & 0.76 & 0.92 \\ & 1 & 0.95 & 0.77 & 0.76 & 0.76 \\ & & 1 & 0.94 & 0.78 & 0.77 \\ & & & 1 & 0.95 & 0.77 \\ & & & & 1 & 0.96 \\ & & & & & 1 \end{pmatrix}$$

Following the idea of testing significance of regression coefficients, we apply a test for significance of an edge between two variables. We test if the posterior edge probability is greater than 0.5 under a 95% credibility interval. If this is true, then the edge is statistically significant, otherwise there is no link between the variables.

By comparing the results with the data-generating process, we observe that fixing a threshold $\tau = 0$ or 1 produces an accurate description of the data-generating process. According to the scale provided by Kass and Raftery (1995), using $\tau = 0$ leads to accepting edges that are not worth more than a bare mention. However, fixing $\tau = 1$ produces strong evidence against the null network. When $\tau = 2$, we observe from the results that the algorithm overestimates the number of links in the network therefore yielding an unsatisfactory representation of the links. Based on these results, we consider $\tau = 1$ as a robust logarithm Bayes factor threshold for our empirical application, which is presented in the next section.

4. APPLICATION

In this section we apply our proposed model to the estimation of the systemic credit risk of the largest listed banks of the Eurozone. We consider the Eurozone because of the changes in progress in this area, where the ECB starts the takeover of the supervision of the largest banks on November 1, 2014. It is a gradual process that aims at replacing the previous fragmented supervision and regulation (between 17 member states) into a unified one, with common rules and practices. It thus seems timely and important to focus the analysis on the banks that belong to this area, with the aim of contributing to identifying the most contagious institutions, at the super-national level.

4.1. Data

In this research we consider only large banks, whose total assets are greater than 30bn euro, and that are included in the ECB comprehensive assessment review. As we aim for an approach that integrates market information with bank-specific data, we consider only publicly listed banks, for which market data is available. Finally, in the case of a banking group with more entities that satisfy the above criteria, we consider only the controlling entity. The complete list of the 45 considered financial institutions is in Table 1, with the corresponding ticker code acronyms. To complete the information, Table 1 contains, besides bank names and their codes, their prevalent country, whether they are Systemically Important Financial Institutions (SIFIs) either at the global or at the domestic level, their total assets from the last available balance sheet (in thousands of euro), the year over year variation of their equity returns (in percentage points), and the corresponding standard deviation of the returns, also in percentage points.

From Table 1, note that 12 of the 17 Eurozone countries have at least one bank with total assets greater than 30bn euro. The five missing countries are Estonia, Luxembourg, Malta, Slovakia, and Slovenia.

Apart from the missing (small) countries, Table 1 underlines a remarkable difference in the banking structure of the different countries, which is to be remembered when interpreting the obtained results. Some countries have a large number of banks in the sample: Italy, the third populated country, has 12, Spain, the fourth populated country, has 7. Others have instead a limited number of banks: Germany, the most populated country, has only 5, and so does France, the second populated country. Note that Greece, a small country, with the most troubled Eurozone economy, has 4. For comparison purposes, the ECB assessment sample (that includes unlisted banks) contains a total of 130 banks in the Eurozone, among which there are 24 German banks, 13 French, 15 Italian, 16 Spanish and 4 Greek ones. Looking at country representation note that our sample of

Table 1. Banks Bankscope – Listed and Supervised by the European Central Bank.

Bank Name	Code	Country	T.Assets	SIFI	YOY	Stdev
Deutsche Bank	DBK	Germany	2012329	G	3.96	4.00
BNP Paribas	BNP	France	1907290	G	26.94	3.59
Credit Agricole	ACA	France	1842361	G	37.42	.39
Banco Santander	SAN	Spain	1269628	G	10.90	3.60
Societe Generale	GLE	France	1250696	G	44.17	4.96
ING Groep	INGA	Netherland	1168632	G	24.86	4.08
UniCredit	UCG	Italy	926827	G	41.21	4.99
Intesa Sanpaolo	ISP	Italy	673472	D	30.72	4.45
Banco Bilbao Vizcaya Argentaria	BBVA	Spain	637785	G	32.08	3.82
Commerzbank	CBK	Germany	635878	G	–48.65	7.28
Natixis-BPCE Group	KN	France	528370	G	37.39	4.74
Dexia	DEXB	Belgium	357210	G	–133.50	22.46
Caixabank	CABK	Spain	348294	–	25.86	5.08
Bankia	BKIA	Spain	282310	–	–469.33	32.13
KBC Group	KBC	Belgium	256886	D	72.92	4.16
CIC Credit Mutuel Group	CC	France	235732	–	33.69	2.02
Banca Monte dei Paschi di Siena	BMPS	Italy	218882	–	2.39	6.29
Erste Group Bank	EBS	Austria	213824	D	24.34	4.52
Banco de Sabadell	SAB	Spain	161547	–	–3.57	6.21
Banco Popular Espanol	POP	Spain	157618	–	–44.69	12.19
Bank of Ireland	BIR	Ireland	148146	D	91.20	5.80
Raiffeisen Bank International	RBI	Austria	136116	D	–25.59	3.91
Unione di Banche Italiane	UBI	Italy	132433	–	38.72	5.30
Banco Popolare	BP	Italy	131921	–	7.77	6.13
Allied Irish Banks	AIB	Ireland	122516	D	53.60	8.68
National Bank of Greece	ETE	Greece	104798	D	–186.01	25.86
BancoComercial Portugues	BCP	Portugal	89744	D	25.95	5.92
BancoEspirito Santo	BES	Portugal	83690	–	37.05	7.37
Wustenrot & Wurttembergische	WUW	Germany	77192	–	–4.68	1.89
Mediobanca	MB	Italy	72841	–	29.96	5.55
Piraeus Bank	TPEIR	Greece	70406	D	–146.26	21.33
Eurobank Ergasias	EUROB	Greece	67653	D	–7.49	19.93
Banca popolare Emilia Romagna	BPE	Italy	61637	–	30.46	5.77
Alpha Bank	ALPHA	Greece	58357	D	–121.81	18.04
Bankinter	BKT	Spain	58165	–	83.17	5.68
Banca Popolare di Milano	PMI	Italy	52475	–	10.74	6.06
Banca Carige	CRG	Italy	49325	–	–18.23	7.07
Aareal Bank	ARL	Germany	45734	–	37.33	4.34
Pohjola Bank Oyj	POH1S	Finland	44623	–	14.87	3.38
Banco BPI	BPI	Portugal	44564	D	22.54	6.02
Bank of Cyprus	CPB	Cyprus	33762	–	–60.01	10.77
IKB Deutsche Industriebank	IKB	Germany	31593	–	59.85	8.36
Banca Popolare di Sondrio	BPSO	Italy	32349	–	–3.35	3.74
Credito Emiliano	CE	Italy	30748	–	30.57	4.50
Credito Valtellinese	CVAL	Italy	29896	–	–11.50	5.68

G, global systemically important bank; D, domestically important bank; T., Asset: total assets; YOY: year on year returns; Stdev: standard deviations.

banks, which does represent about a third of all banks in the ECB list, is overrepresented in terms of Greek banks (it contains them all) and Italian banks (has about 80% of them), and underrepresented in terms of German banks (has about 20% of them). The above bias is the result of the structure of the banking systems in the different countries, and especially of medium-sized banks, that typically operate at a local (subnational) level: while most such banks, in Italy (and in Greece), are listed, in Germany this is rarely the case, with other countries lying in between.

The market and financial data that we have considered has been obtained from Bankscope, a comprehensive database for individual banks across the world provided by the private company Bureau Van Dijk. In particular, the information that we use covers a year of price stock data, just before the start of the ECB supervision, with observations on a weekly basis. Precisely, our observation period goes from October 22, 2012 to October 18, 2013. From the Bankscope database, we have extracted the weekly closing stock prices for each bank, P_t and, then, transformed them into weekly returns, defined as: $R_t = \log(P_t/P_{t-1})$, where t is a week in the last year and $t-1$ the week that precedes it.

4.2. Convergence Diagnostics

As discussed in the previous section, we choose $\tau = 1$ as a threshold value for the Bayes factor. We run 2,200 Gibbs iterations with 200 as burn-in. Our MCMC simulations are implemented in Matlab. We have verified posterior convergence of the algorithm by diagnosing the complexity of the network. We have then monitored convergence using potential scale reduction factor (PSRF) of Gelman and Rubin (1992). The approach divides the sampled chains into three parts to compute within-chain and between-chain covariances to test whether all the chains converge to the same posterior distribution. The chain is said to have properly converged if PSRF $\leq$ 1.2. The results from our convergence diagnostics, shown in Fig. 1 indicates that all our simulations converged with a PSRF smaller than 1.02.

4.3. Results

We now present the results from the application of Bayesian graphical models to the weekly log returns of the 45 considered banks. We applied to such data both a (nonhierarchical) Bayesian graphical model, and our

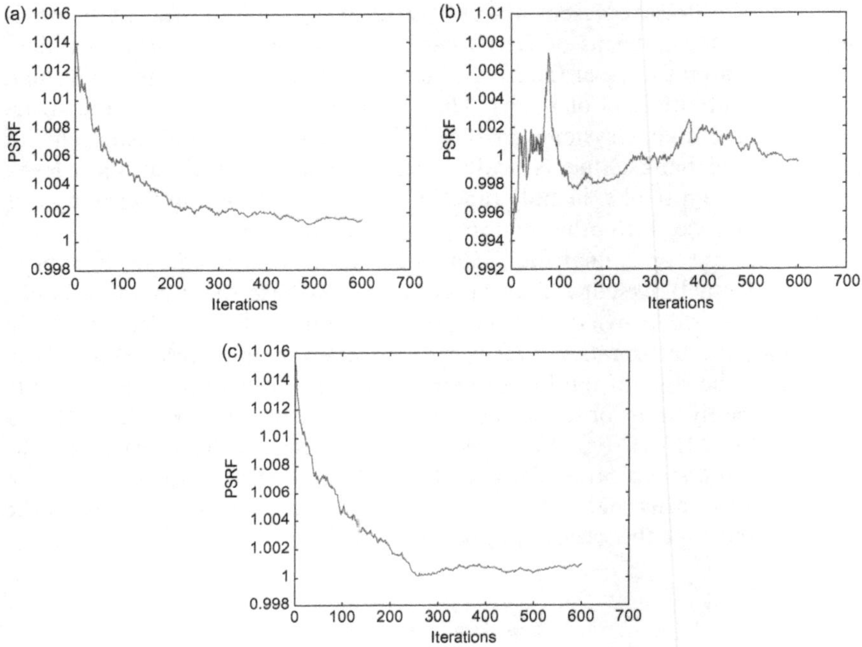

Fig. 1. Evolution of the Potential Scale Reduction Factor (PSRF) for Convergence Diagnostics of the Networks: (a) Non-hierarchical Network; (b): Hierarchical Network (between Countries); (c): Hierarchical Network (between Banks).

proposed Bayesian hierarchical models that enlarges the state space of each bank log return variables with country-specific log return variables, obtained averaging the returns of the banks in each country. This inclusion, if significant, is expected to lower the connections between banks, and especially cross-border ones, as connections between banks may be replaced by connections between countries.

Fig. 2(a) shows the unconditional graphical network model between banks with the highest a posteriori probability, in the nonhierarchical model, and Fig. 2(b) the same best model, conditional on the country-to-country relationships, in the hierarchical model.

From Fig. 2(a) note that the model is highly interconnected, and it is rather difficult to interpret. The only apparent message conveyed is about banks that appear "peripheral," with a lower degree and a lower centrality

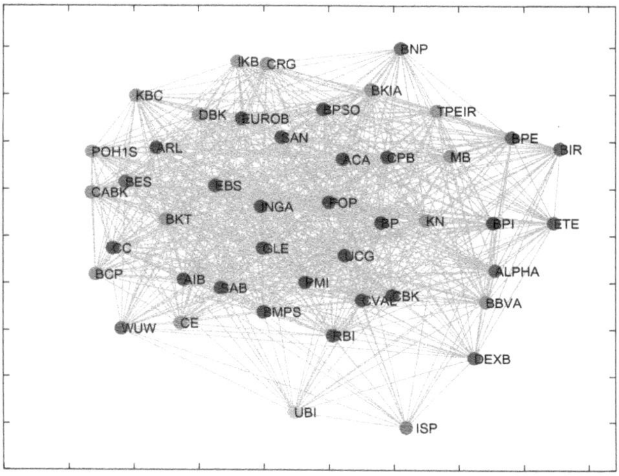

(a) Non-hierarchical Network (Total Links : 711)

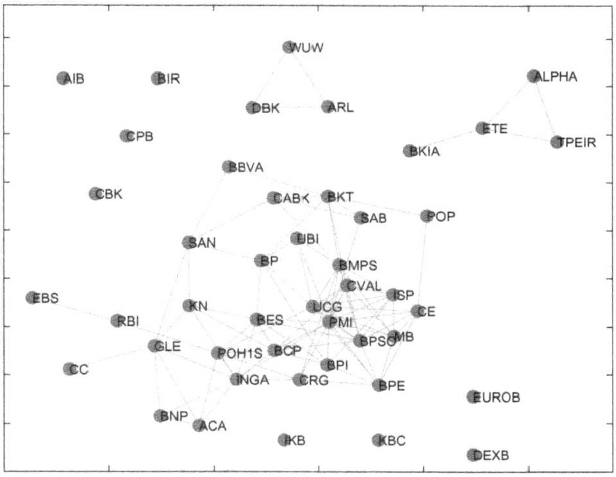

(b) Hierarchical Network (Total Links : 95)

Fig. 2. Comparison of Interbank Connections from (a): a Non-hierarchical Network and (b): a Hierarchical Network. Banks Are Represented Using Their Bank Codes (See Table 1 for Bank Names).

measures. Such banks range from the troubled Greek banks Alpha Bank and Piraeus Bank, and Dexia bank, to medium size banks such as Wustenrot and Wurttenbergische in Germany, Pohjola Bank in Finland, UBI in Italy, and two internationally oriented SIFIs such as Banco de Bilbao and BNP. These banks, because of their lower centrality, seem to appear less contagious/subject to contagion.

Fig. 2(b), instead presents a much clearer picture: the total number of links drop from 711 to 95. There is a remarkable difference between the unconditional results in Fig. 2(a) and those obtained conditionally on a country effect, in Fig. 2(b). The most evident difference is the lower degree of connectivity between banks, especially at the cross-border level.

This means that the country effect explains a lot of the comovements between the returns of the banks. Table 3, that contains, for each bank, the centrality measure and the total number of connections, gives further evidence in this direction: indeed, most banks have connections mainly with banks of their country, and this is a further confirmation of the fact that share prices of banks heavily depend on the risk of the countries in which they operate. This also explains why the most central and connected banks in Table 3 are Italian banks, which are more represented in the sample. To the other side of the spectrum are German banks that appear disconnected from other countries' banks. Similarly, Greek banks form their own circle of troubled banks, together with the Spanish Bankia.

Besides the existence of a country effect, Table 3 emphasizes that some Systemically Important Financial Institutions (SIFI in Table 1) are potentially more contagious than others, as they are more connected. This is the case of Unicredit, heavily linked to most Italian banks; of ING, strongly interconnected at an international level; and, although to a lesser extent, for the French banks BPCE, Credit Agricole, and Societe General and the Spanish Banco Santander. Other SIFIs, including the two German ones Deutsche bank and Commerzbank, as well as BNP Paribas, are less central.

We now examine in detail the inter country linkages estimated by the hierarchical model. Fig. 3 shows the between country graphical network that is learned (model averaged) from the data, and Table 2 gives the corresponding centrality measures. To aid interpretation, we have added a marker for the estimated sign of the partial correlation found: positive or negative.

From Fig. 3 and Table 2, notice that country financial systems are not much connected, confirming the image of Europe as a "market with financial frictions."

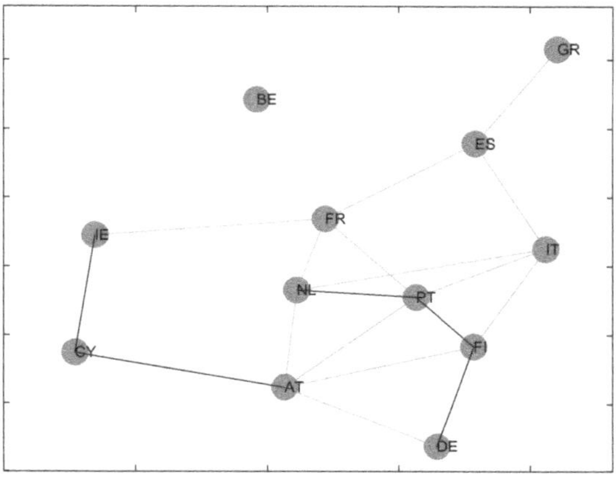

Fig. 3. Intercountry Connections from the Hierarchical Network (Total Links : 19). *AT* stands for Austria, *BE* for Belgium, *CY* for Cyprus, *DE* for Germany, *ES* for Spain, *FI* for Finland, *FR* for France, *GR* for Greece, *IE* for Ireland, *IT* for Italy, *NL* for the Netherlands and *PT* for Portugal.

Table 2. Intercountry Centrality Measures.

Rank	Code	Country	Eig.V	Degree	Num. Banks	YOY	Stdev
1	PT	Portugal	0.4497	5	5	5.24	4.64
2	FI	Finland	0.4148	5	1	24.34	4.52
3	AT	Austria	0.3714	5	7	15.45	3.80
4	FR	France	0.3695	5	3	11.21	6.68
5	NL	Netherland	0.3692	4	12	59.85	8.36
6	IT	Italy	0.3412	4	1	−10.39	9.49
7	DE	Germany	0.1895	2	5	16.04	4.92
8	ES	Spain	0.1818	3	2	−61.51	10.98
9	CY	Cyprus	0.1178	2	1	44.17	4.96
10	IE	Ireland	0.1174	2	2	16.18	4.63
11	GR	Greece	0.0438	1	4	−24.67	11.65
12	BE	Belgium	0	0	2	24.16	4.00

In particular, due to the still persisting strong crisis that generates high credit loss impairments, southern European countries, such as Greece, Spain, and Italy are connected with each other. Similarly, so are stronger economies such as Germany, Austria, and Finland. France act as a gate between troubled southern countries and stronger ones, along with the

Netherlands and Portugal whose banks have a high international exposure. Finally, Belgium, Ireland, and Cyprus, smaller economies, follow very specific paths: Belgium, disconnected, because of the very different behavior of its two banks: Dexia, under restructuring, and KBC, healthier; Cyprus, which went through a dramatic financial crisis, in the last year; Ireland, whose banks have gone through a year of gains, that followed years of crisis.

Indeed, looking at Table 2, the most contagious countries seem France, Portugal, Finland, Austria, and Netherlands. While the French economy is indeed a gate between southern and Northern European economies, the other centralities can be explained by the multinational activities of the banks of the related countries.

Having estimated the country-to-country connections, we now look in detail the interdependencies between banks, that emerge from Fig. 2(b). As already remarked, Table 3 gives the corresponding numeric description. In particular, it gives, for each bank, its eigenvalue centrality measure, and the total number of links in the conditional model. Such total number is split between the within country links and the cross-border links of each bank.

We first focus on the within country links that are estimated by the Bayesian model averaged hierarchical graphical model. Fig. 4 extracts from Fig. 2(b) the bank relationships that are learned from the data, within each country.

A first remark is that, conditionally on the country-to country relationships, the number of links between banks, within country, decreases from 96 to 58.

The results in Fig. 4 can be read off jointly with the column of Table 3 that shows the within country links, country by country. We focus on the largest ones.

Italy

Italian banks seem to rotate around the two largest banks: Unicredit, and Intesa San Paolo, as well as with the investment bank Mediobanca. Among the other banks the most central ones are the cooperative territorial banks such as Popolare di Milano, Banca Popolare di Sondrio, Credito Valtellinese, and Banca Popolare Emilia Romagna, indeed linked with each other. Note that the two most troubled banks, Monte dei Paschi di Siena and Banca Carige, have different centrality: the former is more connected and, therefore, can spread its distress among more neighbors. We also

Table 3. Interbank Centrality Measures from the Hierarchical Network.

	Code	Bank Name	Country	Eig.V	Degree		
					Tot	W-C	A-C
1	PMI	Banca Popolare di Milano	Italy	0.3995	14	9	5
2	UCG	UniCredit	Italy	0.3314	11	10	1
3	BPSO	Banca Popolare di Sondrio	Italy	0.3052	9	7	2
4	CVAL	Credito Valtellinese	Italy	0.2587	8	6	2
5	BMPS	Banca Monte dei Paschi di Siena	Italy	0.2557	9	7	2
6	ISP	Intesa Sanpaolo	Italy	0.2552	7	6	1
7	MB	Mediobanca	Italy	0.2482	7	7	0
8	BPE	Banca popolare Emilia Romagna	Italy	0.2324	7	6	1
9	BCP	Banco Comercial Portugues	Portugal	0.2289	8	2	6
10	BES	Banco Espirito Santo	Portugal	0.2145	8	2	6
11	CE	Credito Emiliano	Italy	0.1999	6	5	1
12	BPI	Banco BPI	Portugal	0.1879	6	2	4
13	INGA	ING Groep	Netherland	0.1751	9	0	9
14	POH1S	Pohjola Bank Oyj	Finland	0.1490	7	0	7
15	UBI	Unione di Banche Italiane	Italy	0.1437	4	4	0
16	CRG	Banca Carige	Italy	0.1375	5	3	2
17	BP	Banco Popolare	Italy	0.1325	6	2	4
18	BKT	Bankinter	Spain	0.1081	7	4	3
19	KN	Natixis-BPCE Group	France	0.0921	5	1	4
20	SAB	Banco de Sabadell	Spain	0.0732	3	2	1
21	CABK	Caixabank	Spain	0.0620	4	3	1
22	ACA	Credit Agricole	France	0.0530	4	2	2
23	GLE	Societe Generale	France	0.0525	6	4	2
24	SAN	Banco Santander	Spain	0.0461	5	2	3
25	POP	Banco Popular Espanol	Spain	0.0396	2	1	1
26	BNP	BNP Paribas	France	0.0360	3	2	1
27	BBVA	Banco Bilbao Vizcaya Argentaria	Spain	0.0198	2	2	0
28	RBI	Raiffeisen Bank International	Austria	0.0195	2	1	1
29	CC	CIC Credit Mutuel Group	France	0.0067	1	1	0
30	EBS	Erste Group Bank	Austria	0.0025	1	1	0
31	BIR	Bank of Ireland	Ireland	0	0	0	0
32	ALPHA	Alpha Bank	Greece	0	2	2	0
33	EUROB	Eurobank Ergasias	Greece	0	0	0	0
34	TPEIR	Piraeus Bank	Greece	0	2	2	0
35	ETE	National Bank of Greece	Greece	0	3	2	1
36	BKIA	Bankia	Spain	0	1	0	1
37	DBK	Deutsche Bank	Germany	0	2	2	0
38	CBK	Commerzbank	Germany	0	0	0	0
39	DEXB	Dexia	Belgium	0	0	0	0
40	KBC	KBC Group	Belgium	0	0	0	0
41	CPB	Bank of Cyprus	Cyprus	0	0	0	0
42	WUW	Wustenrot & Wurttembergische	Germany	0	2	2	0
43	ARL	Aareal Bank	Germany	0	2	2	0
44	IKB	IKB Deutsche Industriebank	Germany	0	0	0	0
45	AIB	Allied Irish Banks	Ireland	0	0	0	0

Tot, total degree; W-C, within country degree; A-C, across country degree.

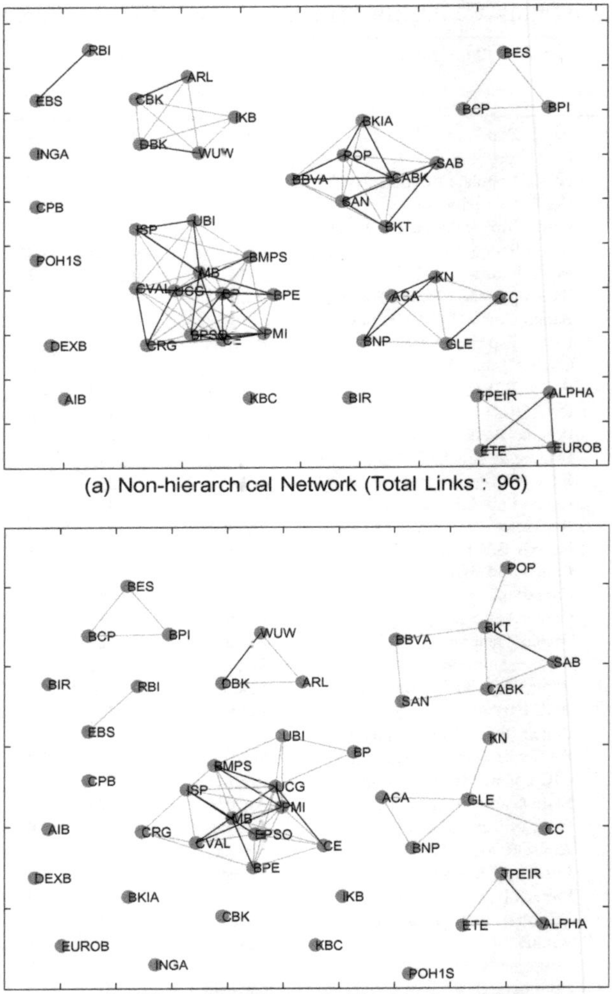

(a) Non-hierarchical Network (Total Links : 96)

(b) Hierarchical Network (Total Links : 58)

Fig. 4. Comparison of Within-Country Connections from (a): a Non-hierarchical Network and (b): a Hierarchical Network. Banks Are Represented Using Their Bank Codes (See Table 1 for Bank Names).

underline that a remarkable aspect of the model is that it seems to be able to capture even regional (subnational) links between banks, expression of the fact that banks may share the same lenders and, therefore, similar risks. This is the case for CVAL and BPSO, both operating mainly in the Lombardy region; BP and UBI, both operating in the north center-east region; BPE and CE both operating in the Emilia Romagna region.

Spain

In the case of Spain, the territorial banks Bankinter and Caixabank are the most central, within country. The two large banks Santander and Banco de Bilbao are less central, but related to each other. Finally, Bankia is disconnected from the others, due to its state led restructuring. Considerations on regional dependences can be drawn for Spanish banks in a similar fashion to Italiian ones.

France

In France the Systematically Important Institution Societe Generale acts as a gate between the territorial bank groups BPCE and Credit Mutuel and the larger groups: BNP Paribas and Credit Agricole.

Germany

In Germany the Systematically Important Deutsche Bank is connected to both Aareal Bank and Wustenrot, while Commerzbank and IKB are isolated, with their respectively bad and good performances.

Greece

All Greek banks but Eurobank (involved in a restructuring plan) are related with each other, and this reflects the fact that they reflect a common deep financial crisis of the country.

From the above discussion, the existence of a country effect, that determines statistical correlations and dependencies between banks, is clear. From an economical viewpoint such effect can be explained as follows. On the asset side, banks acts as lenders and, thus, if the economy is doing badly, so do banks, as enterprises do not repay the given credit and/or go default. In addition, countries with high public debts typically persuade banks to buy the related government securities. When the economy is badly performing, the value of such bond decreases. On the liability side, depositors from troubled countries typically put less money in banks or withdraw it. Furthermore, when a bank does badly, the government may intervene, and capitalize it by buying a relevant portion of shares. Because of this interlinkage between banks and their countries, we expect stock prices of banks to be highly determined by a country risk effect, which may increase the interconnections between banks in troubled countries, such as southern European ones.

We now look at bank to bank connections, that are estimated by our model, across countries, rather than within each country. Fig. 5 shows the across country between bank network that is learned from the data, extracted from Fig. 2(b).

The results in Fig. 5 can be read off jointly with the column of Table 3 that shows the across country links.

We remind the reader that, on the basis of the assumptions of the proposed hierarchical model, the less a country is connected, the less likely the presence of significant cross-border relationships for its banks. To exemplify, a cross-border correlation for a German (or an Irish) bank is less likely to remain in the model than a similar (in size) cross-border correlation for a French (or a Portuguese) bank, as France (and Portugal) are much more central than Germany (and Belgium) in the country-to-country network.

From Fig. 5, notice that the number of connections significantly decreases, from 615 to 37, much more than what do the within country connections.

From Fig. 5 and Table 3, the most connected banks, at the cross-border level, are either banks in small countries, such as ING, Pohjola, Banco Commercial Portugues, and Banco Espirito Santo, or multinational SIFIs such as Unicredit and Societe Generale. Notice also the high centrality of territorial banks such as Banca Popolare di Milano, Banco Popolare, Banco BPI, and BPCE, presumably as they have relevant portions of shares in the hand of international investment funds, that look at them with common portfolio based strategies.

Another interesting remark that can be drawn from Fig. 5 concerns the banks with no cross-border connections: the Greek and the German banks, whose country, as already seen, is disconnected from the others; the Irish

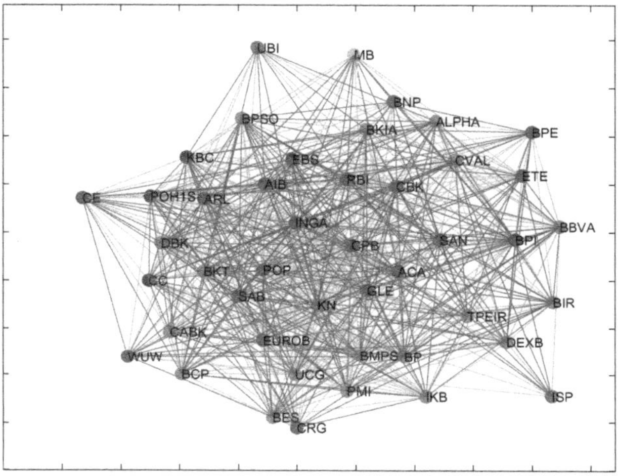

(a) Non-hierarchical Network (Total Links : 615)

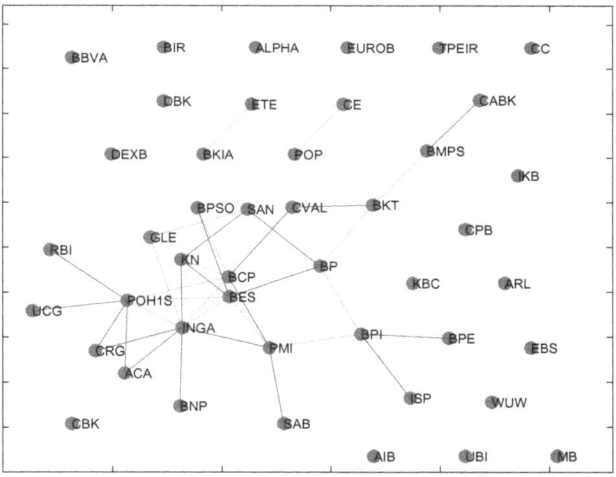

(b) Hierarchical Network (Total Links : 37)

Fig. 5. Comparison of Interbank Connections across Countries from (a): a Non-hierarchical Network and (b): a Hierarchical Network. Banks Are Represented Using Their Bank Codes (See Table 1 for Bank Names). The Colour of Links Indicate the Signs of the Partial Correlations: Red for Negative and Blue for Positive.

banks which have no specific connections beyond the aggregate country ones and, finally, the Bank of Cyprus, Bankia, and the Belgian banks.

A final remark concerns the comparison between SIFIs, whose contagion is the most dangerous because of their asset size. The most central ones appear Unicredit, ING, and the French banks Societe Generale, BPCE, and Credit Agricole. Banco Santander, BNP Paribas, and Banco Bilbao are connected, but less than the previous ones; Commerzbank, Dexia, and Deutsche Bank are the least connected.

4.4. Sensitivity Analysis

For sensitivity analysis purposes, we have run our computational algorithm with different log Bayes factor thresholds, $\tau=0$ and $\tau=2$. Table 4 shows the result of the total links of both hierarchical and non-hierarchical networks for the different Bayes factor thresholds.

Our results, summarized in Table 4 show that the results do not differ considerably from previous ones but the corresponding networks are, respectively, less or more interconnected. For example, in terms of the between country network, the estimated present links drop from 19 to 12 for $\tau=0$, and increase to 46 for $\tau=2$. In terms of specific relationships, we remark that the substance of what we found and discussed for the threshold of $\tau=1$, chosen according to what presented in the Simulation section, remains stable with other choices of the threshold.

Table 4. Sensitivity Analysis with Different Log Bayes Factor Thresholds.

Type	Network	Bayes Factor Threshold and Total Links		
		$\tau=0$	$\tau=1$	$\tau=2$
Nonhierarchical	Over-all inter banks	611	711	867
	Within-country	85	96	106
	Interbank across country	526	615	761
	SIFI	44	45	51
Hierarchical	Over-all inter banks	59	95	343
	Within-country	41	58	93
	Interbank across country	18	37	250
	SIFI	8	11	28
	Inter-countries	12	19	46

5. CONCLUSIONS

Financial network models are a useful tool to model interconnectedness and systemic risks in financial systems. Such models are essentially descriptive, and based on highly correlated networks.

The motivation of this paper is to provide a stochastic framework for financial network models, aimed at a more parsimonious and more realistic representation.

The paper contains two main research contributions in this direction. First we have introduced Bayesian Gaussian graphical models in the field of systemic risk modelling, thus estimating the adjacency matrix of a network in a robust and coherent way. Following this approach, (Bayesian) confidence intervals for the centrality measures can be easily obtained, from the simulated output. Second, we have proposed a hierarchical Bayesian graphical model that can usefully decompose dependencies between financial institutions into correlations into between countries and correlations between institutions, within and/or across countries.

We have applied our proposed methods to the largest Eurozone banks, with the aim of identifying central institutions, more subject to contagion of or, conversely, whose failure could result in further distress or breakdowns in the whole system.

Our results show that, in the transmission of the perceived default risk, there is a strong country effect, which reflects the weakness and the strength of the underlying economies. Besides the country effect, the most central banks appear the large international ones, especially if from a relatively small country.

A methodological extension of the paper may assume using non uniform priors over the different graphical structures. This will lead to penalizing overly complicated models. Although this may introduce a selection bias, it may be help model selection when large networks are being considered.

Another extension of the paper considered directed, rather than undirected, graphical Gaussian models. Doing so we may be able to provide "causal" directions of systemic risks and not only symmetric ones. We remark, however, that this extension will increase substantially the number of possible models to be considered, at the expense of computational efficiency of the model selection procedure.

From an applied viewpoint, in the paper we have considered only stock market data and not balance sheet data. This because the latter is typically backward-looking, whereas market data incorporate future expectations.

We believe, however, that the two sources of information are complementary and should be combined in a methodological extension of our work, possibly also considering "soft" information, which may be "forward looking", such as that present in social network data.

ACKNOWLEDGMENT

We acknowledge financial support from the Italian ministry PRIN MISURA project: Multivariate models for risk assessment. We also thank the referee for useful comments and discussion that have helped improving the content of the paper.

REFERENCES

Acharya, V. V., Pedersen, L. H., Philippon, T., & Richardson, M. (2010). *Measuring systemic risk*. Working Paper No. 1002, Federal Reserve Bank of Cleveland, Cleveland, OH.

Adrian, T., & Brunnermeier, M. K. (2010). *CoVaR*. Staff Reports No. 348. Federal Reserve Bank of New York, New York, NY.

Ahelegbey, D. F., Billio, M., & Casarin, R. (2014). Bayesian graphical models for structural vector autoregressive processes. *Journal of Applied Econometrics* (Forthcoming).

Altman, E. I. (1968). Financial ratios, discriminant analysis and the prediction of corporate bankruptcy. *The Journal of Finance*, *23*(4), 589–609.

Arena, M. (2005). Bank failures and bank fundamentals: A comparative analysis of LATIN AMERICA and East Asia during the nineties using bank-level data. *Journal of Banking and Finance*, *32*, 299–310.

Barigozzi, M., & Brownlees, C. (2013). *Nets: Network estimation for time series*. Working Paper. Social Science Research Network.

Basel Committee on Banking Supervision. (2011). *Basel iii: A global regulatory framework for more resilient banks and banking systems*. Technical report, Bank for International Settlements, Basel, Switzerland.

Billio, M., Getmansky, M., Lo, A. W., & Pelizzon, L. (2012). Econometric measures of connectedness and systemic risk in the finance and insurance sectors. *Journal of Financial Economics*, *104*(3), 535–559.

Brownlees, C. T., & Engle, R. (2011). *Volatility, correlation and tails for systemic risk measurement*. Technical report. New York University, New York, NY.

Brunnermeier, M. K., & Oehmke, M. (2012). *Bubbles, financial crises, and systemic risk*. NBER Working Papers 18398. National Bureau of Economic Research, Inc., Cambridge, MA.

Carvalho, C., Massam, H., & West, M. (2007). Simulation of hyper-inverse Wishart distributions in graphical models. *Biometrika*, *94*, 647–659.

Carvalho, C. M., & West, M. (2007). Dynamic matrix-variate graphical models. *Bayesian Analysis*, *2*, 69–98.

Casella, G., & Robert, C. P. (2004). *Monte Carlo statistical methods*. New York, NY: Springer Verlag.

Castrén, O., & Kavonius, I. K. (2009). *Balance sheet interlinkages and macro-financial risk analysis in the euro area*. Working Paper No. 1124. European Central Bank, Frankfurt, Germany.

Chen, H., Cummins, J. D., Viswanathan, K. S., & Weiss, M. A. (2013). Systemic risk and the interconnectedness between banks and insurers: An econometric analysis. *Journal of Risk and Insurance*, forthcoming.

Cole, R. A., & Gunther, J. W. (1998). Predicting bank failures: A comparison of on-and off-site monitoring systems. *Journal of Financial Services Research*, *13*(2), 103–117.

Cont, R., Moussa, A., & Santos, E. (2013). Network structure and systemic risk in banking systems. In J.-P. Fouque & J. A. Langsam (Eds.), *Handbook on systemic risk* (pp. 327–368). New York: Cambridge University Press.

Corander, J., & Villani, M. (2006). A Bayesian approach to modelling graphical vector autoregressions. *Journal of Time Series Analysis*, *27*(1), 141–156.

Cowell, R. G., Dawid, A. P., Lauritzen, S. L., & Spiegelhalter, D. J. (1999). *Probabilistic networks and expert systems. Statistics for engineering and information science*. New York, NY: Springer Verlag.

Davis, E. P., & Karim, D. (2008). Comparing early warning systems for banking crises. *Journal of Financial Stability* *4*(2), 89–120.

Dawid, A. P., & Lauritzen, S. L. (1993). Hyper Markov laws in the statistical analysis of decomposable graphical models. *The Annals of Statistics*, *21*, 1272–1317.

De Lisa, R., Zedda, S., Vallascas, F., Campolongo, F., & Marchesi, M. (2011). Modelling deposit insurance scheme losses in a Basel 2 framework. *Journal of Financial Services Research*, *40*(3), 123–141.

Edwards, D. (1990). Hierarchical interaction models (with discussion). *Journal of the Royal Statistical Society, Series B*, *52*, 3–20; 51–72.

Fantazzini, D., & Maggi, M. (2012). Computing reliable default probabilities in turbulent times. In *Rethinking valuation and pricing models* (pp. 241–255). London: Academic Press.

Frydenberg, M., & Lauritzen, S. L. (1989). Decomposition of maximum likelihood in mixed graphical interaction models. *Biometrika*, *76*(3), 539–555.

Furfine, C. H. (2003). Interbank exposures: Quantifying the risk of contagion. *Journal of Money, Credit and Banking*, *35*(1), 111–128.

Gai, P., & Kapadia, S. (2009). *A network model of super-systemic crises*. Documentos de trabajo del Banco Central. Banco Central de Chile, Santiago, Chile.

Gelman, A., & Rubin, D. B. (1992). Inference from iterative simulation using multiple sequences, (with discussion). *Statistical Science*, *7*, 457–511.

Georg, C.-P., & Poschmann, J. (2010). *Systemic risk in a network model of interbank markets with central bank activity*. Jena Economic Research Papers No. 2010-033. Jena, Germany.

Giudici, P., & Castelo, R. (2003). Improving Markov chain Monte Carlo model search for data mining. *Machine Learning*, *50*(1–2), 127–158.

Giudici, P., & Green, P. J. (1999). Decomposable graphical Gaussian model determination. *Biometrika*, *86*, 785–801.

Guo, J., Levina, E., Michailidis, G., & Zhu, J. (2011). Joint estimation of multiple graphical models. *Biometrika*, *98*(1), 1–15.

Guo, J., Levina, E., Michailidis, G., & Zhu, J. (2013). Estimating heterogeneous graphical models for discrete data with an application to roll call voting. *Annals of Applied Statistics*, forthcoming.

Gup, B. E. (1998). *Bank failures in the major trading countries of the world: Causes and estimation*. Westport, CT: Quorum Books.

Gyftodimos, E., & Flach, P. A. (2002). Hierarchical bayesian networks: A probabilistic reasoning model for structured domains. In Proceedings of the ICML-2002 workshop on development of representations, University of New South Wales, Sydney (pp. 23–30).

Halaj, G. (2013). *Optimal asset structure of a bank–bank reactions to stressful market conditions*. Working Paper Series 1533. European Central Bank, Frankfurt, Germany.

Hensman, J., Lawrence, N. D., & Rattray, M. (2013). Hierarchical Bayesian modelling of gene expression time series across irregularly sampled replicates and clusters. *BMC Bioinformatics*, *14*(1), 1–12.

Huang, X., Zhou, H., & Zhu, H. (2011). Systemic risk contributions (BIS Papers chapters). In Bank for International Settlements (Ed.), *Macroprudential regulation and policy* (Vol. 60, pp. 36–43). Bank for International Settlements, Basel, Switzerland.

Idier, J., Lamé, G., & Mésonnier, J. S. (2013). *How useful is the marginal expected shortfall for the measurement of systemic exposure? A practical assessment*. Working Papers 348. Banque de France, Paris, France.

Kanno, M. (2013). *Credit migration forecasting and correlation between business and credit cycles*. Technical report, Kanagawa University, Rokkakubashi, Japan.

Kass, R. E., & Raftery, A. E. (1995). Bayes factors. *Journal of the American Statistical Association*, *90*(430), 773–795.

Kenny, G., Kostka, T., & Masera, F. (2013). *Can macroeconomists forecast risk? Event-based evidence from the Euro area SPF*. Working Paper Series 1540. European Central Bank, Frankfurt, Germany.

Klomp, J., & Haan, J. d. (2012). Banking risk and regulation: Does one size fit all? *Journal of Banking and Finance*, *36*(12), 3197–3212.

Koopman, S. J., Lucas, A., & Schwaab, B. (2011). Modeling frailty-correlated defaults using many macroeconomic covariates. *Journal of Econometrics*, *162*(2), 312–325.

Lauritzen, S. L. (1996). *Graphical models*. Oxford: Oxford University Press.

Lauritzen, S. L., & Wermuth, N. (1989). Graphical models for associations between variables, some of which are quantitative and some qualitative. *Annals of Statistics*, *17*, 31–57.

Liu, J. (1994). The collapsed Gibbs sampler in Bayesian computations with applications to a gene regulation problem. *Journal of the American Statistical Association*, *89*(427), 958–966.

Madigan, D., & York, J. (1995). Bayesian graphical models for discrete data. *International Statistical Review*, *63*(2), 215–232.

Mare, D. S. (2012). *Contribution of macroeconomic factors to the prediction of small bank failures*. Working Paper. Social Science Research Network.

Markose, S., Giansante, S., Gatkowski, M., & Shaghaghi, A. R. (2010). *Too interconnected to fail: Financial contagion and systemic risk in network model of CDS and other credit enhancement obligations of US banks*. Economics Discussion Papers No. 683, University of Essex, Essex, pp. 21–50.

Merton, R. C. (1974). On the pricing of corporate debt: The risk structure of interest rates. *Journal of Finance*, *2*, 449–471.

Nier, E., Yang, J., Yorulmazer, T., & Alentorn, A. (2007). Network models and financial stability. *Journal of Economic Dynamics and Control, 31*(6), 2033–2060.

Pearl, J. (1988). *Probabilistic reasoning in intelligent systems: Networks of plausible inference.* San Mateo, CA: Morgan Kaufmann Publishers.

Peltonen, T. A., Scheicher, M., & Vuillemey, G. (2013). *The network structure of the CDS market and its determinants.* Working Paper Series 1583. European Central Bank, Frankfurt, Germany.

Resti, A., & Sironi, A. (2007). *Risk management and shareholders' value in banking: From risk measurement models to capital allocation policies* (Vol. 417). London: Wiley.

Roberts, G., & Sahu, S. (1997). Updating schemes, covariance structure, blocking and parametrisation for the Gibbs sampler. *Journal of the Royal Statistical Society, 59*, 291–318.

Rose, P. S., & Kolari, J. W. (1985). Early warning systems as a monitoring device for bank condition. *Quarterly Journal of Business and Economics, 24*(1), 43–60.

Roth, M. (1994). Too big to fail and the instability of the banking system: Some insights from foreign countries. *Business Economics, 4*, 43–49.

Segoviano, M. A., & Goodhart, C. A. (2009). *Banking stability measures.* IMF Working Papers 09/4. International Monetary Fund, Washington, DC.

Sharpe, W. F. (1964). Capital asset prices: A theory of market equilibrium under conditions of risk. *Journal of Finance, 19*(3), 425–442.

Sinkey, J. F. (1975). A multivariate statistical analysis of the characteristics of problem banks. *The Journal of Finance, 30*(1), 21–36.

Tam, K. Y., & Kiang, M. Y. (1992). Managerial applications of neural networks: The case of bank failure predictions. *Management Science, 38*(7), 926–947.

Vasicek, O. A. (1984). *Credit valuation.* Technical report. KMV corporation, Moody's, New York.

Vazquez, F., & Federico, P. (2012). *Bank funding structures and risk: Evidence from the global financial crisis.* Working Paper. Social Science Research Network.

Wang, H., & Li, S. Z. (2012). Efficient Gaussian graphical model determination under g-Wishart prior distributions. *Electronic Journal of Statistics, 6*, 168–198.

Wermuth, N., & Lauritzen, S. L. (1990). On substantive research hypotheses, conditional independence graphs and graphical chain models (with discussion). *Journal of the Royal Statistical Society, Series B, 52*(1), 21–50.

Whittaker, J. (1990). *Graphical models in applied multivariate statistics.* Chichester: Wiley.

Zhang, D., Wells, M. T., Turnbull, B. W., Sparrow, D., & Cassano, P. A. (2005). Hierarchical graphical models. *Journal of the American Statistical Association, 100*(471), 719–727.

PARALLEL CONSTRAINED HAMILTONIAN MONTE CARLO FOR BEKK MODEL COMPARISON

Martin Burda

Department of Economics, University of Toronto, Toronto, Canada

ABSTRACT

The BEKK GARCH class of models presents a popular set of tools for applied analysis of dynamic conditional covariances. Within this class the analyst faces a range of model choices that trade off flexibility with parameter parsimony. In the most flexible unrestricted BEKK the parameter dimensionality increases quickly with the number of variables. Covariance targeting decreases model dimensionality but induces a set of nonlinear constraints on the underlying parameter space that are difficult to implement. Recently, the rotated BEKK (RBEKK) has been proposed whereby a targeted BEKK model is applied after the spectral decomposition of the conditional covariance matrix. An easily estimable RBEKK implies a full albeit constrained BEKK for the unrotated returns. However, the degree of the implied restrictiveness is currently unknown. In this paper, we suggest a Bayesian approach to estimation of the BEKK model with targeting based on Constrained Hamiltonian Monte Carlo (CHMC). We take advantage of suitable parallelization of the

Bayesian Model Comparison
Advances in Econometrics, Volume 34, 155–179

ISSN: 0731-9053/doi:10.1108/S0731-905320140000034008

problem within CHMC utilizing the newly available computing power of multi-core CPUs and Graphical Processing Units (GPUs) that enables us to deal effectively with the inherent nonlinear constraints posed by covariance targeting in relatively high dimensions. Using parallel CHMC we perform a model comparison in terms of predictive ability of the targeted BEKK with the RBEKK in the context of an application concerning a multivariate dynamic volatility analysis of a Dow Jones Industrial returns portfolio. Although the RBEKK does improve over a diagonal BEKK restriction, it is clearly dominated by the full targeted BEKK model.

Keywords: Dynamic conditional covariances; Markov chain Monte Carlo; GPU acceleration

JEL classifications: C11; C15; C32; C63

1. INTRODUCTION

Modeling time series volatility dynamics is essential in many areas of economics and finance, such as risk management, portfolio selection, and derivative pricing. The generalized autoregressive conditional heteroskedasticity (GARCH) model has been used as a key tool of applied volatility analysis for nearly two decades. The core GARCH specification asserts that the best predictor of the variance in the next period is a weighted average of the long-run average variance, the variance predicted for the current period, and the new information in the current period. Such an updating rule acts as a description of adaptive learning behavior and can be thought of as Bayesian updating (Engle, 2001).

In a recent paper by Caporin and McAleer (2012) the BEKK GARCH[1] model and the dynamic conditional correlation (DCC) model (Engle, 2002) are singled out as the "two most widely used models of conditional covariances and correlations" in the class of multivariate GARCH models. Nonetheless, in the full unrestricted version of either model type, the number of parameters increases relatively quickly with the model dimensionality, such as with the number of assets in a portfolio.

Parameter parsimony can be achieved by imposition of restrictions on the full model (Billio & Caporin, 2009; Ding & Engle, 2001; Engle, Shephard, & Sheppard, 2009). However, doing so is traded off with model flexibility. For example, Burda and Maheu (2012)show that restricted versions of the BEKK, the scalar and diagonal models, are clearly dominated by the full BEKK version in terms of model fit. Indeed, such parameter restrictions generally operate on the parameters driving the model dynamics.

An alternative way of reducing dimensionality is via so-called *targeting* which imposes a structure on the model intercepts based on sample information. Targeting yields model constraints in a structured way so that its implied long-run solution of the covariance (or correlation) ensures its independence from any of the parameters driving the model dynamics. Hence, targeting differs from merely setting a subset of parameters to zero as in the scalar and diagonal models. Estimation under covariance targeting was originally proposed by Engle and Mezrich (1996) as a two-step estimation procedure, whereby the unconditional covariance matrix of the observed process is estimated in the first step and the remaining parameters are estimated in the second step. However, Aielli (2011) shows that the advantage of targeting as a tool for controlling the curse of dimensionality does not hold for DCC models, as the sample correlation is an inconsistent estimator of the long-run correlation matrix. Caporin and McAleer (2012) also argue that targeting can be rigorously applied only to BEKK and its use in DCC models inappropriate.

In a recent paper, Noureldin, Shephard, and Sheppard (2014) introduce a new class of multivariate volatility models, the rotated BEKK (RBEKK) within the general family of Rotated ARCH (RARCH) models. The conditional covariance matrix is first subjected to spectral decomposition and then a BEKK model is applied to the rotated returns. The RBEKK is relatively easy to estimate under targeting when a simple model such as the diagonal BEKK is used in the second stage. In contrast, in the unrotated full BEKK model, covariance targeting implies a set of model constraints that are nonlinear in parameters and "extremely complicated, except for the scalar case" (Caporin & McAleer, 2012, p. 742).

However, the RBEKK class of models is not equivalent to the full BEKK model specification. Instead, the RBEKK constitutes a constrained version of the BEKK, with or without covariance targeting. At present the degree of restrictiveness of RBEKK relative to full BEKK is unknown.[2] Indeed, in the frequentist literature, estimation and model comparison of the full BEKK especially in the case with targeting presents a significant challenge due to the highly nonlinear set of implied parameter constraints that ensure positive definiteness of the conditional covariance matrix.

In this paper, we suggest a Bayesian approach to estimation and model comparison of the full BEKK model with targeting relative to the corresponding RBEKK class based on Constrained Hamiltonian Monte Carlo (CHMC) (Neal, 2011). CHMC is a recent statistical method that deals very effectively with complicated nonlinear model constraints in the context of relatively high-dimensional problems and computationally costly log-likelihoods. CHMC is particularly useful in cases where it is difficult to accurately approximate the log-likelihood surface around the current parameter draw in real time necessary for obtaining sufficiently high acceptance probabilities in importance sampling (IS) methods, such as for recursive models in finance. Complicated constraints pose further challenge for IS-based methods. In such situations one would typically resort to random walk (RW) style sampling that is fast to run and does not require the knowledge of the properties of the underlying log-likelihood. However, RW mechanisms can lead to very slow exploration of the parameter space with high autocorrelations among draws which would require a prohibitively large size of the Markov chain to be obtained in implementation to achieve satisfactory mixing and convergence. CHMC combines the advantages of sampling that is relatively cheap with RW-like intensity but superior parameter space exploration.

CHMC falls within a general class of Markov chain Monte Carlo (MCMC) methods that can be used under both the Bayesian and the frequentist paradigm, applied to posterior densities or directly to model likelihoods without prior information (Chernozhukov & Hong, 2003). We elaborate in detail on the CHMC procedure, its application to the BEKK and RBEKK with targeting, and its usefulness in applicability to a wider class of problems of similar nature.

We further describe parallel implementation of CHMC suitable for use on multi-core CPUs, or Graphical Processing Units (GPUs). Offloading computationally intensive and suitably parallelized implementation steps to GPUs readily available in present-day workstations has been shown to speed up serial code execution by orders of magnitude (Aldrich, Fernández-Villaverde, Gallant, & Rubio Ramirez, 2011; Geweke & Durham, 2012; Morozov & Mathur, 2012), even though to date, the adoption of GPU computing technology in economics and econometrics has been relatively slow compared to other fields (Casarin, Grassi, Ravazzolo, & van Dijk, 2013). Due to the popularity of BEKK type models in practice, we believe the model comparison evidence provided herein will serve as a useful reference for applied analysts.

2. BEKK GARCH

The BEKK class of multivariate GARCH models was introduced by Engle and Kroner (1995). Their specification was sufficiently general to allow the inclusion of special factor structures (Bauwens, Laurent, & Rombouts, 2006). Here we focus on the BEKK with all orders set to one, which is the predominantly used version in applications (Silvennoinen & Teräsvirta, 2009). Let r_t be a $N \times 1$ vector of asset returns with $t = 1,\ldots,T$ and denote the information set as $\mathcal{F}_{t-1} = \{r_1,\ldots,r_{t-1}\}$. Under the BEKK model the returns follow

$$r_t | \mathcal{F}_{t-1} \sim NID(0,\Sigma_t) \tag{1}$$

$$\Sigma_t = CC' + Ar_{t-1}r'_{t-1}A' + B\Sigma_{t-1}B' \tag{2}$$

Σ_t is a positive definite $N \times N$ conditional covariance matrix of r_t given information at time $t-1$, C is a lower triangular matrix, and A and B are $N \times N$ matrices. We maintain a Gaussian assumption and a zero intercept for simplicity. The total number of parameters in this model is $N(N+1)/2 + 2N^2$, collected into the vector $\theta = (vech(C)', vech(A)', vech(B)')'$ with dimension $d = \dim(\theta)$. The diagonal BEKK results from the restriction on A and B to be diagonal matrices, with $d = N(N+1)/2 + 2N$, and in the scalar BEKK $A = \alpha I_N$, $B = \beta I_N$, $C = \psi I_n$, with $d = 3$.

2.1. Covariance Targeting

The BEKK model with targeting is defined as

$$\Sigma_t = (\overline{\Sigma} - A\overline{\Sigma}A' - B\overline{\Sigma}B') + Ar_{t-1}r'_{t-1}A' + B\Sigma_{t-1}B' \tag{3}$$

where

$$E[\Sigma_t] = E[\Sigma_{t-1}] = \overline{\Sigma}$$
$$E[r_{t-1}r'_{t-1}] = \overline{\Sigma}$$

There are two types of constraints on the model that need to be taken into account. The first type, covariance stationarity constraints, are generally simple to impose, as shown by Engle and Kroner (1995). The second type, positive definiteness of the conditional covariance matrices, is satisfied by construction in the full BEKK version, but becomes difficult to deal with in the targeted case. CM note that one way in which positive definiteness of Σ_t can be guaranteed is by imposing positive definiteness on $\overline{\Sigma} - A\overline{\Sigma}A' - B\overline{\Sigma}B'$ but note that such constraint is complicated. In a Bayesian MCMC scheme the constraint can be assessed by checking invertibility of Σ_t for each t and taking corrective action in the sampling procedure if the constraint is violated. A Bayesian approach is favorable also on inference grounds. In general the occurrence of the constrained likelihood argmax on the parameter space boundary cannot be ruled out a priori. Furthermore, the log-likelihood surface can be highly asymmetric, with the degree of asymmetry increasing in higher dimensions. Due to both of these features, the accuracy of QMLE asymptotic inference, based on a Fisher information matrix that is symmetric by construction, is questionable. Bayesian MCMC-based procedures, including CHMC presented in the next section, provide an alternative estimation and inference approach for either constraint case, unfettered by the aforementioned likelihood irregularities.

2.2. *Rotated BEKK*

In a recent paper Noureldin et al. (2014) proposed a so-called rotated BEKK (RBEKK) within a class of rotated ARCH models. The idea is to transform the original data by rotation and then subject the rotated data to a BEKK model specification. Since the rotated residuals are orthogonal a suitable parametrization is the diagonal BEKK which makes the model implementation simple even under covariance targeting. Using the spectral decomposition, let

$$E[\Sigma_t] \equiv \overline{H} = P\Lambda P'$$

where P is a matrix of eigenvectors and the eigenvalue matrix Λ is diagonal with nonnegative elements. The symmetric square root of $\overline{H}$ is

$$\overline{H}^{1/2} = P\Lambda^{1/2}P'$$

Letting

$$r_t = \overline{H}^{1/2} e_t$$

the rotated returns are defined as

$$e_t = \overline{H}^{-1/2} r_t$$
$$= P\Lambda^{-1/2} P' r_t$$

where $Var[e_t] = I_d$, the d-dimensional identity matrix. The conditional covariance of the rotated returns is $Var[e_t|\mathcal{F}_{t-1}] = G_t$, where $\Sigma_t = \overline{H}^{1/2} G_t \overline{H}^{1/2}$. Any dynamic covariance model for Σ_t can also be applied to model the dynamics of G_t. The resulting RBEKK model is thus specified as

$$G_t = (I_d - AA - BB') + Ae_t e'_{t-1} A' + BG_{t-1} B' \quad (4)$$

with $G_0 = I_d$. Noureldin et al. (2014) consider the scalar, diagonal, common persistence (CP), and orthogonal parameter matrix BEKK specifications for the model (4). In their analysis, the diagonal model outperforms the other specifications. They note that the diagonal model in Eq. (4) implies a complete BEKK for the unrotated returns r_t. However, such model is only a restricted version of the fully flexible BEKK model in Eq. (3). Since Noureldin et al. (2014) have not performed a model comparison of Eq. (4) with Eq. (3), the relative degree of restrictiveness of Eq. (4) has been left as an open question which we answer in this paper using recent advances in Bayesian MCMC methodology and parallel computing power.

3. CONSTRAINED HAMILTONIAN MONTE CARLO

Hamiltonian (or Hybrid) Monte Carlo (HMC) is a class of MCMC methods featuring multi-step distant proposals whose path follows the evolution fo Hamiltonian dynamics. HMC has its roots in the physics literature where it was introduced for simulating molecular dynamics (Duane, Kennedy, Pendleton, & Roweth, 1987). It has since become popular in a number of application areas including statistical physics

(Akhmatskaya, Bou-Rabee, & Reich, 2009; Gupta, Kilcup, & Sharpe, 1988), computational chemistry (Tuckerman, Berne, Martyna, & Klein, 1993), or a generic tool for Bayesian statistical inference (Beskos, Pillai, Roberts, Sanz-Serna, & Stuart, 2013; Ishwaran, 1999; Liu, 2004; Neal, 1993, 2011). HMC is most applicable to situations when a good importance sampler is not available or practical to implement and one would thus typically need to rely on RW sampling. However, HMC has been shown to yield samples far more efficient than obtained by the RW Metropolis–Hastings (M–H) mechanism (Neal, 2011; Rasmussen, 2003).

HMC and its constrained version, CHMC, applies to a general class of models, nesting the BEKK, that is parametrized by a Euclidean vector $\theta \in \Theta$ for which all information in the sample is contained in the model likelihood $\pi(\theta)$ assumed known up to an integrating constant. Formally, this class of models can be characterized by a family $\mathscr{P}_\theta$ of probability measures on a measurable space $(\Theta, \mathscr{B})$ where $\mathscr{B}$ is the Borel σ-algebra. The purpose of MCMC methods is to formulate a Markov chain on the parameter space Θ for which, under certain conditions, $\pi(\theta) \in \mathscr{P}_\theta$ is the invariant (also called "equilibrium" or "long-run") distribution. The Markov chain of draws of θ can be used to construct simulation-based estimates of the required integrals, and functionals $h(\theta)$ of θ that are expressed as integrals. These functionals include objects of interest for inference on θ such as quantiles of $\pi(\theta)$.

The Markov chain sampling mechanism specifies a method for generating a sequence of random variables $\{\theta_r\}_{r=1}^{R}$, starting from an initial point θ_0, in the form of conditional distributions for the draws $\theta_{r+1}|\theta_r \sim G(\theta_r)$. Under relatively weak regularity conditions (Robert & Casella, 2004), the average of the Markov chain converges to the expectation under the stationary distribution

$$\lim_{R \to \infty} \frac{1}{R} \sum_{r=1}^{R} h(\theta_r) = E_\pi[h(\theta)]$$

A Markov chain with this property is called ergodic. As a means of approximation we rely on large but finite number of draws $R \in \mathbb{N}$ which the analyst has the discretion to select in applications.

$G(\theta_r)$ can be obtained from a given (economic) model and its corresponding likelihood $\pi(\theta)$. Typically, $\pi(\theta)$ has a complicated form which precludes direct sampling in which case the M–H principle is usually employed for $\theta_{r+1}|\theta_r$ from $G(\theta_r)$ (see Chib & Greenberg, 1995 for a detailed overview). Suppose we have a proposal-generating density $q(\theta^*_{r+1}|\theta_r)$ where θ^*_{r+1} is a proposed state given the current state θ_r of the Markov chain. The M–H principle stipulates that θ^*_{r+1} be accepted as the next state θ_{r+1} with the acceptance probability

$$\alpha(\theta_r, \theta^*_{r+1}) = \min\left[\frac{\pi(\theta^*_{r+1})q(\theta_r|\theta^*_{r+1})}{\pi(\theta_r)q(\theta^*_{r+1}|\theta_r)}, 1\right] \tag{5}$$

otherwise $\theta_{r+1} = \theta_r$. Then the Markov chain satisfies the so-called detailed balance condition

$$\pi(\theta_r)q(\theta^*_{r+1}|\theta_r)\alpha(\theta_r, \theta^*_{r+1}) = \pi(\theta^*_{r+1})q(\theta_r|\theta^*_{r+1})\alpha(\theta^*_{r+1}, \theta_r)$$

which is sufficient for ergodicity. $\alpha(\theta^*_{r+1}, \theta_r)$ is the probability of the move $\theta_r|\theta^*_{r+1}$ if the dynamics of the proposal-generating mechanism were to be reversed. The proposal-generating density $q(\theta^*_{r+1}|\theta_r)$ can be chosen to be sampled easily even though $\pi(\theta)$ may be difficult or expensive to sample from. The popular Gibbs sampler arises as a special case when the M–H sampler is factored into conditional densities. The proposal draws $\theta^*_{r+1}|\theta_r$ from $q(\theta^*_{r+1}|\theta_r)$ in Eq. (5) are generated in one step.

In contrast, HMC uses a whole sequence of proposal steps whereby the last step in the sequence becomes the proposal draw. This facilitates efficient exploration of the parameter space with the resulting Markov chain. The proposal sequence is constructed using difference equations of the law of motion yielding high acceptance probability even for distant proposals. The parameter space Θ is augmented with a set of independent auxiliary stochastic parameters $\gamma \in \Gamma$ that fulfill a supplementary role in the proposal algorithm, facilitating the directional guidance of the proposal mechanism. The proposal sequence takes the form $\{\theta^k_r, \gamma^k_r\}^L_{k=1}$ starting from the current state $(\theta_r, \gamma_r) = (\theta^0_r, \gamma^0_r)$ and yielding a proposal $(\theta_r, \gamma_r) = (\theta^L_r, \gamma^L_r)$. The detailed balance is then satisfied using the acceptance probability

$$\alpha(\theta_r, \gamma_r; \theta^*_{r+1}, \gamma^*_{r+1}) = \min\left[\frac{\pi(\theta^*_{r+1}, \gamma^*_{r+1})q(\theta_r, \gamma_r|\theta^*_{r+1}, \gamma^*_{r+1})}{\pi(\theta_r, \gamma_r)q(\theta^*_{r+1}, \gamma^*_{r+1}|\theta_r, \gamma_r)}, 1\right] \tag{6}$$

In CHMC, constraints are incorporated into the HMC proposal mechanism via "hard walls" representing a barrier against which the proposal sequence, simulating a particle movement, bounces off elastically.

Constraints thus do not provide grounds for proposal rejection, eliminating any associated redundancies. Heuristically, the constraint is checked at each step of the proposal sequence and if it is violated then the trajectory of the sequence is reflected off the hard wall posed by the constraint. This facilitates efficient exploration of the parameter space even in the presence of highly complex parameter constraints. We provide a synthesis of the technical principles of CHMC in a generally accessible form in Appendix A.

We use a parallelized version of CHMC for the BEKK likelihood to facilitate feasibility of implementation in relatively high dimensions ($d = 200$). Although the BEKK model is formulated as intrinsically recursive, evaluation of conditionally independent objects can take full advantage of the computational speedups afforded by multi-core and GPU architecture of present-day computers. In particular, the positive definiteness constraint on Σ_t in Eq. (3) is verified independently for each t by checking the positiveness of the eigenvalues of Σ_t. Similarly, computation of the matrix inverse Σ_t^{-1}, the determinant $|\Sigma_t|$, and the term $Ar_{t-1}r'_{t-1}A'$ in evaluating the BEKK log-likelihood are evaluated independently for each t, so that only the last term $B\Sigma_{t-1}B'$ in Eq. (3) is left for sequential recursion. Further details on the practical implementation on multi-core and GPU architectures using the Fortran compiler are given in Appendix B.

4. APPLICATION AND MODEL COMPARISON

The model comparison is performed in the context of an empirical application with close-to-close daily returns data on 10 stocks[3] in the Dow Jones Industrial Index (DJIA). The sample period is February 1, 2008–December 31, 2013, total of $T = 1,510$ observations, covering the recent financial crisis. The source of the data is Yahoo! Finance which is accessible online. We use close prices adjusted for dividends and splits. The time series of the daily returns data are plotted in Fig. 1. Summary statistics are provided in Table C1. The sample mean for all series is close to 0 and skewness is small. The associated parameter dimensionality for the portfolio is 200 for the targeted BEKK, 20 for a diagonal BEKK specification, and 20 for the RBEKK.

For identification, the diagonal elements of C and the first element of both A and B are restricted to be positive (Engle & Kroner, 1995), but otherwise our prior on all other parameters is diffuse. In the

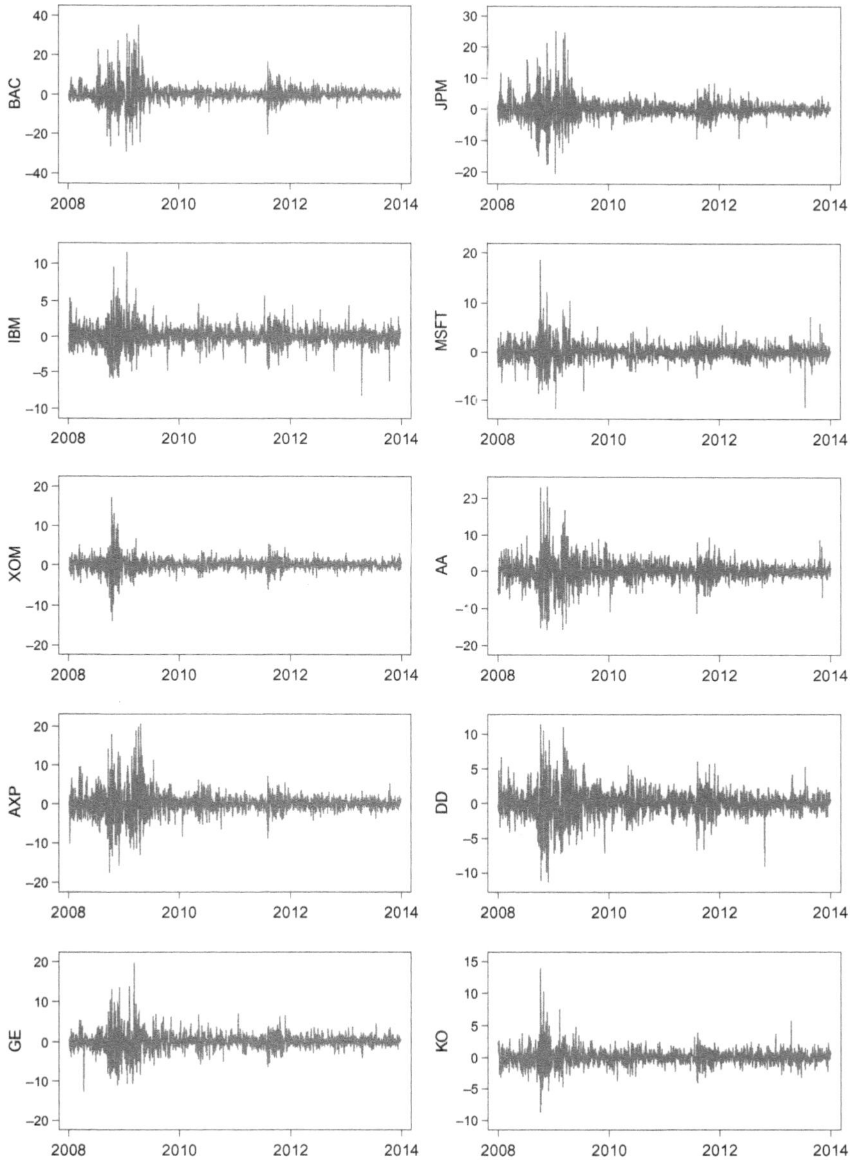

Fig. 1. Time Series of Daily Returns Data.

implementation, we utilize the analytical expressions for the gradient of the BEKK log-likelihood from (Hafner & Herwartz, 2008). The initial Σ_1 in the full BEKK recursion is set to the sample covariance.

For model comparison we use two criteria. First, we compute the predictive log scores for the full BEKK and the RBEKK, each with covariance targeting, integrating across samples from the posterior predictive densities conditional on data $y_{-h},\ldots,y_t$ for each $t=1,\ldots,T$, where $y_{-h},\ldots,y_0$ represents a burn-in sample. The period $t=1,\ldots,T$ spans over the last 120 data points in the sample time series, which corresponds to about half a year of daily trading data. The technical details of the implementation are described below. The full BEKK scored the value of −1,743.8 while the RBEKK scored −1,792.2 on this measure, attesting to a better predictive ability of the former model.

Second, we use the same criterion as in Noureldin et al. (2014), albeit in a Bayesian setting, that is we compare the 1-step predictive ability of BEKK, diagonal BEKK, and RBEKK (with second-stage diagonal BEKK), each with covariance targeting. From the model parameters of the rotated returns e_t in the RBEKK we calculate the implied model parameters and likelihood for the original returns r_t as in Noureldin et al. (2014). The likelihood for r_t can be calculated using a transformation of random variables with the Jacobian $\frac{\partial r_t}{\partial e_t}=P\Lambda^{1/2}P'$ and its determinant $|P\Lambda^{1/2}P'|=|\Lambda^{1/2}|$ where the second equality follows from P being orthogonal. Hence, for a time series of length T we have $\ln L_r=\ln L_e+\frac{T}{2}\ln|\Lambda|$ where $\ln L_r$ and $\ln L_e$ denote the log-likelihoods for r_t and e_t, respectively. Following Noureldin et al. (2014) we also perform a margin-copula predictive likelihood decomposition in order to assess the fit of multivariate volatility models by separating the gains/losses due to the forecast variances and correlations. Let $\ln L_i=\sum_{t=1}^{T}\ln f(r_{i,t}|\mathcal{F}_{t-1})$ be the marginal log-likelihood for the ith asset. The copula log-likelihood is then given by $\ln L_r-\sum_{i=1}^{N}\ln L_i$ where $\ln L_r$ is the model's joint log-likelihood.

The results are based on Markov chains of 10,000 draws following a 1,000 burn-in section. The output was obtained using an optimized Fortran code run on two 2.53 GHz 4-core Xeon X5550 CPUs with 48 GB of RAM per node each containing two NVIDIA M2070 (Fermi) GPUs each with 6 GB of RAM. We use the rule of thumb target acceptance rate of about 70% (Beskos, Pillai, Roberts, Sanz-Serna, & Stuart, 2010), with 20 steps in each proposal sequence.

The wallclock run time is about 11.5 hours the diagonal RBEKK, and about 36 hours for the full BEKK. As a point of comparison, for the full BEKK model, a serial execution takes about 567 hours, OpenMP (without

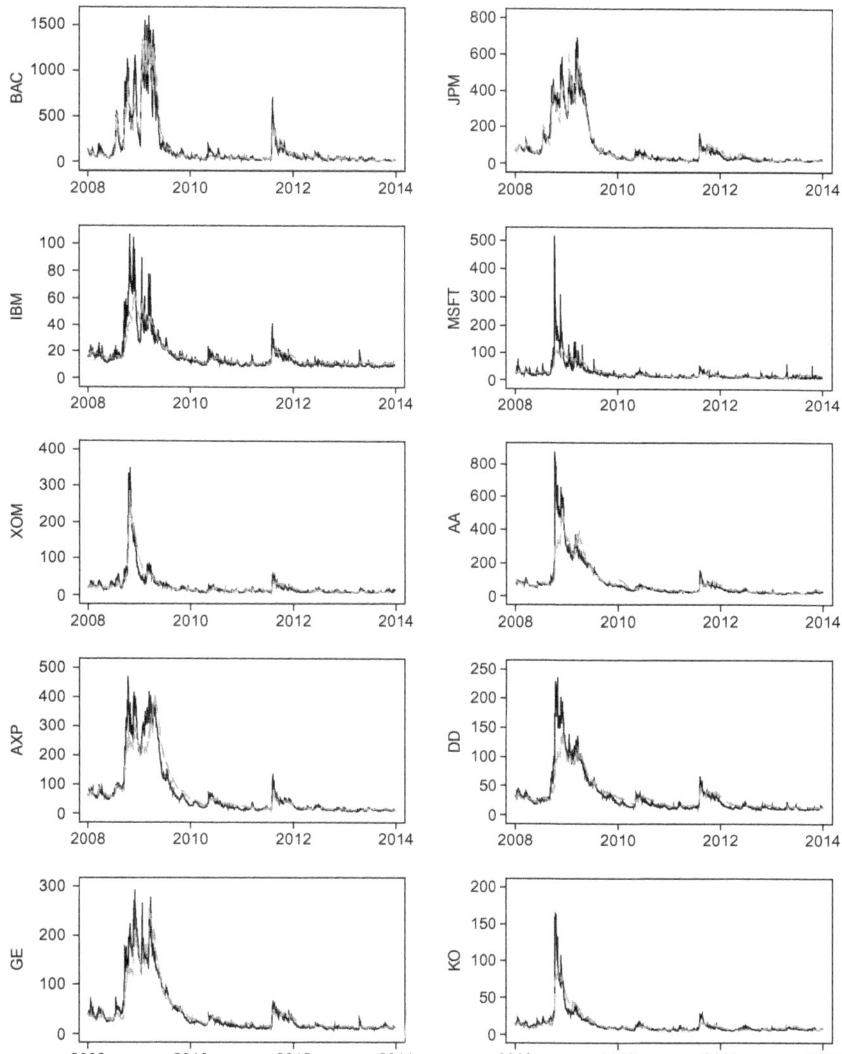

Fig. 2. Predicted Variances. In each graph, the solid line corresponds to the full BEKK model and the dashed line to the RBEKK model.

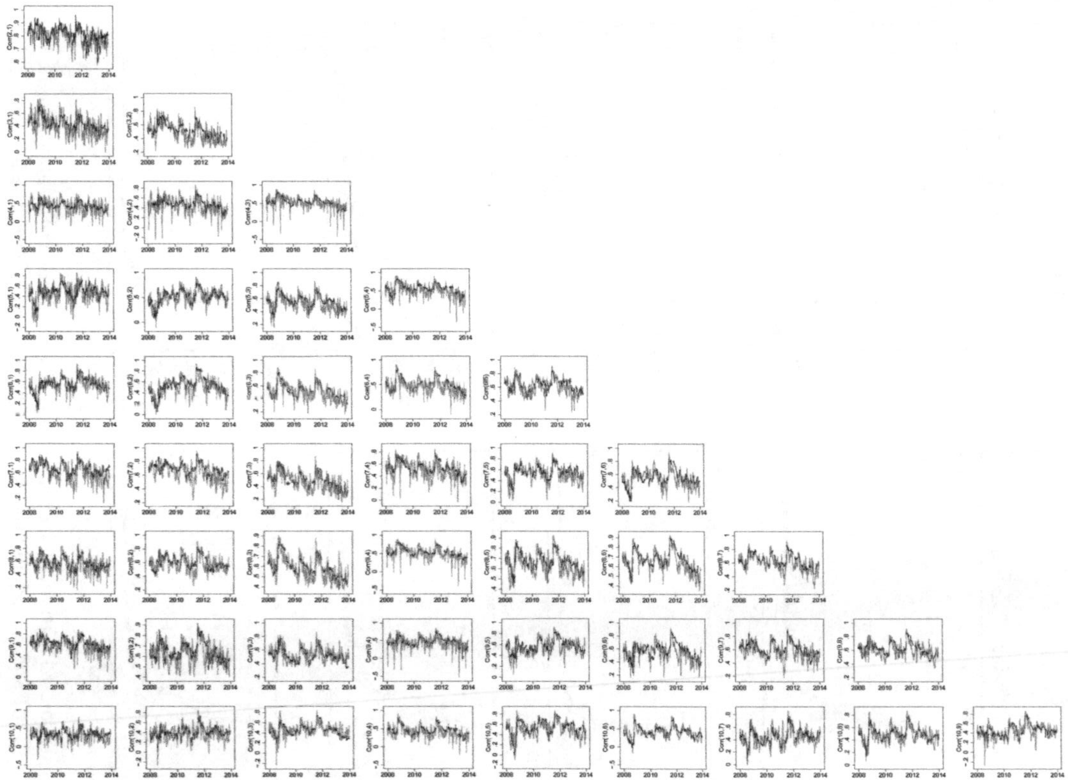

Fig. 3. Predicted Conditional Correlations. In each graph, the solid line corresponds to the full BEKK model and the dashed line to the RBEKK model.

OpenACC compilation) takes about 562 hours, and OpenACC (without OpenMP compilation) takes about 38 hours. The timing is inevitably implementation-specific. We sought to make maximum use of GPU parallelization wherever possible while the remaining code stayed intrinsically serial (recursive) which is reflected in the relatively high speed-up due to the GPU unit and relatively low speed-up due to the multi-CPU unit.

The model comparison results are reported in Table C2 for $N=10$. We report the minimum and maximum values of posterior means of draws of parameter matrices A and B, posterior means of margin-copula draws, and the corresponding means of the log-likelihood values (posterior values under a diffuse prior). The RBEKK improves in terms of predictive ability over the diagonal BEKK. However, the RBEKK is shown to be dominated by the full BEKK model and hence we contend that the latter should be the model of choice whenever sufficient computational power allows for its implementation.

The posterior means of predicted dynamic variances are shown in Fig. 2 and dynamic conditional correlations in Fig. 3. In each case, the full BEKK model adjusts more readily to abrupt deviations than the RBEKK.

5. CONCLUSIONS

In this paper we suggest an effective approach for estimation and inference on the full BEKK GARCH model with targeting and compare its predictive ability with the RBEKK model. The approach is based on parallelized CHMC, a recent statistical technique devised to handle nonlinear constraints in the context of relatively costly and irregular likelihoods. We show, in the context of an application to multivariate dynamic volatility analysis of a Dow Jones Industrial returns portfolio, that the RBEKK is dominated by the full targeted BEKK, although the former presents an improvement over a diagonal BEKK restriction.

NOTES

1. The model was proposed by Engle and Kroner (1995) who note that the acronym comes from an earlier version of their paper, which synthesized the work on multivariate ARCH models by Yoshi Baba, Rob Engle, Dennis Kraft, and Ken Kroner.

2. Noureldin et al. (2014) only provide a model comparison of several members of the RBEKK class with OGARCH, GOGARCH, and the diagonal BEKK, but not the full BEKK with or without targeting.

3. These are: Bank of America Corp. (BAC), JP Morgan Chase & Co. (JPM), International Business Machines (IBM), Microsoft (MSFT), Exxon Mobil Corp. (XOM), Alcoa (AA), American Express Inc. (AXP), DuPont (DD), General Electric (GE), The Coca Cola Co. (KO).

4. In the physics literature, θ denotes the position (or state) variable and $-\ln\pi(\theta)$ describes its potential energy, while γ is the momentum variable with kinetic energy $\gamma' M^{-1}\gamma/2$, yielding the total energy $H(\theta,\gamma)$ of the system, up to a constant of proportionality. M is a constant, symmetric, positive-definite "mass" matrix.

5. In the physics literature, the Hamiltonian dynamics describe the evolution of (θ,γ) that keeps the total energy $H(\theta,\gamma)$ constant.

6. For further details on GPU computing and its applications, see, e.g., the NVidia Cuda Zone at https://developer.nvidia.com/get-started-parallel-computing.

ACKNOWLEDGMENTS

I would like to thank John Maheu for valuable comments. This work was made possible by the facilities of the Shared Hierarchical Academic Research Computing Network (SHARCNET: www.sharcnet.ca) and it was supported by grants from the Social Sciences and Humanities Research Council of Canada (SSHRC: www.sshrc-crsh.gc.ca).

REFERENCES

Aielli, G. P. (2011). *Dynamic conditional correlation: On properties and estimation.* SSRN Working Paper. Retrieved from http://papers.ssrn.com/sol3/papers.cfmabstract id=1507743

Akhmatskaya, E., Bou-Rabee, N., & Reich, S. (2009). A comparison of Generalized Hybrid Monte Carlo methods with and without momentum flip. *Journal of Computational Physics*, *228*(6), 2256–2265.

Aldrich, E. M., Fernández-Villaverde, J., Gallant, A. R., & Rubio Ramirez, J. F. (2011). Tapping the supercomputer under your desk: Solving dynamic equilibrium models with graphics processors. *Journal of Economic Dynamics and Control*, *35*, 386–393.

Bauwens, L., Laurent, S., & Rombouts, J. V. K. (2006). Multivariate GARCH models: A survey. *Journal of Applied Econometrics*, *21*, 79–109.

Beskos, A., Pillai, N. S., Roberts, G. O., Sanz-Serna, J. M., & Stuart, A. M. (2010). The acceptance probability of the Hybrid Monte Carlo method in high-dimensional problems. *AIP Conference Proceedings*, *1281*(1), 23–27.

Beskos, A., Pillai, N. S., Roberts, G. O., Sanz-Serna, J. M., & Stuart, A. M. (2013). Optimal tuning of the hybrid Monte Carlo algorithm. *Bernoulli*, *19*(5A), 1501–1534.

Billio, M., & Caporin, M. (2009). A generalized dynamic conditional correlation model for portfolio risk evaluation. *Mathematics and Computers in Simulation, 79*(8), 2566–2578.

Burda, M., & Maheu, J. M. (2012). Bayesian adaptively updated Hamiltonian Monte Carlo with an application to high-dimensional BEKK GARCH models. *Studies in Nonlinear Dynamics & Econometrics, 17*(4), 345–372.

Caporin, M., & McAleer, M. (2012). Do we really need both BEKK and DCC? A tale of two multivariate GARCH models. *Journal of Economic Surveys, 26*(4), 736–751.

Casarin, R., Grassi, S., Ravazzolo, F., & van Dijk, H. K. (2013). *Parallel sequential Monte Carlo for efficient density combination: The DeCo Matlab toolbox*. Discussion Paper, Tinbergen Institute.

Chernozhukov, V., & Hong, H. (2003). An MCMC approach to classical estimation. *Journal of Econometrics, 115*(3), 293–346.

Chib, S., & Greenberg, E. (1995). Understanding the Metropolis–Hastings algorithm. *American Statistician, 49*(4), 327–335.

Ding, Z., & Engle, R. (2001). Large scale conditional covariance matrix modeling, estimation and testing. *Academia Economic Papers, 29*, 157–184.

Duane, S., Kennedy, A., Pendleton, B., & Roweth, D. (1987). Hybrid Monte Carlo. *Physics Letters B, 195*(2), 216–222.

Engle, R. (2001). GARCH 101: The use of ARCH/GARCH models in applied econometrics. *Journal of Economic Perspectives, 15*(4), 157–168.

Engle, R. F. (2002). Dynamic conditional correlation: A simple class of multivariate generalized autoregressive conditional heteroskedasticity models. *Journal of Business and Economic Statistics, 20*, 339–350.

Engle, R. F., & Kroner, K. F. (1995). Multivariate simultaneous generalized ARCH. *Econometric Theory, 11*(1), 122–150.

Engle, R. F., & Mezrich, J. (1996). GARCH for groups. *Risk, 9*(8), 36–40.

Engle, R. F., Shephard, N., & Sheppard, K. (2009). *Fitting vast dimensional time-varying covariance models*. Retrieved from http://ssrn.com/abstract=1354497

Geweke, J., & Durham, G. (2012). *Massively parallel sequential Monte Carlo for Bayesian inference*. Working Paper. Available at SSRN: http://dx.doi.org/10.2139/ssrn.1964731.

Gupta, R., Kilcup, G., & Sharpe, S. (1988). Tuning the Hybrid Monte Carlo algorithm. *Physical Review D, 38*(4), 1278–1287.

Hafner, C. M., & Herwartz, H. (2008). Analytical quasi maximum likelihood inference in multivariate volatility models. *Metrika, 67*, 219–239.

Ishwaran, H. (1999). Applications of Hybrid Monte Carlo to generalized linear models: Quasicomplete separation and neural networks. *Journal of Computational and Graphical Statistics, 8*, 779–799.

Leimkuhler, B., & Reich, S. (2004). *Simulating hamiltonian dynamics*. Cambridge: Cambridge University Press.

Liu, J. S. (2004). *Monte Carlo strategies in scientific computing*. Springer Series in Statistics. New York: Springer.

Morozov, S., & Mathur, S. (2012). Massively parallel computation using graphics processors with application to optimal experimentation in dynamic control. *Computational Economics* 40(2), 151–182. doi:10.1007/s10614-011-9297-4

Neal, R. M. (1993). Probabilistic inference using Markov chain Monte Carlo methods. Technical Report CRG-TR-93-1, Department of Computer Science, University of Toronto.

Neal, R. M. (2011). MCMC using Hamiltonian dynamics. In S. Brooks, A. Gelman, G. Jones, & X.-L. Meng (Eds.), *Handbook of Markov Chain Monte Carlo*. Boca Raton, FL: Chapman & Hall.

Noureldin, D., Shephard, N., & Sheppard, K. (2014). Multivariate rotated ARCH models. *Journal of Econometrics 179*(1), 16–30. doi:10.1016/j.jeconom.2013.10.003. Retrieved from http://www.sciencedirect.com/science/article/pii/S0304407613002078

Rasmussen, C. E. (2003). Gaussian processes to speed up Hybrid Monte Carlo for expensive Bayesian integrals. *Bayesian Statistics*, *7*, 651–659.

Robert, C. P., & Casella, G. (2004 *Monte Carlo statistical methods* (2nd ed.). New York, NY: Springer.

Silvennoinen, A., & Teräsvirta, T. (2009). Modeling multivariate autoregressive conditional heteroskedasticity with the double smooth transition conditional correlation GARCH model. *Journal of Financial Econometrics.*, *7*(4), 373–411.

Tuckerman, M., Berne, B., Martyna, G., & Klein, M. (1993). Efficient molecular dynamics and Hybrid Monte Carlo algorithms for path integrals. *The Journal of Chemical Physics*, *99*(4), 2796–2808.

APPENDIX A: STATISTICAL PROPERTIES OF CONSTRAINED HAMILTONIAN MONTE CARLO

In this section we provide the stochastic background for CHMC. This synthesis is based on previously published material (Burda & Maheu, 2012; Neal, 2011; and references therein). However, the bulk of literature presenting HMC methods does so in terms of the physical laws of motion based on preservation of total energy in the phase-space. Here we take a fully stochastic perspective familiar to the applied econometrician. The CHMC principle is thus presented in terms of the joint density over the augmented parameter space leading to a Metropolis acceptance probability update.

A.1. CHMC Principle

Consider a vector of parameters of interest $\theta \in \mathbb{R}^d$ distributed according to the density kernel $\pi(\theta)$. Let $\gamma \in \mathbb{R}^d$ denote a vector of auxiliary parameters with $\gamma \sim \Phi(\gamma; 0, M)$ where Φ denotes the Gaussian distribution with mean vector 0 and covariance matrix M, independent of θ. M can be either set to the identity matrix or to a covariance matrix estimated by the inverse of the Fisher information matrix around the mode, as we do in our implementation. Denote the joint density of (θ, γ) by $\pi(\theta, \gamma)$. Denote the constraint on the parameter space by a density kernel $\exp(C_r(\theta))$ where

$$C_r(\theta) = \begin{cases} 0 \text{ if the constraint is satisfied} \\ C_r > 0 \text{ s.t. } \lim_{r \to \infty} C_r = \infty \text{ if the constraint is violated} \end{cases}$$

Then the negative of the logarithm of the joint density of (θ, γ) is given by the constrained Hamiltonian equation[4]

$$H(\theta, \gamma) = -C_r(\theta) - \ln \pi(\theta) + \frac{1}{2}\ln((2\pi)^d |M|) + \frac{1}{2}\gamma' M^{-1} \gamma \tag{A.1}$$

CHMC is formulated in the following steps:

1. Draw an initial auxiliary parameter vector $\gamma_r^0 \sim \Phi(\gamma; 0, M)$;
2. Transition from (θ_r, γ_r) to $(\theta_r^L, \gamma_r^L) = (\theta_{r+1}^*, \gamma_{r+1}^*)$ according to the constrained Hamiltonian dynamics;
3. Accept $(\theta_{r+1}^*, \gamma_{r+1}^*)$ with probability

$$\alpha(\theta_r, \gamma_r; \theta^*_{r+1}, \gamma^*_{r+1}) = \min[\exp(-H(\theta^*_{r+1}, \gamma^*_{r+1}) + H(\theta^0_r, \gamma^0_r)), 1] \tag{A.2}$$

otherwise keep (θ_r, γ_r) as the next MC draw.

The constraints on the parameter space Θ are reflected in Step 2. We will now describe each step in detail.

Step 1 provides a stochastic initialization of the system akin to a RW draw in order to make the resulting Markov chain $\{(\theta_r, \gamma_r)\}_{r=1}^R$ irreducible and aperiodic (Ishwaran, 1999). In contrast to RW, this so-called refreshment move is performed on the auxiliary variable γ as opposed to the original parameter of interest θ, setting $\theta^0_r = \theta_r$. The initial refreshment draw of γ^0_r is equivalent to a Gibbs step on the parameter space of (θ, γ) accepted with probability 1. Since it only applies to γ, it will leave the target joint distribution of (θ, γ) invariant and subsequent steps can be performed conditional on γ^0_r (Neal, 2011).

Step 2 constructs a sequence $\{\theta^k_r, \gamma^k_r\}_{k=1}^L$ according to the Hamiltonian dynamics starting from the current state (θ^0_r, γ^0_r), with the newly refreshed γ^0_r, and setting the last member of the sequence as the CHMC new state proposal $(\theta^*_{r+1}, \gamma^*_{r+1}) = (\theta^L_r, \gamma^L_r)$. The transition from (θ^0_r, γ^0_r) to (θ^L_r, γ^L_r) via the proposal sequence $\{\theta^k_r, \gamma^k_r\}_{k=1}^L$ taken according to the discretized Hamiltonian dynamics (A.5)–(A.7) is fully deterministic, placing a Dirac delta probability mass $\delta(\theta^k_r, \gamma^k_r) = 1$ on each (θ^k_r, γ^k_r) conditional on (θ^0_r, γ^0_r). The CHMC acceptance probability in Eq. (A.2) is specified in terms of the difference between the Hamiltonian (A.1) evaluated at the initial (θ^0_r, γ^0_r) and at the proposal $(\theta^*_{r+1}, \gamma^*_{r+1})$ The role of the Hamiltonian dynamics is to ensure that the acceptance probability (A.2) for $(\theta^*_{r+1}, \gamma^*_{r+1})$ is kept close to 1. This corresponds to maintaining the difference $-H(\theta^*_{r+1}, \gamma^*_{r+1}) + H(\theta^0_r, \gamma^0_r)$ close to zero throughout the sequence $\{\theta^k_r, \gamma^k_r\}_{k=1}^L$. This property of the transition from (θ_r, γ_r) to $(\theta^*_{r+1}, \gamma^*_{r+1})$ can be achieved by conceptualizing θ and γ as functions of continuous time t and specifying their evolution using the Hamiltonian dynamics equations[5]

$$\frac{d\theta_i}{dt} = \frac{\partial H(\theta, \gamma)}{\partial \gamma_i} = [M^{-1}\gamma]_i \tag{A.3}$$

$$\frac{d\gamma_i}{dt} = -\frac{\partial H(\theta, \gamma)}{\partial \theta_i} = \nabla_{\theta_i} \ln \pi(\theta) \tag{A.4}$$

for $i=1,\ldots,d$. For any discrete time interval of duration s, Eqs. (A.3)–(A.4) define a mapping T_s from the state of the system at time t to the state at time $t+s$. The differential Eqs. (A.3)–(A.4) are generally solved by numerical methods, typically the Stormer–Verlet (or leapfrog) numerical integrator (Leimkuhler & Reich, 2004). For each step in constructing the proposal sequence $\{\theta_r^k, \gamma_r^k\}_{k=1}^L$, CHMC discretizes the Hamiltonian dynamics (A.3)–(A.4) as follows: for some small $\varepsilon \in \mathbb{R}$ first take a half-ε step in γ

$$\gamma(t+\varepsilon/2)=\gamma(t)+(\varepsilon/2)\nabla_\theta \ln\pi(\theta(t)) \tag{A.5}$$

then take a full ε step in θ

$$\theta(t+\varepsilon)=\theta(t)+\varepsilon M^{-1}\gamma(t+\varepsilon/2) \tag{A.6}$$

check the constraint at $\theta(t+\varepsilon)$ for each dimension i of θ, if for any i the constraint is violated then set $\gamma_i(t+\varepsilon/2)=-\gamma_i(t+\varepsilon/2)$ reversing the proposal dynamics and take further steps in θ_i until the constraint is satisfied, then finish with another half-ε step in γ

$$\gamma(t+\varepsilon)=\gamma(t+\varepsilon/2)+(\varepsilon/2)\nabla_\theta \ln\pi(\theta(t+\varepsilon)) \tag{A.7}$$

Intuitively, the proposal trajectory bounces off the "walls" given by the constraint. Since a move in γ only occurs conditional on θ for which the constraint is satisfied in which case $C_r(\theta)=0$, and the move in θ only depends on $\partial H(\theta,\gamma)/\partial\gamma_i$ which does not involve $C_r(\theta)$, the exact functional form of $C_r(\theta)$ is inconsequential as long as it is differentiable in θ to define valid Hamiltonian dynamics.

From the statistical perspective, γ plays the role of an auxiliary variable that parametrizes (a functional of) $\pi(\theta)$ providing it with an additional degree of flexibility to maintain the acceptance probability close to one for every k. Even though $\ln\pi(\theta_r^k)$ can deviate substantially from $\ln\pi(\theta_r^0)$, resulting in favorable mixing for θ, the additional terms in γ in Eq. (A.1) compensate for this deviation maintaining the overall level of $H(\theta_r^k, \gamma_r^k)$ close to constant over $k=1,\ldots,L$ when used in accordance with Eqs. (A.5)–(A.7), since $\frac{\partial H(\theta,\gamma)}{\partial\gamma_i}$ and $\frac{\partial H(\theta,\gamma)}{\partial\theta_i}$ enter with the opposite signs in Eqs. (A.3)–(A.4). In contrast, without the additional parametrization with γ, if only $\ln\pi(\theta_r^k)$ were to be used in the proposal mechanism as is the case in RW style samplers, the M–H acceptance probability would often drop to zero relatively quickly.

Step 3 applies a Metropolis correction to the proposal $(\theta^*_{r+1}, \gamma^*_{r+1})$. In continuous time, or for $\varepsilon \to 0$, Eqs. (A.3)–(A.4) would keep $-H(\theta^*_{r+1}, \gamma^*_{r+1}) + H(\theta_r, \gamma_r) = 0$ exactly resulting in $\alpha(\theta_r, \theta^*_{r+1}) = 1$ but for discrete $\varepsilon > 0$, in general, $-H(\theta^*, \gamma^*) + H(\theta, \gamma) \neq 0$ necessitating the Metropolis step.

The system (A.5)–(A.7) is time reversible and symmetric in (θ, γ), which implies that the forward and reverse transition probabilities $q(\theta^L_r, \gamma^L_r | \theta^0_r, \gamma^0_r)$ and $q(\theta^0_r, \gamma^0_r | \theta^L_r, \gamma^L_r)$ are equal: this simplifies the M–H acceptance ratio in Eq. (6) to the Metropolis form $\pi(\theta^*_{r+1}, \gamma^*_{r+1})/\pi(\theta^0_r, \gamma^0_r)$. From the definition of the Hamiltonian $H(\theta, \gamma)$ in Eq. (A.1) as the negative of the log-joint densities, the joint density of (θ, π) is given by

$$\pi(\theta, \gamma) = \exp[-H(\theta, \gamma)] = \pi(\vartheta)((2\pi)^d |M|)^{-1/2} \exp\left(-\frac{1}{2}\gamma' M^{-1} \gamma\right) \tag{A.8}$$

Hence, the Metropolis acceptance probability takes the form

$$\begin{aligned} \alpha(\theta_r, \gamma_r; \theta^*_{r+1}, \gamma^*_{r+1}) &= \min\left[\frac{\pi(\theta^*_{r+1}, \gamma^*_{r+1})}{\pi(\theta^0_r, \gamma^0_r)}, 1\right] \\ &= \min[\exp(-H(\theta^*_{r+1}, \gamma^*_{r+1}) + H(\theta^0_r, \gamma^0_r)), 1] \end{aligned}$$

The expression for $\alpha(\theta_r, \gamma_r; \theta^*_{r+1}, \gamma^*_{r+1})$ shows, as noted above, that the CHMC acceptance probability is given in terms of the difference of the Hamiltonian equations $H(\theta^0_r, \gamma^0_r) - H(\theta^*_{r+1}, \gamma^*_{r+1})$. The closer can we keep this difference to zero, the closer the acceptance probability is to one. A key feature of the Hamiltonian dynamics (A.3)–(A.4) in Step 2 is that they maintain $H(\theta, \gamma)$ constant over the parameter space in continuous time conditional on $H(\theta^0_r, \gamma^0_r)$ obtained in Step 1, while their discretization (A.5)–(A.7) closely approximates this property for discrete time steps $\varepsilon > 0$ with a global error of order ε^2 corrected by the Metropolis update in Step 3.

APPENDIX B: PARALLELIZATION

At present there are many ways in which algebraic operations can be computed in parallel. The purpose of this section is to motivate the ease of implementation of parallelizable code for researchers who are interested in computation speedup but have so far worked with serial code only. The current computer architecture for parallelization can be broadly divided into two categories: multi-core CPUs and GPUs. Multi-core CPUs are typically faster (in the range of 2.2–4.7 GHz) and have direct memory access but the numbers of cores in one node is limited (typically 8–32 cores). GPUs are slower (around 1 GHz) but the number of cores within one node is in the order of hundreds. Distributed memory access may pose a wallclock computing time bottleneck for GPUs, depending on the application.[6] Multi-core CPUs are therefore suited for relatively few intensive operations while GPUs are more beneficial for evaluating a very large number of simple algebraic operations.

For our parallel CHMC code implementation we used the Fortran Intel compiler on a multi-core GPU Unix machine. Starting with a serial code, we parallelized loops with independent operations within the loop by adding a single line of so-called OpenMP directives (starting with `!$omp`) in front of loops to be run in parallel on multi-core CPUs, and two lines of so-called OpenACC directives interfacing CUDA (starting with `!$acc`) in front of loops to be run in parallel on GPUs. Examples of pseudo-code are as follows:

Multi-core CPUs:

```
!$omp parallel do default(shared) private(j)
do j = 1, T
invert (H (:, :, j))

    ⋮

end do
```

GPU acceleration:

```
!$acc data copy(theta) present(returns_data)
!$acc kernels loop
do j = 1, T
evaluate Ar_{t-1}r'_{t-1}A'

    ⋮

end do
```

OpenMP and OpenACC directives can be freely combined in the same code. In the Fortran PGI compiler the parallelization is activated when the code is compiled with the flags -openmp and -acc, respectively. When compiled without these flags, the directives are ignored by the compiler since the exclamation character "!" represents the comment-out sign in Fortran, and a serial code is generated as a result. The parallelized code is fully portable across different platforms (Intel, AMD; NVidia, GeForce).

APPENDIX C: APPLICATION AND MODEL COMPARISON RESULTS

Table C1. Summary Statistics.

	Mean	SD	Skewness
AA	−0.019	3.307	0.253
AXP	0.090	3.011	0.507
BAC	0.048	4.609	0.741
DD	0.063	2.073	−0.165
GE	0.025	2.386	0.348
IBM	0.057	1.521	0.189
JPM	0.087	3.400	0.879
KO	0.040	1.346	0.891
MSFT	0.032	1.964	0.466
XOM	0.030	1.761	0.612

Table C2. Model Comparison: $N = 10$.

Parameters	DBEKK	RBEKK		BEKK	
		Diag	*Off-diag*	*Diag*	*Off-diag*
min(A)	0.127	0.065	−0.043	0.037	−0.098
max(A)	0.186	0.308	0.522	0.332	0.175
min(B)	0.980	0.951	−0.012	0.919	−0.090
max(B)	0.988	0.994	0.011	1.001	0.093
LL decomposition					
Margin (BAC)	−3,832	−3,802		−3,771	
Margin (JPM)	−3,478	−3,461		−3,433	
Margin (IBM)	−2,616	−2,614		−2,568	
Margin (MSFT)	−2,968	−2,963		−2,911	
Margin (XOM)	−2,607	−2,592		−2,553	
Margin (AA)	−3,595	−3,581		−3,542	
Margin (AXP)	−3,314	−3,303		−3,249	
Margin (DD)	−2,988	−2,979		−2,927	
Margin (GE)	−3,092	−3,086		−3,041	
Margin (KO)	−2,334	−2,337		−2,299	
Copula	4,364	4,350		4,473	
Total LL	−26,463	−26,373		−25,826	

FACTOR SELECTION IN DYNAMIC HEDGE FUND REPLICATION MODELS: A BAYESIAN APPROACH

Guillaume Weisang

Graduate School of Management, Clark University, Worcester, MA, USA

ABSTRACT

In this paper, I propose an algorithm combining adaptive sampling and Reversible Jump MCMC to deal with the problem of variable selection in time-varying linear model. These types of model arise naturally in financial application as illustrated by a motivational example. The methodology proposed here, dubbed adaptive reversible jump variable selection, differs from typical approaches by avoiding estimation of the factors and the difficulties stemming from the presence of the documented single factor bias. Illustrated by several simulated examples, the algorithm is shown to select the appropriate variables among a large set of candidates.

Keywords: Tracking problem; hedge fund replication; alternative beta; Kalman filter; factor selection; model selection

JEL classifications: C1; C11; C32; C52; G11

Bayesian Model Comparison
Advances in Econometrics, Volume 34, 181–222

ISSN: 0731-9053/doi:10.1108/S0731-905320140000034009

This paper deals with the problem of variable selection in a time-varying linear model. Motivation for this paper comes from a typical financial problem: selecting the risk factors for the estimation of the time-varying risk exposures of an unobservable portfolio, such as it happens, for example, in hedge fund replication. The methodology proposed here differs from typical approaches to this problem by avoiding the estimation of risk factors and the difficulties stemming from the presence of the documented single factor bias (e.g., Brown, 1989). Instead, I propose and develop a Bayesian model comparison approach based on combining a reversible-jump MCMC (RJ-MCMC) (Green, 1995) with adaptive sampling (Nott & Kohn, 2005) to explore the posterior distribution of the models. The resulting algorithm, dubbed adaptive reversible jump variable selection, is illustrated on several examples and is shown to select the appropriate variables among a large set of candidates.

The paper proceeds as follows. Section 1 provides a detailed exposition of the motivation behind the formulation of the problem considered in these pages. Section 2 develops the proposed approach, while Section 3 reports the results of several simulated examples. Finally, Section 4 presents a discussion the algorithm's performance before concluding in Section 5.

1. MOTIVATION

As of January 2014, the hedge fund industry represented upward of US $2.63 trillion in assets under management,[1] and its performance compared to standard asset classes and markets over the last two decades (especially during bear markets and crises) have made it very attractive to institutional investors seeking diversification and craving its time-varying correlation to traditional investments. However, regulatory frameworks around the world sometime prevent, and often substantially limit, direct investments into speculative financial vehicles like hedge funds by fiduciary investors on the ground of sound risk management practices. As a result, hedge fund replication, either as risk management exercise or as a financial engineering process has become more relevant than ever to provide institutional investors with the tools to take advantage of the growth and the characteristics of this industry. We can formalize the hedge fund replication problem in the following way.

1.1. The Hedge Fund Investor's Problem

An (institutional) investor would like to invest in a managed portfolio at time 0 over a horizon T. Typically, the evolution of this managed portfolio is only partially observable in the sense that its exact composition is unknown but the investor has periodically access to a summary of its performance through its periodic returns $\{r_\tau^{\mathrm{HF}}\}_{\tau=1}^{T}$. This investment opportunity is represented by the values $\{W_\tau^{\mathrm{HF}}\}_{\tau=0}^{T}$ of her share of the managed portfolio over the investment period. Instead, she must resort to a *replicating portfolio* represented by $\{W_\tau^{\mathrm{Clone}}\}_{\tau=0}^{T}$. This replicating portfolio can either be a real investment prospect or be used to estimate the sources of the risk attached to a current investment in our investor's portfolio.

The investor's problem is then to minimize the expected cost $\mathbb{E}\| W_T^{\mathrm{HF}} - W_T^{\mathrm{Clone}}\|_2^2$ at horizon T of using the clone portfolio. One can show that

$$\arg\min_{\pi} E_\pi \| W_T^{\mathrm{HF}} - W_T^{\mathrm{Clone}}\|_2^2 = \arg\min_{\pi} \sum_{\tau=1}^{T} \mathrm{Var}_\pi (r_\tau^{\mathrm{HF}} - r_\tau^{\mathrm{Clone}})$$

where $\{r_\tau^{\mathrm{Clone}}\}_{\tau=0}^{T}$ represent the returns of the clone portfolio over the investment horizon $[0,T]$. The investment policy π represents the sequence of investment decisions made at the beginning of each period regarding the composition of the clone portfolio and it verifies the following assumptions

$$\mathbb{E}_\pi (r_\tau^{\mathrm{HF}} - r_\tau^{\mathrm{Clone}}) = 0 \quad \forall \tau$$

$$\mathrm{Cov}_\pi (r_s^{\mathrm{HF}} - r_s^{\mathrm{Clone}}, r_\tau^{\mathrm{HF}} - r_\tau^{\mathrm{Clone}}) = 0 \quad \forall s \neq \tau$$

This particular investor must therefore minimize the *tracking error* $\mathrm{Var}_\pi (r_\tau^{\mathrm{HF}} - r_\tau^{\mathrm{Clone}})$ for every period τ, $\tau = 1,\ldots, T$.

1.2. The Factor Approach

Each observation r_τ^{HF} informs us in some sense about the nature of the unobserved composition of the managed portfolio over period τ since $r_\tau^{\mathrm{HF}} = \sum_{i \in I} \omega_{i\tau} r_{i\tau}$ where I represents the set of assets composing the managed portfolio, $r_{i\tau}$ represents the return of asset i over period τ, and $\omega_{i\tau}$

represents the weight of asset i in the managed portfolio. However, the universe of assets that can comprise the managed portfolio is typically not restricted to any asset class, will include every asset available on financial markets, as well as derivative securities that depend in a nonlinear fashion on the value of other investment assets, and can even contain instruments that are not available to the investor.

A standard approach – e.g., Arbitrage Pricing Theory (Ross, 1976) or the extension of Sharpe's style analysis (Sharpe, 1992) for performance assessment to hedge fund replication (Fung & Hsieh, 1997) – is to summarize the sources of variations in the returns of all assets, in the investment universe $\mathcal{U}$, using a linear factor structure. That is, there exists a set of risk factors $\{F_j\}_{j=1,\ldots,m}$ such that

$$\forall i \in \mathcal{U}, \forall \tau \quad r_{i\tau} = \alpha_i + \sum_{j=1}^{m} \beta_{ij} F_{j\tau} + \xi_{i\tau} \quad \text{with} \quad \mathbb{E}(\xi_{i\tau} | F_{1\tau} \ldots F_{m\tau}) = 0 \tag{1}$$

The return r_τ^{HF} can therefore be written as

$$r_\tau^{\mathrm{HF}} = \alpha_\tau^{\mathrm{HF}} + \sum_{j=1}^{m} w_{j\tau}^{\mathrm{HF}} F_{j\tau} + \varepsilon_\tau \tag{2}$$

such that

$$\alpha_\tau^{\mathrm{HF}} = \sum_{i \in I} \alpha_i \omega_{i\tau}, \quad w_{j\tau}^{\mathrm{HF}} = \sum_{i \in I} \omega_{i\tau} \beta_{ij}, \quad \text{and} \quad \varepsilon_\tau = \sum_{i \in I} \omega_{i\tau} \xi_{i\tau} \tag{3}$$

Contrary to the assumption behind traditional hedge fund replication approaches (e.g., Fung & Hsieh, 1997), the objective of the investor in Section 1.1 is to get access to the evolution through time of the factor loadings $\{w_{j\tau}^{\mathrm{HF}}\}$ – also called risk exposures in the financial literature – rather than the fund's "alpha," $\alpha_\tau^{\mathrm{HF}}$, often considered to be a proxy measure for the skill of the portfolio manager.

1.3. LEQG Criterion and Bayesian Filtering

We assume the following linear quadratic Gaussian (LQG) state-space model for the *unobservable* dynamic of the managed portfolio

$$\begin{cases} \mathbf{w}_{\tau+1}^{\mathrm{HF}} = \mathbf{T}_{\tau}\mathbf{w}_{\tau}^{\mathrm{HF}} + \nu_{\tau+1} \\ r_{\tau}^{\mathrm{HF}} \quad = \mathbf{f'}_{\tau}\mathbf{w}_{\tau}^{\mathrm{HF}} + \eta_{\tau} \end{cases} \tag{4}$$

with $\nu_\tau \sim NIID(\mathbf{0}_{m\times 1}, \mathbf{Q}_\tau)$ and $\eta_\tau \sim NIID(0,S_\tau)$, and $\mathbf{f}_\tau = (F_{1\tau},\ldots,F_{m\tau})'$ is the vector of returns of the factors influencing the evolution of the managed portfolio's returns at time τ. $\mathbf{T}_\tau$, the transition matrix of the state vector $\mathbf{w}_\tau^{\mathrm{HF}}$, represents the deterministic part of the evolution of managed portfolio's exposure. It is a known quantity at time τ. The Gaussian assumptions made here can be relaxed so long as the noises $\boldsymbol{n}$ and η retain distributions from the exponential family, as explained below.

While the investor of Section 1.1 minimizes with respect to the investment policy π the expected cost $\mathbb{E}_\pi[\mathbb{C}]$ where $\mathbb{C} = \| W_T^{\mathrm{HF}} - W_T^{\mathrm{Clone}} \|_2^2$, Whittle (1996) shows that the investor's problem is encompassed in a more general linear exponential quadratic Gaussian (LEQG) formulation in which an investor minimizes the expected exponential *disutility* function $e^{-\theta\mathbb{C}}$, where θ is a risk sensitivity parameter. One then minimizes the scaled cost criterion

$$\chi_\pi(\theta) = -\theta^{-1}\log\left[E(e^{-\theta\mathbb{C}})\right] = \mu - \theta\frac{v}{2} + o(v) \tag{5}$$

where μ and v are respectively the expected value and the variance of $\mathbb{C}$ under policy π. The two formulations, LGQ and LEQG, are equivalent for small values of θ. In fact, when $\theta = 0$, the risk neutral case, LEQG reduces to LQG. When $\theta > 0$, it corresponds to an increasingly risk-seeking behavior of the investor; while for $\theta < 0$, the investor is increasingly risk averse.

From the statistical point-of-view, the investor must estimate the state vector $\mathbf{w}_\tau^{\mathrm{HF}}$ for every period $\tau = 1,\ldots,T$. Once estimated, the value $\hat{\mathbf{w}}_\tau^{\mathrm{HF}}$ of the state vector is then used as an input into the policy decision made at τ in order to minimize the cost over the period $[\tau+1,T]$. Define the discrepancy $\mathbb{D}(r_\tau^{\mathrm{HF}}, \mathbf{w}_\tau^{\mathrm{HF}})$ such that the joint distribution of $(r_\tau^{\mathrm{HF}}, \mathbf{w}_\tau^{\mathrm{HF}})$ is proportional to $e^{-\mathbb{D}(r_\tau^{\mathrm{HF}}, \mathbf{w}_\tau^{\mathrm{HF}})}$. Whittle (1996) demonstrates a certainty equivalence principle for the LEQG formulation. That is, at $0 < \tau \leq T$, the optimal value of the control input u_τ ($\pi = \{u_\tau\}_{\tau=1}^T$) minimizing the cost $\mathbb{C}$ is just the optimal value obtained if $\mathbf{w}_\tau^{\mathrm{HF}}$ were known but with $\mathbf{w}_\tau^{\mathrm{HF}}$ replaced by the estimate $\hat{\mathbf{w}}_\tau^{\mathrm{HF}} (= \mathbf{w}_\tau^{\mathrm{Clone}})$ given by minimizing the discrepancy $\mathbb{D}$ over the past history $[0,\tau]$. However, there is in general no separation between estimation of the state vector $\mathbf{w}^{\mathrm{HF}}$ and control of cost $\mathbb{C}$ as control costs (past and futures) affect the value of the estimates $\{\hat{\mathbf{w}}_\tau^{\mathrm{HF}}\}_{\tau=1}^T$ and noise statistics $\mathbf{Q}_\tau$ and H_τ affect the control rule.

In the risk-neutral case ($\theta = 0$), minimization of the discrepancy (for estimation of the state vector) over the pasts observations, and minimization of cost (for control) over the future are in fact separate. In this case, one can "solve" for $\mathbf{w}_{\tau}^{HF}$ in Eq. (4) by applying the Kalman filter. In the more general case ($\theta \neq 0$), the investor's problem is solved by Bayesian filtering, usually involving simple modifications of the Kalman filter's recursion.

Replication exercises similar to the one presented in Section 1 have been treated repeatedly in the Economics and the Finance literatures (Corielli & Marcellino, 2006; Glabadanidis & Zolotoy, 2013). In particular, in the case of hedge fund replication, traditional approaches consist in extending Sharpe's (1992) Style Analysis using a determined set of factors and OLS methodology (possibly on rolling windows) to extract the factor loadings (e.g., Amenc, Géhin, Martellini, & Meyfredi, 2007; Amenc, Martellini, Meyfredi, & Ziemann, 2010; Fung & Hsieh, 2007). Roncalli and Teiletche (2008) proposed using Kalman filtering to extract the time-varying factor loadings and demonstrated its superiority to OLS-based methodologies on several examples. Roncalli and Weisang (2011) formalized hedge fund replication with the aim of extracting the fund's "betas" as the tracking problem (4). However, the efficiency of a replication methodology depends on the set of factors used both in terms of minimization of a tracking error criterion or the ability to interpret and make sense in an investment or economic context of the results of the replication methodology. The purpose of this paper is to propose a methodology whose goal is to select the appropriate set of factors for the tracking problem (4). The short exposition in Section 1.3 suggests that replication of the managed portfolio should be done eventually using Bayesian filtering on the set of selected factors. Since our concern is only to identify the factors, we will assume a risk-neutral investor and therefore use Kalman filtering. We leave to the reader the care of implementing the replication strategy using their own risk aversion parameters leading to the use of modified versions of the filtering algorithm.

1.4. A Word on Latent Factor Estimation and Single Factor Bias

Using Monte-Carlo simulations, Brown (1989) demonstrated that any empirical solution to the estimation of the *K* equally important latent factors in an economy – as can be hypothesized in the APT – will lead to the false inference of one important "factor" identified as the return on an equally weighted market index. In the decades that followed, most of econometrics developments in the identification and estimation of factors in factor models has focused on the asymptotic theory behind factor estimation – that is, when the number N of assets in the economy and the number of observations T are considered to grow infinitively large (e.g., Bai, 2003; Bai & Ng, 2002, 2006; Forni, Hallin, Lippi, & Reichlin, 2000, 2004, 2005; Forni & Lippi, 2001; Hallin & Liška, 2007) – unfortunately not addressing the finite sample estimation problem brought to light by Brown. While Brown's (1989) paper was widely noticed and cited, it appears its consequences on traditional factor estimation, using for example Principal Component Analysis (e.g., Darolles & Mero, 2011) had not been fully realized until Harding (2007) among others brought to light the connection with Random Matrix Theory (RMT). In a short note, Harding (2008) explains what RMT tells us about the distribution of the eigenvalues of the latent factors' covariance matrix[2] (the usual starting point in a latent factor estimation) and what direct consequences we can draw for their estimation.

Using the same example as Brown (1989), Harding (2008) illustrates that in a factor economy with $N=50$ to $N=200$ assets assumed to have K equally important factors (where K is left for estimation) over $T=80$ time periods – these sample sizes are similar to typical situations as sketched in Section 1.1 – methodologies based on traditional principal component analysis cannot estimate beyond the first factor and its eigenvalue and even that is done with significant bias. To identify correctly all the factors (with $K=10$), the minimum amount of data required is $T_{\min} \approx 40,000$ (that's over 110 years of daily data or over 3,300 years of monthly data!). More recent results show that, under the assumption of "weak factors," such as is likely the case in the example cited above, the principal component estimators of factors and factor loadings are inconsistent (Onatski, 2012). Onatski (2012) provides asymptotic formulae to estimate the amount of inconsistency and the parameters of the asymptotic distributions of factors and factor loadings that seem to work well empirically even in the case of fairly small sample sizes N and T in the case of *static* factor models.

The consequences for factor identification and estimation, hence for model selection in our setting, are not difficult to derive. Even if one could accurately select the correct number of factors using an information criterion (e.g., Onatski, 2009, 2010), estimating more than the first factor is difficult unless you possess an extremely large amount of data. Unfortunately, it is unrealistic to hope to identify effectively these factors and in the meantime assume model stability (i.e., constant factor loadings). The approach adopted here avoids this difficulty. Its underlying idea is to explore the posterior distribution of the possible models using observable variables. The goal is not to estimate and infer the factors' structure behind all hedge fund returns, but on the contrary to identify the best models with the objective of forecasting (cf. Section 1.3).

2. METHODOLOGY

2.1. Bayesian Model Selection, Posterior Distribution, and Bayes Factors

A Bayesian *variable dimension model* is defined as a collection of models ($m=1,\ldots,M$), $\mathcal{M}_m=\{f(\cdot|\boldsymbol{\theta}_m);\boldsymbol{\theta}_m\in\Theta_m\}$, associated with a collection of priors on the parameters of these models, $\pi_m(\boldsymbol{\theta}_m)$ and a prior distribution on the indices of these models, $\{\varrho(m), m=1,\ldots,M\}$ (Robert & Casella, 2004). The vector $\boldsymbol{\theta}_m$ (of length n_m) represents the parameters of the model $\mathcal{M}_m$ and is an element of the parameter space Θ_m. The complete parameter space Θ of a Bayesian variable dimension model is given by $\Theta=\cup\{m\}\times\Theta_m$. The more concise notation $\pi(m,\boldsymbol{\theta}_m)=\varrho(m)\pi_m(\boldsymbol{\theta}_m)$ is used to describe the prior distribution of the models. As noted by Robert and Casella (2004), one can immediately see the advantage of this Bayesian representation of the problem from an inferential point-of-view. Indeed, once the prior and the model are defined, the model selected from the data $\mathbf{z}$ is determined from the posterior probabilities

$$p(\mathcal{M}_i|\mathbf{z})=\frac{\varrho(i)\int_{\Theta_i}f_i(z|\boldsymbol{\theta}_i)\pi_i(\boldsymbol{\theta}_i)d\boldsymbol{\theta}_i}{\sum_j\varrho(j)\int_{\Theta_j}f_j(z|\boldsymbol{\theta}_j)\pi_j(\boldsymbol{\theta}_j)d\boldsymbol{\theta}_j}$$

by either taking the largest $p(\mathcal{M}_i|\mathbf{z})$ or using *model averaging* through

$$\sum_j p(\mathcal{M}_j|\mathbf{z})\int_{\Theta_j}f_j(z|\boldsymbol{\theta}_j)\pi_j(\boldsymbol{\theta}_j|\mathbf{z})d\boldsymbol{\theta}_j$$

where $\mathbf{z}$ denotes the observed dataset, as a predictive distribution. Model comparison and hypothesis testing are handled by means of the Bayes factor defined by the ratio of the posterior odds of two competing models to their prior odds, and given by the ratio of the marginalized likelihoods $B_{ij}=p(\mathbf{z}|\mathcal{M}_i)/p(\mathbf{z}|\mathcal{M}_j)$. B_{ij} provides evidence in favor of $\mathcal{M}_i$ against $\mathcal{M}_j$ if $B_{ij}>1$; conversely, B_{ij} provides evidence against $\mathcal{M}_i$ in favor of $\mathcal{M}_j$ if $B_{ij}<1$. Further review and discussion of the Bayes factor, including cutoff point tables for hypothesis testing, are provided by Kass and Raftery (1995).

The problem of model selection is therefore straightforward in the Bayesian context if one has an analytical expression of the marginal likelihood of model $\mathcal{M}_i$ given by $\int_{\Theta_i} f_i(z|\boldsymbol{\theta}_i)\pi_i(\boldsymbol{\theta}_i)d\boldsymbol{\theta}_i$ and the normalizing constant $\sum_j \varrho(j)\int_{\Theta_j} f_j(z|\boldsymbol{\theta}_j)\pi_j(\boldsymbol{\theta}_j)d\boldsymbol{\theta}_j$. Often, the marginal likelihood can be calculated analytically when choosing appropriate conjugate prior distributions on the parameters $\boldsymbol{\theta}_i$ of the model. The normalizing constant can be calculated over all possible models or estimated by means of a Markov chain if the number of possible models is large. For the problem of factor selection in the model described in Section 1.3, we can first expect that in practical situations the number of possible factors is large, and hence calculation of the normalizing constant is computationally intractable. Second, an analytical expression of the marginal likelihood is in general not available. Therefore, one must resort to simulating draws from the posterior distribution on the model space to approximate both the normalizing constant of the model's posterior distribution and the marginal likelihoods. Green (1995) provides a generic algorithm which ensures the ergodicity and convergence of such a Markov chain. Approximations of the marginal likelihoods and the Bayes factors can be obtained by using either standard Monte Carlo or Laplace–Metropolis estimators through processing of the posterior simulations after conditioning on $\mathcal{M}_i$ (cf. Kass & Raftery, 1995; Lewis & Raftery, 1997).

2.2. *Adaptive RJ-MCMC Variable Selection*

Green (1995) introduced a Metropolis–Hastings sampling algorithm moving between parameter subspaces of differing dimensionality for the purpose of model determination by supplying a general and constructive formulation for trans-dimensional jumps between $\boldsymbol{\theta}_i$ in model $\mathcal{M}_i$ and $\boldsymbol{\theta}_j$ in model $\mathcal{M}_j$. Given that the chain is currently in state $(\mathcal{M}_i, \boldsymbol{\theta}_i)$, a new value for the chain $(\mathcal{M}_j, \boldsymbol{\theta}_j)$ is proposed from some proposal distribution $q(\boldsymbol{\theta}_i, d\boldsymbol{\theta}_j)$, which is then subsequently accepted or rejected. Green (1995) showed that, if $\pi(d\boldsymbol{\theta}_i)q(\boldsymbol{\theta}_i, d\boldsymbol{\theta}_j)$ is absolutely continuous with respect to a symmetric measure $\xi(d\boldsymbol{\theta}_i, d\boldsymbol{\theta}_j)$ on $\Theta \times \Theta$, and if $g_m(\boldsymbol{\theta}_i, \boldsymbol{\theta}_j)$ is the Radon-Nikodym derivative of $\pi(d\boldsymbol{\theta}_i)q(\boldsymbol{\theta}_i, d\boldsymbol{\theta}_j)$ with respect to ξ, then detailed balance of the Markov chain is preserved if we accept the proposed new state with the probability $\rho_m(\boldsymbol{\theta}_i, \boldsymbol{\theta}_j)$ in the usual Metropolis–Hastings form: $\rho_m(\boldsymbol{\theta}_i, \boldsymbol{\theta}_j) = \min(1, g_m(\boldsymbol{\theta}_j, \boldsymbol{\theta}_i)/g_m(\boldsymbol{\theta}_i, \boldsymbol{\theta}_j))$ (see also Robert & Casella, 2004).

Green's (1995) insight was that, in a move from a model $\mathcal{M}_i$ to a model $\mathcal{M}_j$ with $n_i < n_j$, one can supplement each of the spaces Θ_i and Θ_j with adequate artificial spaces in order to create a *bijection* between them. For instance, one generates a random vector $\mathbf{V}$ of length $n_j - n_i$ consisting of variables drawn from some proposal density $\varphi(\cdot)$ with joint density $\varphi_{n_j - n_i}(\mathbf{v}) = \prod_{\ell=1}^{n_j - n_i} \varphi(v_\ell)$. Having generated $\mathbf{V}$, the move from $\boldsymbol{\theta}_i$ to $\boldsymbol{\theta}_j = f_{i,j}(\boldsymbol{\theta}_j, \mathbf{V})$ is proposed where the jump transformation $f_{i,j} : \Theta_i \times \mathbb{R}^{n_j - n_i} \mapsto \Theta_j$ denotes a bijection, mapping the current state of the chain together with the generated random vector to a point in a higher dimensional space. One then poses

$$g_m(\boldsymbol{\theta}_i, \boldsymbol{\theta}_j) = \pi(i, \boldsymbol{\theta}_i | \mathbf{z}) \pi_{ij} \varphi_{n_j - n_i}(\mathbf{v}) \tag{6}$$

$$g_m(\boldsymbol{\theta}_j, \boldsymbol{\theta}_i) = \pi(j, \boldsymbol{\theta}_j | \mathbf{z}) \pi_{ji} \left| \frac{\partial f_{i,j}(\boldsymbol{\theta}_i, \mathbf{v})}{\partial(\boldsymbol{\theta}_i, \mathbf{v})} \right| \tag{7}$$

since the move from $\mathcal{M}_j$ to $\mathcal{M}_i$ is deterministic in this case[3] and π_{ij} denotes the probability of choosing a move from model $\mathcal{M}_i$ to model $\mathcal{M}_j$. This move is then accepted with probability

$$\rho_m(\boldsymbol{\theta}_i, \boldsymbol{\theta}_j) = \min\left(1, (\text{likelihood ratio}) \frac{\pi(j, \boldsymbol{\theta}_j)}{\pi(i, \boldsymbol{\theta}_i)} \frac{\pi_{ji}}{\pi_{ij} \varphi_{n_j - n_i}(\mathbf{v})} \left| \frac{\partial f_{i,j}(\boldsymbol{\theta}_i, \mathbf{v})}{\partial(\boldsymbol{\theta}_i, \mathbf{v})} \right| \right) \tag{8}$$

where the likelihood ratio is the ratio of the likelihoods of model $\mathcal{M}_j$ to model $\mathcal{M}_i$.

In Eq. (4), the problem of model and variable selection consists in choosing the actual factors to which the managed portfolio is exposed. The model selection process thus entails choosing both *the number and the nature* of the factors. Hence, if we note $\mathcal{F}$ the set of factors to choose from, and $\boldsymbol{\theta}_{\mathcal{F}}$ a binary latent vector of length $m*$, the cardinal number of $\mathcal{F}$, indicating which of the factors are included in the model, we see that the prior distribution ϱ on the models can be expressed as

$$\varrho(i) = \varrho(m_i, \boldsymbol{\theta}_{\mathcal{F}}) = P\left(\boldsymbol{\theta}_{\mathcal{F}} \middle| \sum_{\ell=1}^{m*} \theta_{\mathcal{F},\ell} = m_i \right) \times p(m_i) \tag{9}$$

where m_i is the number of factors in the model. A natural prior for $\boldsymbol{\theta}_{\mathcal{F}}$ is to choose $\prod_{i=1}^{m*} \mathscr{B}_{p_i(\boldsymbol{\theta}_{\mathcal{F}})}(\gamma)$ where $\mathscr{B}_{p_i(\boldsymbol{\theta}_{\mathcal{F}})}(\gamma) = p_i(\boldsymbol{\theta}_{\mathcal{F}})^{\gamma}(1 - p_i(\boldsymbol{\theta}_{\mathcal{F}}))^{1-\gamma}$ is a Bernoulli distribution with parameter $p_i(\boldsymbol{\theta}_{\mathcal{F}}) \in (0, 1)$. This parameter can either be considered fixed or can be adapted depending on the current state of the chain (Nott & Kohn, 2005). Adaptive MCMC techniques have been developed to increase the efficiency of the sampling scheme and the precision of the approximation of the posterior density for a given amount of computation time.

The class of adaptive proposal densities proposed by Nott and Kohn (2005) adapt the parameter $p_i(\boldsymbol{\theta}_{\mathcal{F}})$ using a best linear predictor based on past iterations of the chain to approximate the full marginal distribution $\pi(\gamma_i = 1 | \mathbf{g}_{-i} = \boldsymbol{\theta}_{\mathcal{F},-i})$ of component γ_i. The algorithm is started with a fixed $\mathbf{p}$ and adaptation is started only after a sufficient number of iterations of the chain and updated periodically. After N_{iter} iterations of the chain past the burn-in period, the adaptation scheme is

$$p_i(\boldsymbol{\theta}_{\mathcal{F}}) = \min[\max(\hat{\theta}_{\mathcal{F},i} - d_{i,i}^{-1}\mathbf{w}_{-i}^{(i)\prime}(\boldsymbol{\theta}_{\mathcal{F},-i} - \hat{\boldsymbol{\theta}}_{\mathcal{F},-i}), \delta), 1 - \delta] \tag{10}$$

where the subscript $-i$ designs a vector from which the ith element has be removed, $\hat{\theta}_{\mathcal{F},i}$ is the estimated mean of the ith component of $\boldsymbol{\theta}_{\mathcal{F}}$, $\mathbf{w}^{(i)}$ is the ith column of the inverse of the estimated ridge covariance matrix $\mathbf{W}^{-1} = \text{Cov}(\boldsymbol{q}_{\mathcal{F}}^{(1)}, \ldots, \boldsymbol{q}_{\mathcal{F}}^{(N_{\text{iter}})}) + \lambda I$, and $d_{i,i}$ its ith diagonal element. δ is a design parameter ensuring a probability. The same values $\delta = \lambda = 0.01$ as used by Nott and Kohn (2005) are used in the following.

Adaptation has to be done with care as we leave the Markovian framework of Green (1995) to ensure the convergence of the chain to the correct density. The theoretical convergence of a combination of RJ-MCMC and adaptive proposal sampling to the posterior distribution of the models $p(\mathcal{M}|\mathbf{z})$ is ensured by Theorem 1 in Nott and Kohn (2005) which describes the conditions upon which introducing adaptive sampling in a general MCMC sampling scheme on a finite sample space will not perturb the convergence toward the stationary distribution. Theorem 2 in Nott and Kohn (2005) is an ergodic theorem which allows for estimation of expectations over a sample path of the chain with respect to its stationary distribution. Conditioned on $\mathcal{M}_i$, the sub-chains are Markov chains since the proposal densities $q(\theta, \theta' | \mathcal{M}_i)$ are *not* adaptive.

A pseudo-code version of the algorithm is provided in Algorithm 1. T_{adpt} represents the number of iterations between adaptations of the sampling parameters of the model composition's vector $\boldsymbol{\theta}_{\mathbf{f}}$.

Algorithm 1. Adaptive Reversible-Jump MCMC Variable Selection

for i from 1 to $N_{sim}+N_b$ **do**
  $j \leftarrow \mathscr{U}(\{1,\ldots,m^*\})$ ▷ Draw one factor to consider
  $\boldsymbol{\theta}_{\mathcal{F},\text{prop}} \leftarrow \boldsymbol{\theta}_{\mathcal{F}}$, $\boldsymbol{\theta}_{\mathcal{F},\text{prop},j} \leftarrow \mathscr{B}_{p_j(\boldsymbol{\theta}_{\mathcal{F}})}$ ▷ Bernoulli experiment
  Jump.Type←MoveType ($\boldsymbol{\theta}_{\mathcal{F},\text{prop}}$) ▷ Choose RJ-MCMC move type
  ($\boldsymbol{\theta}_{\text{prop}} \leftarrow$ Proposal ($\boldsymbol{\theta}_m$, $\boldsymbol{\theta}_{\mathcal{F},\text{prop}}$, Jump.Type) ▷ Propose new parameters
  $l_{\text{prop}} \leftarrow$ ModelLikelihood ($\boldsymbol{\theta}_{\text{prop}}$) ▷ Calculate Model Likelihood l_{prop}
  $\alpha \leftarrow$ AcceptProb $(\boldsymbol{\theta}_{\mathcal{F}}^{(i-1)}, \boldsymbol{\theta}^{(i-1)}, \ell^{(i-1)}, \boldsymbol{\theta}_m, \boldsymbol{\theta}_{\mathcal{F},\text{prop}}, \ell_{\text{prop}})$
  $u \leftarrow \mathscr{U}(0,1)$ ▷ MH Step
  if $u<\alpha$ **then**
    $(\boldsymbol{\theta}_{\mathcal{F}}^{(i)}, \boldsymbol{\theta}^{(i)}) \leftarrow (\boldsymbol{\theta}_{\mathcal{F},\text{prop}}, \boldsymbol{\theta}_{\text{prop}})$
  else $(\boldsymbol{\theta}_{\mathcal{F}}^{(i)}, \boldsymbol{\theta}^{(i)}) \leftarrow (\boldsymbol{\theta}_{\mathcal{F}}^{(i-1)}, \boldsymbol{\theta}^{(i-1)})$
  end if
  if $(i>N_b)$ and $(i-N_b \bmod T_{adpt})=0$ **then** ▷ Adaptive Sampling
    update $p_j(\boldsymbol{\theta}_{\mathcal{F}})$ according to (10)
  end if
end for

2.3. *Moment Condition*

One of the major difficulties in using RJ-MCMC comes from designing the reversible jumps of the Markov chain between models. In model (4) and in the context of factor selection described in Section 2.2 , this question takes the form of jumps between different compositions of the vector $\boldsymbol{\theta}_{\mathcal{F}}$. Even when a jump is between models that differ only by one component, the reversibility condition renders the design difficult. Questions like what should be the initial weight $w_{0,m+1}$ and variance parameters should be given to a newly included factor must be answered hand-in-hand with the reverse question what should be done with the initial weight and variance parameters of a factor removed from the current composition.

In order to address this question, the algorithm proposed here was derived using a slightly modified version of model (4) where the transition matrix $\mathbf{T}_\tau$ is assumed to be time-invariant, and the processes $\{\mathbf{w}_\tau^{\text{HF}}\}_{\tau=0}^T$ and $\{r_\tau^{\text{HF}}\}_{\tau=1}^T$ are assumed to be weakly stationary. Hence, the modified model is given by

$$\begin{cases} \mathbf{w}_{\tau+1}^{\text{HF}} = \mathbf{T}\,\mathbf{w}_\tau^{\text{HF}} + \nu_{\tau+1} \\ r_\tau^{\text{HF}} = \mathbf{f}'_\tau \mathbf{w}_\tau^{\text{HF}} + \eta_\tau \end{cases} \tag{11}$$

where $\nu_\tau \sim NIID(\mathbf{0}_{m\times 1}, \mathbf{Q})$ and $\eta_\tau \sim NIID(0,S)$ and $\mathbf{f}_\tau$ are as given in Eq. (4). Using Eq. (11), the weak stationarity assumption on the state vector allows to express the variance Σ_τ of the state vector as $\Sigma = \mathbf{T}\Sigma\mathbf{T}' + \mathbf{Q}$. Furthermore, since the deterministic part of the evolution of the state vector is here unknown, we make the simplifying assumption that $\mathbf{T} = \lambda \mathbf{I}_m$ with $|\lambda| < 1$. Thus, $\Sigma = (1-\lambda^2)^{-1}\mathbf{Q}$.

In Eq. (11), only the process $\{r_\tau^{\mathrm{HF}}\}_{\tau=1}^{T}$ is observed, and the reversible jumps can therefore be designed to preserve its sample variance. The complete derivation of this moment condition on $\mathrm{Var}(r_\tau^{\mathrm{HF}})$, provided in Appendix A, is straightforward if tedious. Under the assumption of weak stationarity,

$$\mathrm{Var}(r_\tau^{\mathrm{HF}}) = \mathrm{Tr}(\Sigma\Sigma_F) + \mathbf{f}'\Sigma\mathbf{f} + S \tag{12}$$

Denoting $\mathbf{Q} = \{q_{ij}\}_{i,j=1,\ldots,m}$, $\mathbf{f} = (\mu_1, \ldots, \mu_m)'$, and $\Sigma_F = \{\sigma_{ij}\}_{i,j=1,\ldots,m}$, the moment condition (12) can be expressed in a scalar form. Indeed, given that

$$\mathrm{Tr}(\Sigma\Sigma_F) = \frac{1}{1-\lambda^2}\sum_{i=1}^{m}\sum_{j=1}^{m} q_{ij}\sigma_{ij} \tag{13}$$

and

$$\mathbf{f}'\Sigma\mathbf{f} = (1-\lambda^2)^{-1}\sum_{i=1}^{m}\sum_{j=1}^{m} q_{ij}\mu_i\mu_j \tag{14}$$

we find

$$(1-\lambda^2)\mathrm{Var}(r_\tau^{\mathrm{HF}}) = \sum_{i=1}^{m}\sum_{j=1}^{m} q_{ij}(\mu_i\mu_j + \sigma_{ij}) + (1-\lambda^2)S \tag{15}$$

2.4. Reversible Jumps and Bijective Transformations

For simplicity's sake, only jumps between models that are "close" together in the sense that the dimension of the state-vector changes by one or less are considered.

Birth/Death. In a "Birth" or "Death" move, a factor is randomly selected to be added to or deleted from the model.

Split/Combine. In a "Split" or "Combine" move, the exposition to a randomly selected factor, already present in the model, is either "split " (respectively "combined") with another randomly selected factor that is to be included in the model (respectively dropped out of the model).

Fixed dimensional move. This type of move is tantamount to a traditional Metropolis–Hastings move in a fixed dimensional setting for the Markov chain of the current model's parameters. We can distinguish between two different types of fixed dimensional moves: a "covariance" move where only the matrix **Q** is changed; and a "Long Term Memory" move where only the memory coefficient λ is changed.

Even though *Birth/Death* and *Split/Combine* moves seem very close in spirit, they are in fact fundamentally different. While the *Birth* and *Death* moves are simply conceived, they may result in "wild" jumps around the support of the models' posterior distribution, especially since the calibration of the jump proposal distributions is very difficult in this case (cf. Richardson & Green, 1997; Robert & Casella, 2004). The *split* and *combine* moves on the other hand are constructed to be more "local" to the models' posterior distribution in the sense that only the parameters associated with the two factors involved are modified. Both *Birth/Death* and *Split/Combine* moves are designed to preserve moment condition (15).

Only the case where **Q**, hence $\sum$, is diagonal are considered.[4] In that case, Eq. (15) simplifies to

$$(1-\lambda^2)\mathrm{Var}(r_\tau^{\mathrm{HF}}) = \sum_{i=1}^{m} q_{ii}(\mu_i^2 + \sigma_{ii}) + (1-\lambda^2)S \tag{16}$$

and the contribution of the ith factor to the second moment of the returns' process is given by $q_{ii}(\mu_i^2 + \sigma_{ii})$.

Let's denote by s the current state of the Markov chain and by $\dagger$ the equivalent quantities in the proposed new state. Eq. (17) translates into the following equality that must be preserved by each new proposed state

$$\sum_{i=1}^{m(s)} q_{ii}(s)(\mu_i^2 + \sigma_{ii}) = \sum_{i=1}^{m^\dagger} q_{ii}^\dagger((\mu_i^\dagger)^2 + \sigma_{ii}^\dagger) \tag{17}$$

Assume for notational convenience and without loss of generality that the new (respectively removed) factor in the proposed state will always be last in the state vector, and that when split (respectively combined) it is with the second to last factor.[5]

Split move. Eq. (17) then simplifies into the following

$$q_{m(s)m(s)}(s) = q^{\dagger}_{m(s)m(s)} + q^{\dagger}_{m^{\dagger}m^{\dagger}} \frac{(\mu^{\dagger}_{m^{\dagger}m^{\dagger}})^2 + \sigma^{\dagger}_{m^{\dagger}m^{\dagger}}}{\mu^2_{m(s)m(s)} + \sigma_{m(s)m(s)}} \tag{18}$$

If one generates $u \sim (0, 1)$, one can pose

$$q^{\dagger}_{m(s)m(s)} = uq_{m(s)m(s)}(s) \tag{19}$$

and

$$q^{\dagger}_{m^{\dagger}m^{\dagger}} = (1-u)q_{m(s)m(s)}(s) \frac{\mu^2_{m(s)m(s)} + \sigma_{m(s)m(s)}}{(\mu^{\dagger}_{m^{\dagger}m^{\dagger}})^2 + \sigma^{\dagger}_{m^{\dagger}m^{\dagger}}} \tag{20}$$

Combine move. Similarly to the *Split* move, one can define the reverse move by posing

$$q^{\dagger}_{m^{\dagger}m^{\dagger}} = q_{m^{\dagger}m^{\dagger}}(s) + q_{m(s)m(s)}(s) \frac{\mu^2_{m(s)m(s)} + \sigma_{m(s)m(s)}}{(\mu^{\dagger}_{m^{\dagger}m^{\dagger}})^2 + \sigma^{\dagger}_{m^{\dagger}m^{\dagger}}} \tag{21}$$

$$q^{\dagger}_{m(s)m(s)} = \frac{q_{m^{\dagger}m^{\dagger}}(s)}{q_{m^{\dagger}m^{\dagger}}(s) + q_{m(s)m(s)}(s) \frac{\mu^2_{m(s)m(s)} + \sigma_{m(s)m(s)}}{\left(\mu^{\dagger}_{m^{\dagger}m^{\dagger}}\right)^2 + \sigma^{\dagger}_{m^{\dagger}m^{\dagger}}}} \tag{22}$$

noting that in the case where a factor is removed $m^{\dagger} = m(s) - 1$. The second equality above is needed only to complete the bijection but does not enter in the new proposed state.

Birth move. Eq. (17) can be rewritten into

$$\sum_{i=1}^{m(s)} (q_{ii}(s) - q^{\dagger}_{ii})(\mu_i^2 + \sigma_{ii}) = q^{\dagger}_{m^{\dagger}m^{\dagger}} \left((\mu^{\dagger}_{m^{\dagger}m^{\dagger}})^2 + \sigma^{\dagger}_{m^{\dagger}m^{\dagger}}\right) \tag{23}$$

For the purpose of the *Birth* move, I will assume for $i = 1, \ldots, m(s)$

$$(q_{ii}(s) - q_{ii}^{\dagger})\frac{\mu_{ii}^2 + \sigma_{ii}}{(\mu_{m^\dagger m^\dagger}^{\dagger})^2 + \sigma_{m^\dagger m^\dagger}^{\dagger}} = \frac{q_{m^\dagger m^\dagger}^{\dagger}}{m(s)} \tag{24}$$

If one generates $v \sim \Gamma^{-1}(\theta_1, \theta_2)$, then

$$\forall i = 1,\ldots, m(s) \quad q_{ii}^{\dagger} = q_{ii}(s) - \frac{v}{m(s)}\frac{(\mu_{m^\dagger m^\dagger}^{\dagger})^2 + \sigma_{m^\dagger m^\dagger}^{\dagger}}{\mu_{ii}^2 + \sigma_{ii}} \tag{25}$$

$$q_{m^\dagger m^\dagger}^{\dagger} = v \tag{26}$$

and we must have

$$\forall i = 1,\ldots, m(s) \quad v < q_{ii}(s)m(s)\frac{\mu_{ii}^2 + \sigma_{ii}}{(\mu_{m^\dagger m^\dagger}^{\dagger})^2 + \sigma_{m^\dagger m^\dagger}^{\dagger}} \tag{27}$$

Death move. Similarly to the *Birth* move, the reverse move is given by posing

$$q_{ii}^{\dagger} = q_{ii}(s) + \frac{q_{m(s)m(s)}}{m^{\dagger}}\frac{\mu_{ii}^2 + \sigma_{ii}}{\mu_{m(s)m(s)}^2 + \sigma_{m(s)m(s)}} \tag{28}$$

The bijective transformations of the moves are then defined as (cf. Section 2.2)

$$Birth: \begin{pmatrix} \mathbf{w}_0^{(1)}(s) \\ \vdots \\ \mathbf{w}_0^{(m(s))}(s) \\ u_1 \\ q_{11}(s) \\ \vdots \\ q_{m(s)m(s)}(s) \\ u_2 \end{pmatrix} \rightarrow \begin{pmatrix} (1-u_1)\mathbf{w}_0^{(1)}(s) \\ \vdots \\ (1-u_1)\mathbf{w}_0^{(m(s))}(s) \\ u_1 \\ q_{11}(s) - \frac{u_2}{m(s)}\frac{(\mu_{m^\dagger m^\dagger}^{\dagger})^2 + \sigma_{m^\dagger m^\dagger}^{\dagger}}{\mu_{11}^2 + \sigma_{11}} \\ \vdots \\ q_{m(s)m(s)}(s) - \frac{u_2}{m(s)}\frac{(\mu_{m^\dagger m^\dagger}^{\dagger})^2 + \sigma_{m^\dagger m^\dagger}^{\dagger}}{\mu_{m(s)m(s)}^2 + \sigma_{m(s)m(s)}} \\ u_2 \end{pmatrix} = \begin{pmatrix} \mathbf{w}_0^{(1)^{\dagger}} \\ \vdots \\ \mathbf{w}_0^{(m(s))^{\dagger}} \\ \mathbf{w}_0^{(m^\dagger)^{\dagger}} \\ q_{11}^{\dagger} \\ \vdots \\ q_{m(s)m(s)}^{\dagger} \\ q_{m^\dagger m^\dagger}^{\dagger} \end{pmatrix} \tag{29}$$

$$Split: \begin{pmatrix} \mathbf{w}_0^{(m(s))}(s) \\ u_3 \\ q_{m(s)}(s) \\ u_4 \end{pmatrix} \rightarrow \begin{pmatrix} u_3 \mathbf{w}_0^{(m(s))}(s) \\ (1-u_3)\mathbf{w}_0^{(m(s))}(s) \\ u_4 q_{m(s)m(s)}(s) \\ (1-u_4) q_{m(s)m(s)}(s) \dfrac{\mu_{m(s)}^2 + \sigma_{m(s)m(s)}}{\mu_{m^\dagger}^2 + \sigma_{m^\dagger m^\dagger}} \end{pmatrix} = \begin{pmatrix} \mathbf{w}_0^{(m(s))^\dagger} \\ \mathbf{w}_0^{(m^\dagger)^\dagger} \\ q^\dagger_{m(s)m(s)} \\ q^\dagger_{m^\dagger m^\dagger} \end{pmatrix} \tag{30}$$

where $(u_1, u_2) \sim N\left(0, \frac{1}{2m^\dagger}\right) \times \Gamma^{-1}(\theta_1, \theta_2)$ and $(u_3, u_4) \sim \mathcal{U}(0,1)^2$. The *Death* and *Combine* moves are obviously defined as the inverse of the transformations described in Eqs. (29) and (30).

2.5. Prior Distribution

The following prior distribution for the model's parameters is used at step s

$$\begin{aligned} \pi_{m(s)}(\boldsymbol{\theta}(s)) = (2\pi)^{-\frac{m(s)}{2}} & \frac{\theta_2^{m(s)\theta_1}}{(\Gamma(\theta_1))^{m(s)}} \frac{\Gamma(h_1+h_2)}{\Gamma(h_1)\Gamma(h_2)} \lambda^{h_1-1}(1-\lambda)^{h_2-1} \\ & \times \prod_{i=1}^{m(s)} (q_i(s))^{-\theta_1-1} e^{-\frac{1}{2}\left(\mathbf{w}_0^{(i)}(s) - \frac{1}{m(s)}\right)^2 - \frac{\theta_2}{q_i(s)}} \end{aligned} \tag{31}$$

where $\boldsymbol{\theta}(s) = (\lambda(s), \mathbf{w}_0(s), \mathbf{Q}(s))$. These imply a beta $\mathcal{B}\mathbf{e}(h_1, h_2)$ prior distribution for the long-term memory coefficient $\lambda(s)$; a Normal $N\left(\frac{1}{m(s)}, 1\right)$ prior distribution for the starting value of the state vector $w_0^i(s)$; and an Inverse-Gamma $\mathcal{IG}(\theta_1, \theta_2)$ for $i = 1, \ldots, m(s)$ prior distribution for the variances $q_i(s)$ of the individual transition noise components. The various quantities needed for the calculations of the acceptance probabilities given by Eq. (8) are provided in Appendix B.

For the fixed dimensional moves, in the case of $\mathbf{w}_0$, it is possible to obtain the formula for its full conditional distribution with the help of a Kalman smoother and a simulation smoother (see, e.g., Durbin & Koopman, 2001). Indeed, if $\mathbf{w}_{1|\mathbf{z},\lambda(s),\mathbf{Q}(s)} \sim N(\mathbf{a}_1^\dagger, \mathbf{P}_1^\dagger)$ with $(\mathbf{a}_1^\dagger, \mathbf{P}_1^\dagger)$ given by the Kalman smoother, we have that $\mathbf{w}_{0|\mathbf{w}_1,\mathbf{z},\lambda(s),\mathbf{Q}(s)} \sim N(\mathbf{a}_0^\dagger, \mathbf{P}_0^\dagger)$ with

$$\mathbf{a}_0^\dagger = (\mathbf{P}_0^\dagger)^{-1}[\mathbf{P}_0^{-1}\mathbf{a}_0 + (\mathbf{P}_1^\dagger)^{-1}\mathbf{w}_{1|\mathbf{z},\lambda(s),\mathbf{Q}(s)}]$$

$$(\mathbf{P}_0^{\dagger})^{-1} = \mathbf{P}_0^{-1} + (\mathbf{P}_1^{\dagger})^{-1}$$

Since the full conditional distribution is available, a new initial state $\mathbf{w}_0$ is generated in a Gibbs-like step, instead of a traditional Metropolis–Hastings step.

New values of the covariance matrix are proposed in a random-walk Metropolis step by drawing the diagonal elements of the matrix $\mathbf{Q}$ independently from the prior distribution. The new value $\mathbf{Q}^{\dagger}$ of the covariance matrix is then accepted as the new step in the chain with acceptance probability $\rho_{\text{cov}} = \min\left(1 \frac{\ell_{m(s)}(\mathbf{Q}^{\dagger}|\mathbf{z},\mathbf{w}_0(s),\lambda(s))}{\ell_{m(s)}(\mathbf{Q}(s)|\mathbf{z},\mathbf{w}_0(s),\lambda(s))}\right)$. Similarly, new values for the long-term memory parameter λ are proposed in a random-walk Metropolis step by simply choosing the proposal density equal to the prior distribution. Hence, the new value $\lambda^{\dagger}$ is accepted as $\lambda(s+1)$ in the chain with an acceptance probability given by $\rho_{\text{LT memory}} = \min\left(1, \frac{\ell_{m(s)}(\lambda^{\dagger}|\mathbf{z},\mathbf{w}_0(s),\mathbf{Q}(s))}{\ell_{m(s)}(\lambda(s))|\mathbf{z},\mathbf{w}_0(s),\mathbf{Q}(s)}\right)$.

3. ILLUSTRATIONS

The numerical properties and feasibility of the algorithm presented above are tested using several simulated experiments. The programs were written in the *R* language and rely on the *KFAS* package for the implementation of the filtering and smoothing algorithms.

Four experiments using simulated data were considered. Using observed variables over 239 months, the returns of a fund following each of the following dynamic allocations are simulated

1. US stocks (S&P500) and commodities (GSCI index)
   - $w_\tau^{\text{SP500}} = \sin(2\pi\tau/60)$ and $w_\tau^{\text{GSCI}} = 1 - w_\tau^{\text{SP500}}$
2. US stocks (S&P500 index), commodities (GSCI index), VIX futures and German bonds (GER10Y)
   - $w_\tau^{\text{SP500}} = w_\tau^{\text{GSCI}} = w_\tau^{VIX} = w_\tau^{\text{GER10Y}} = 1/4$
3. US stocks (S&P500 index), commodities (GSCI index), and German bonds (GER10Y)
   - $w_\tau^{\text{SP500}} = \sin(2\pi\tau/60)$, $w_\tau^{\text{GSCI}} = 1 - w_\tau^{\text{SP500}}$,
   - and $w_\tau^{\text{GER10Y}} = 1/2$ if $0 < \tau < 120$ and $w_\tau^{\text{GER10Y}} = 0$ if $\tau \geq 120$
4. European stocks (DJ Eurostoxx 50 Index), Japanese stocks (Topix Index), US small stocks (Russell 2000 index), and FX USD/GBP futures with

- $w_\tau^{\text{sx5e}} = \sin(2\pi\tau/60)$ if $0<\tau<120$ and $w_\tau^{\text{sx5e}} = 0$ if $\tau \geq 120$
- $w_\tau^{\text{tpx}} = 0$ if $0<\tau<120$ and $w_\tau^{\text{tpx}} = \sin(2\pi\tau/60)$ if $\tau \geq 120$
- $w_\tau^{\text{Russell}} = 1/2$ if $0<\tau<125$ and $w_\tau^{\text{Russell}} = 0$ if $\tau \geq 125$
- $w_\tau^{\text{usdgbp}} = 1/2$ if $0<\tau<76$ and $w_\tau^{\text{Russell}} = 0$ if $\tau \geq 76$

To the returns of each simulated fund, some Gaussian measurement noise representing punctual and minor investments not captured by the variables is also added with $\varepsilon_\tau \sim NIID(0,H)$, where $H = 10^{-4}$.

Each experiment was designed to test the algorithm under a different scenario. Experiment 1 considers the case of time-varying factor loadings. Experiment 2 considers the case of four constant and equal factor loadings. Experiment 3 is an evolution of Experiment 1 in which an extraneous factor with constant loading present for half of the interval considered before disappearing. Experiment 4 considers time-varying factors appearing and disappearing during the time interval considered. The nature of the observed factors used in the simulation was also varied in order to alleviate the potential that results are driven by the specific time series used.

The adaptive RJ-MCMC algorithm was then run for each simulated fund on the following set of 11 potential factors

- S&P 500 index,
- DJ EuroStoxx 50/SPX factor,
- Russell 2000/SPX factor,
- TOPIX/SPX factor,
- USD/EUR factor,
- US 10Y Treasury bonds,
- German 10Y Treasury bonds,
- GSCI index,
- USD/JPY factor,
- USD/GPB factor,
- VIX index.

The initial set of factors for the algorithm was set randomly by drawing $m(0)$ factors uniformly from the above list. The initial number of factors $m(0)$ was uniformly drawn from $1,\ldots,m^*$. The values of the hyper-parameters were $p_{\text{Birth}} = p_{\text{Death}} = 1/4$, $p_{\text{Split}} = p_{\text{Combine}} = 3/4$, $p_{\text{Cov}} = p_{\text{LTmem}} = 1/2$, $p_{\text{Port}} = 0$, $\theta_1 = 2.001$, $\theta_2 = 1.05 \times 10^{-2}$, $h_1 = 100$, $h_2 = 2$, $p = (p_1,\ldots,p_{m^*})^{\text{Tau;}} = \frac{1}{2}\mathbf{1}_{1\times m^*}$, $S = 10^{-4}$, and $P_0 = I_{m(0)} - \frac{1}{m(0)}\mathbf{1}_{m(0)\times m(0)}$ Adaptive sampling for the probabilities of including a factor in a proposed model is started 100,000 after the "burn-in" period, and was revised every

100, 000 iterations. Each run of the algorithm took approximately 50 minutes for a chain length of 10^6 iterations.

Summary results for one run of each experiment of the algorithm are provided in Tables 1–4. All summary results are computed after discarding the first 50,000 iterations and using only the first 500,000 chain iterations after "burn-in" period, thinned by a factor 4. Cumulative estimates of the posterior probabilities that a given factor should be included in the underlying model are provided in Figs. 1–3.

Finally, using the factors with the highest posterior probabilities for each example (above 90% after 500,000 iterations of the Markov chains), I compared the Bayesian filtering replication to three common replication methodologies: rolling window Sharpe's Style Analysis (e.g., Fung & Hsieh, 1997; Sharpe, 1992), rolling window OLS estimation (e.g., Darolles & Mero, 2011), and the static optimal portfolio (e.g., Corielli & Marcellino, 2006; Glabadanidis & Zolotoy, 2013). For each example and each set of selected factors, a Kalman filter was estimated using maximum likelihood. Subsequently, the Kalman filter's estimates of the states were used to build a replicating portfolio. Similarly, replicating portfolios were built using states estimates obtained at time τ by, first, unconstrained minimization of the quadratic error over the selected factors (OLS), and second, minimization of the quadratic error over all possible factors subject to positivity constraints and a budget constraint (Sharpe's Style Analysis) on a 24-month rolling window: $[\tau-25; \tau-1]$. As a comparison, the optimal static portfolio, $\Sigma_F^{-1}\sigma_{F,HF}$, over the entire period was also calculated, where $\sigma_{F,HF}$ is the vector of covariance between the fund's returns and the considered factors. For Example 1, the selected factors are S&P 500 and GSCI indexes; for Example 2, all factors except the TOPIX/SPX index are selected; for Example 3, the selected factors are S&P 500 index, US 10Y Treasure bonds, German 10 Treasure bonds, and GSCI index; and, for Example 4, the selected factors are DJ Eurostoxx 50/SPX, Russell 2000/SPX, TOPIX/SPX factors, US 10Y Treasury bonds, and finally the USD/GBP factor. Tracking errors and other useful statistics are reported in Table 6. A graphic of the estimated factor loadings are provided in Fig. 4 for Example 3.

4. DISCUSSION

First, it is worth noting that the algorithm selects the true factors as the most likely factors as measured by their marginal posterior probabilities in the three time-varying examples (cf. Figs. 1–3). In the case of factors with

Table 1. Example 1 – Most Visited Models (At Least 1,000 Iterations of the Chain After Thinning by a Factor 4) – $N_s = 500{,}000$; $N_b = 50{,}000$.

(a) *Laplace–Metropolis estimates of log Bayes factors*

B_{ij}	M_1	M_2	M_3	M_4
M_1	0.0	−102.9	−116.2	−40.9
M_2	102.9	0.0	−13.3	61.9
M_3	116.2	13.3	0.0	75.2
M_4	40.9	−61.9	−75.2	0.0

(b) *Model compositions*

	SP500	DJ EuroStoxx 50	Russell 2000	TOPIX	USD \EUR	US 10Y	GER 10Y	GSCI	USD \JPY	USD \GPB	VIX	# iter. (in %)
M_1	✓							✓				45.4216
M_2	✓		✓					✓				3.4720
M_3	✓					✓		✓				36.7384
M_4	✓		✓			✓		✓				14.3680

Table 2. Example 2 – Most Visited Models (At Least 1000 Iterations of the Chain After Thinning by a Factor 4) – $N_s = 500{,}000$; $N_b = 50{,}000$.

(a) *Laplace–Metropolis estimates of log Bayes factors*

B_{ij}	M_{305}	M_{313}
M_{305}	0	–442
M_{313}	442	0

(b) *Model compositions*

	SP500	DJ EuroStoxx 50	Russell 2000	TOPIX	USD \EUR	US 10Y	GER 10Y	GSCI	USD \JPY	USD \GPB	VIX	# iter. (in %)
M_{45}	✓	✓	✓			✓	✓	✓			✓	0.1000
M_{55}	✓	✓	✓		✓	✓	✓	✓			✓	0.2180
M_{97}	✓	✓			✓		✓	✓	✓		✓	0.1340
M_{101}	✓	✓	✓		✓		✓	✓	✓		✓	0.1640
M_{109}	✓	✓	✓	✓	✓		✓	✓	✓		✓	0.1680
M_{116}	✓	✓	✓			✓	✓	✓	✓		✓	0.1960
M_{127}	✓	✓		✓	✓	✓	✓	✓			✓	0.2560
M_{129}	✓		✓		✓	✓	✓	✓	✓		✓	0.1780
M_{131}	✓	✓	✓		✓	✓	✓	✓	✓		✓	0.8140
M_{138}	✓	✓	✓	✓	✓	✓	✓	✓	✓		✓	0.3740
M_{165}	✓	✓			✓		✓	✓		✓	✓	0.1040
M_{168}	✓	✓	✓		✓		✓	✓		✓	✓	0.1380
M_{182}	✓	✓	✓			✓	✓	✓		✓	✓	0.1520
M_{191}	✓	✓			✓	✓	✓	✓		✓	✓	0.1300
M_{193}	✓		✓		✓	✓	✓	✓		✓	✓	0.1880
M_{195}	✓	✓	✓		✓	✓	✓	✓		✓	✓	0.6300
M_{231}	✓	✓	✓		✓	✓	✓		✓	✓	✓	0.1820
M_{251}	✓	✓		✓		✓	✓	✓	✓	✓	✓	0.4700

(b) *Model compositions*

	SP500	DJ EuroStoxx 50	Russell 2000	TOPIX	USD \EUR	US 10Y	GER 10Y	GSCI	USD \JPY	USD \GPB	VIX	# iter. (in %)
M_{255}	✓	✓	✓		✓		✓	✓	✓	✓	✓	1.0300
M_{259}	✓	✓		✓	✓		✓	✓	✓	✓	✓	0.2580
M_{263}	✓	✓	✓	✓	✓		✓	✓	✓	✓	✓	0.9200
M_{268}	✓		✓			✓	✓	✓	✓	✓	✓	0.1620
M_{270}	✓	✓	✓			✓	✓	✓	✓	✓	✓	0.4240
M_{278}	✓	✓	✓	✓		✓	✓	✓	✓	✓	✓	0.4280
M_{280}	✓				✓	✓	✓	✓	✓	✓	✓	0.1440
M_{282}	✓	✓			✓	✓	✓	✓	✓	✓	✓	0.8720
M_{284}	✓		✓		✓	✓	✓	✓	✓	✓	✓	03760
M_{285}		✓	✓		✓	✓	✓	✓	✓	✓	✓	0.1660
M_{290}	✓	✓		✓	✓	✓	✓	✓	✓	✓	✓	0.6400
M_{292}	✓		✓	✓	✓	✓	✓	✓	✓	✓	✓	0.4040
M_{293}		✓	✓	✓	✓	✓	✓	✓	✓	✓	✓	0.1040
M_{305}	✓	✓	✓		✓	✓	✓	✓	✓	✓	✓	15.6200
M_{313}	✓	✓	✓	✓	✓	✓	✓	✓	✓	✓	✓	68.4700

Table 3. Example 3 – Most Visited Models (At Least 1000 Iterations of the Chain After Thinning by a Factor 4) – $N_s = 500{,}000$; $N_b = 50{,}000$.

(a) *Laplace–Matropolis estimates) of log Bayes factors*

B_{ij}	M_1	M_4	M_6
M_1	0	233	952
M_4	–233	0	719
M_6	–952	–719	0

(b) *Model compositions*

	SP500	DJ EuroStoxx 50	Russell 2000	TOPIX	USD \EUR	US 10Y	GER 10Y	GSCI	USD \JPY	USD \GPB	VIX	# iter. (in %)
M1	✓						✓	✓				3.6987
M3	✓		✓				✓	✓				0.4738
M4	✓					✓	✓	✓				10.6215
M6	✓		✓			✓	✓	✓				90.3169
M7	✓	✓	✓			✓	✓	✓				0.0155
M12	✓					✓	✓	✓		✓		0.0296
M13	✓		✓			✓	✓	✓		✓		0.0956

constant loadings, as in Example 2, the algorithm does not perform as well however. These results are extremely encouraging and are to be celebrated as they demonstrate that the Bayesian approach provides a valid and practical solution to the factor selection problem in this dynamic factor model setting.

In terms of Model selection, model comparison using Bayes factors calculated by post processing of the simulations – i.e., using Laplace–Metropolis estimators of the marginal likelihoods (Lewis & Raftery, 1997) – does not systematically lead to the selection of the true data generating process (DGP). As a result, the algorithm provides some false positives in the sense that it can attribute high posterior probabilities to factors that are not part of the true data generating processes. In Example 1, two additional factors, Russell 2000 and US 10Y Treasury bonds, are included in the model with the largest marginal likelihood (cf. Table 1). In Example 2, all the factors considered end up in the most convincing model ($\mathcal{M}_{313}$)

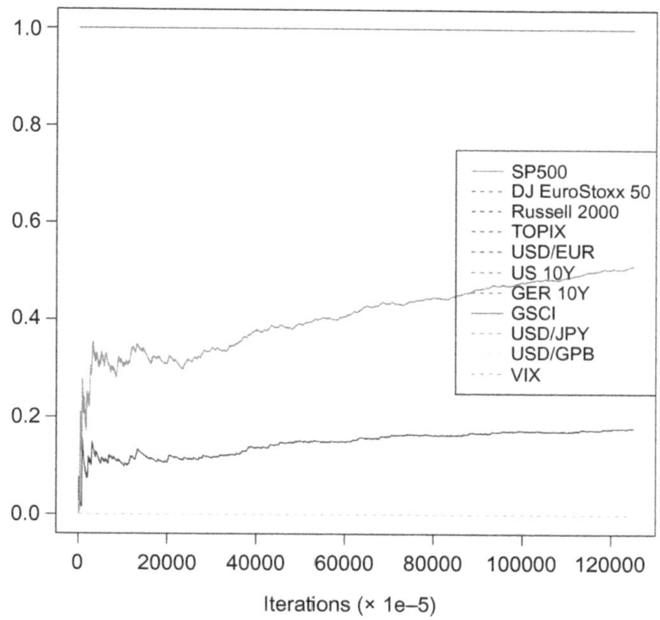

Fig. 1. Example 1 – Factors' Cumulative Posterior Probabilities (After "Burn-In" Period)–True Factors (Continuous) and Erroneous Factors (Dashed).

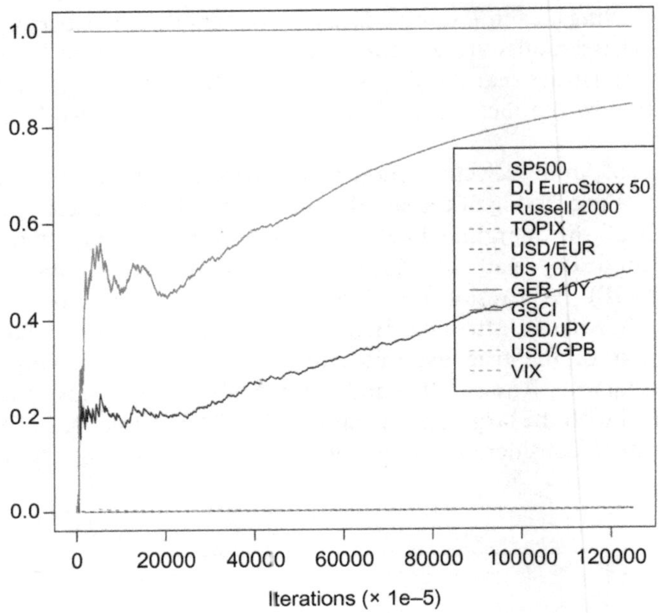

Fig. 2. Example 3 – Factors' Cumulative Posterior Probabilities (After "Burn-In" Period)–True Factors (Continuous) and Erroneous Factors (Dashed).

and with large posterior probabilities (cf. Table 2). In Example 4, one additional factor, US 10Y Treasury bonds, is included in the preferred model (cf. Table 4). While in Example 3, using the Bayes factor leads to selecting the true DGP, the chain spent most iterations in two models with one to two additional factors (cf. Table 3 b), and the posterior probabilities of these additional factors are subsequently large.

There are several possible reasons for these results. First, the algorithm is very slow mixing. The low acceptance probabilities[6] of the MH-step for the jumps are below 10%, and sometime much lower (around 1%). This translates into a slow exploration of the marginal posterior distribution of the models. Ideally, we would like to these values to be around 0.5, which corresponds to a behavior of the chain that is both relatively, fast in its exploration yet still spends some time in each neighborhood. Such low values of the acceptance probabilities for this type of Markov chain is not uncommon however as it is notoriously difficult to calibrate the

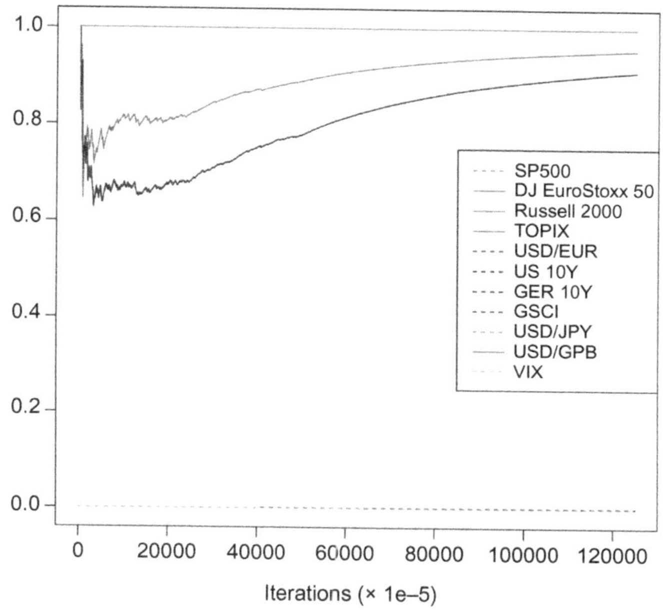

Fig. 3. Example 4 – Factors' Cumulative Posterior Probabilities (After "Burn-In" Period)–True Factors (Continuous) and Erroneous Factors (Dashed).

prior distribution of the models since this information is usually unavailable (Richardson & Green, 1997; Schäfer & Chopin, 2010, 2011). Unfortunately, fixed-dimensional covariance moves also suffer from very low acceptance probabilities suggesting that the posterior distributions of each individual model are not necessarily well explored as well. As a consequence, the post-simulations Laplace–Metropolis estimates of the marginal likelihood may be affected and the Bayes factors are not trustworthy. One possible direction for improving the acceptance probabilities of the covariance move is to resort to Hamiltonian dynamics to this setting (Neal, 2011). This is left to future work however.

Second, a consequence of this slow exploration is the relatively small number of visited models (cf. Table 5). Of the $2^{11} = 2,048$ possible combinations of factors, less than 2% are typically visited in Examples 1, 3, and 4, less than 20% in Example 2. The overlap between the sets of models visited in two runs of the algorithm for the same DGP is not

Table 4. Example 4 – Most Visited Models (At Least 1000 Iterations of the Chain After Thinning by a Factor 4) $-N_s = 500{,}000$; $N_b = 50{,}000$.

(a) *Laplace–Metropolis estimates of log Bayes factors*

M_1	M_3	M_4	M_5	
M_1	0.00	3.33	−43.19	−51.73
M_3	−3.33	0.00	−46.51	−55.05
M_4	43.19	46.51	0.00	−8.54
M_5	51.73	55.05	8.54	0.00

(*b*) *Model compositions*

	SP500	DJ EuroStoxx 50	Russell 2000	TOPIX	USD \EUR	US 10Y	GER 10Y	GSCI	USD \JPY	USD \GPB	VIX	# iter. (in %)
M_1		✓		✓						✓		2.3362
M_3		✓	✓	✓						✓		6.5180
M_4		✓		✓		✓				✓		2.0714
M_5		✓	✓	✓		✓				✓		88.9210
M_6	✓	✓	✓	✓		✓				✓		0.0120
M_7		✓	✓	✓	✓	✓				✓		0.1264
M_8		✓	✓	✓		✓			✓	✓		0.010

Table 5. Typical Number of Visited Models (After Thinning the Chain).

Example	Adapt. Sampling	w/o Adapt. Sampling
1	4	40
2	313	350
3	13	44
4	8	52

always large suggesting that the chains tend to stay in rather local areas and do not explore the complete support of the models posterior distribution. Furthermore, as the adaptation of the proposal densities weighs in on the behavior of the chain, it will naturally have a tendency to remain in a local area and explore around a local maxima of the models' posterior distribution. These phenomena are not uncommon to this kind of algorithms however (Robert & Casella, 2004; Schäfer & Chopin, 2011).

The local nature of the exploration of the posterior distributions of models is actually a feature of the algorithm accentuated by the use of adaptive sampling. Typically, as the number of factors considered could be around several dozens, the number of possible models can reach levels were a complete exploration of the distribution is unfeasible. The rationale for using adaptive sampling is to allow efforts to focus on the appropriate neighborhood of the true model. Table 5 shows that adaptive sampling leads to a reduction of the number of considered models, while not necessarily excluding any relevant factor from consideration. Furthermore, while it is clearly insufficient to reach ideal levels, try-outs with different proposal densities (not reported in this paper) indicate that using adaptive proposals actually helps to increase the acceptance probabilities of jumps between models with relevant factors. However, the local nature of the exploration is also in part responsible for the high marginal posterior probabilities ascribed to the false positives factors. One solution could be to let the adaptation kick in much later in the run of the chain or to increase the number of iterations between adaptive changes in the factors' proposal probabilities. The results of such alterations would most likely only mitigate the problems however.

The case of Example 2 is interesting. With constant factor loadings, the DGP of Example 2 is clearly a special case of model (11) with a static linear model reducing to the measurement equation. In this case, the algorithm cannot discriminate between variables and estimates high posterior probabilities for all factors. When the factor loadings are constant, we are in fact in a situation similar to classic OLS where the coefficient of determination (R^2) of the model increases artificially when including additional irrelevant variables since in this case, the likelihood of the model is inversely related to the sum of squared errors. This is however not very concerning since this special case can be handled more efficiently directly using Bayesian variable selection methods more adapted to this situation (e.g., Nott & Kohn, 2005; Robert & Casella, 2004).

Determining why an additional factor is reported with a high posterior probability by the algorithm could be interesting. It is not clear however if inclusion of the additional factor helps in tracking the second-order moment of the measurement series – Kalman filters have been known in practice to sometime need additional factors for that very reasons (Roncalli & Weisang, 2011). It could also be due to the assumption of a diagonal covariance matrix for the state vector $\mathbf{w}_\tau$ when in practice one can assume that there should be some negative correlation. Finally, it could also be the case that the absence of a constant term in the measurement equation could have an influence on the compositions of the models explored. In Examples 1, 3, and 4 for instance, the most common factor not among the true factor that is found in the compositions of the most visited models is the US 10Y Treasury bonds factor.

In terms of replication performance, it is clear from Table 6 that the Bayesian filtering methodology performs better overall (tracking error, proportion of performance explained, and proportion of performance captured) than the rolling window OLS, the rolling window Sharpe's Style Analysis, or the static optimal replicating portfolio, especially when the managed portfolio has dynamic factor loadings. It is also worth noting that when using the Kalman filter the uncaptured performance is due to the implementation lag of one period rather than the effect of poor estimation. Indeed, as one can see, for example, from Fig. 4, the Kalman filter allows for a close capture of the dynamic factor loadings contrary to other methods presented here, even when additional factors are included (cf. the factor loading of the included US 10Y factor that is not present in the DGP). This last property is predicated however on having the "true" factors included in the set of selected factors, highlighting the need for the model selection algorithm presented in this paper.

Table 6. Replicating Portfolios – Tracking Errors and Statistics for Kalman Filter (KF), 24 Months Rolling Window OLS (OLS), 24 Months Rolling Window Sharpe's Style Analysis (Sharpe SA), and Static Optimal Portfolio (Opt. Port.).

Example	TE[a]				$TE/\sqrt{Var(r_\tau^{HF})}$ (in %)				ΠAB (in %)[b]			
	KF	OLS	Sharpe SA	Opt. Port.	KF	OLS	Sharpe SA	Opt. Port.	KF	OLS	Sharpe SA	Opt. Port.
1	0.011	0.041	0.045	0.045	13	49.05	54.58	54.44	99.88	178.57	76.73	144.12
2	0.0058	0.00	1.1	1.8e-15	0.52	0.00	100.55	0.00	93.60	100.00	13e+85	114.03
3	0.015	0.039	0.083	0.057	13	34.73	74.05	51.24	92.33	178.34	2.85	110.29
4	0.0084	0.021	0.029	0.03	28	70.05	94.13	97.95	96.75	102.50	61.42	153.25

[a] $TE = \sqrt{Var(r_\tau^{Clone} - r_\tau^{HF})}$.

[b] Proportion of the fund's performance replicated by the clone: $\Pi AB = 100\Pi_{\tau=1}^{T}(1 + r_\tau^{Clone})/\Pi_{\tau=1}^{T}(1 + r_\tau^{HF})$.

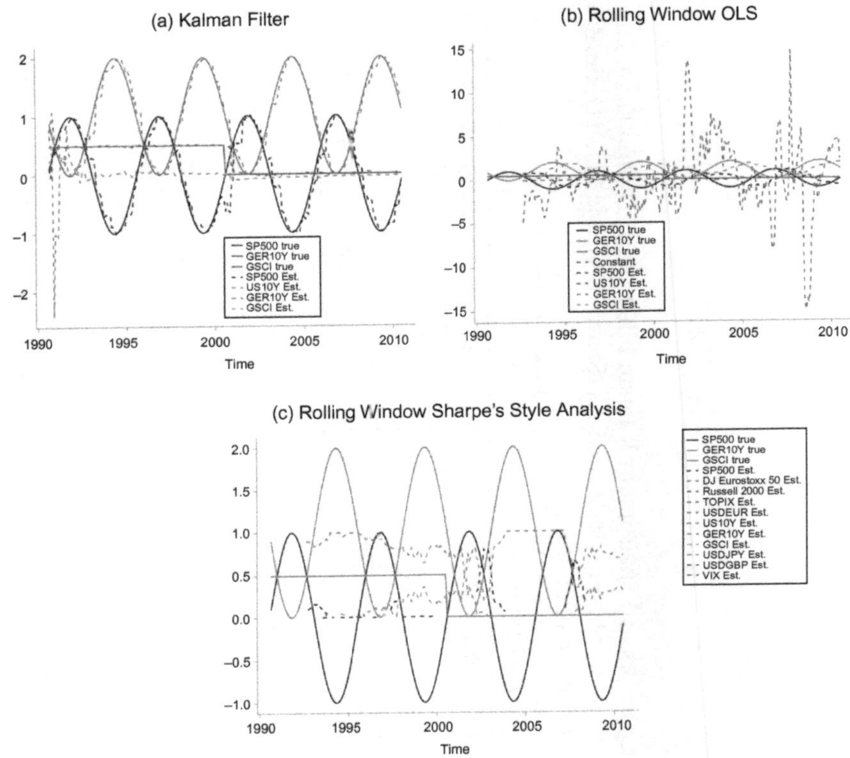

Fig. 4. Example 3 – True (Continuous) and Estimated (Dashed) Factor Loadings.

5. CONCLUSION

In this paper, I have presented an adaptive Reversible Jump MCMC variable selection approach to the problem of factor selection in dynamic factor models. The approach combines adaptive sampling (Nott & Kohn, 2005) to RJ-MCMC (Green, 1995) to explore the posterior distribution of the possible models. It can be applied in settings where a large number of variables are considered and where consideration of every possible model is not an option. If in practice the development and the implementation costs of a reversible jump Markov chain algorithm has likely been a major hurdle to the use of the Bayesian approach in the past, this paper provides the basis for a more wide-spread use of this approach in these dynamic settings.

The results presented here on four simulated examples suggest the algorithm performs successfully. It explores only a small number among all possible model candidates but in a way that avoids irrelevant specifications. The algorithm successfully selects the "true factors" involved in the data generating processes in each examples. The results also suggest that further improvements may be necessary especially in the design of the proposal densities for the covariance moves. One possible direction could be using Hamiltonian dynamics (HMC) to improve the exploration by each subchain of the posterior distribution of a model's parameters. This could lead to better post-simulation estimates of the Bayes factors and would improve tremendously model comparison.

It appears that one potential critique of the approach presented here is the local behavior of the solution provided. One very recent development in the Bayesian variable selection literature could provide a possible avenue for a remedy. Schäfer and Chopin (2011) show that one can use Sequential Monte Carlo methods to implement a "global" exploration of the posterior distribution of the models. This is a very exciting development that I intend to pursue. It is worth noting that algorithms based on particle filters are also more easily amenable to parallelization. Beside the gain in robustness of the solutions, there is therefore also a noticeable improvement in terms of speed of the algorithm.

Another avenue for improvement for this algorithm would be to improve the proposal densities by proposing moves that are more "relevant" by considering non-diagonal Q matrices for the Kalman filter. I have already begun exploring this avenue. Results so far are disappointing however. While it is relatively easy to express the moment condition from which I derived the transformations that govern the model jumps matrix terms to take into consideration the non-diagonal nature of the Q matrix. It is difficult however to find matrix transformations that preserve it, and are easy to calibrate to the problem.

NOTES

1. From Hedge Fund Research's "HFR Global Hedge Fund Industry" year-end 2013 report.

2. The significant import of RMT has been to characterize the distribution of the eigenvalues of the idiosyncratic noise in the factor model – the celebrated Marchenko–Pastur distribution. RMT shows that the sample distribution of these eigenvalues is bounded. One can therefore define the (number of) factors as

corresponding to the (number of) eigenvalues of the sample covariance matrix that lay outside of this bounded support.

3. Green's (1995) algorithm works in the case of more general moves, where both parameter spaces Θ_i and Θ_j are supplemented by artificial spaces to form spaces of equal dimension n. In the general case, $\boldsymbol{\theta}_i$ is completed by drawing $\mathbf{u}_i \sim \varphi_{n-n_i}(\cdot)$ and $\boldsymbol{\theta}_j$ by $\mathbf{v}_j \sim \varphi_{n-n_j}(\cdot)$, so that the mapping between $(\boldsymbol{\theta}_i, \mathbf{u}_i)$ and $(\boldsymbol{\theta}_j, \mathbf{v}_j)$ is a bijection

$$(\boldsymbol{\theta}_j, \mathbf{v}_j) = f_{i,j}(\boldsymbol{\theta}_i, \mathbf{u}_i)$$

In this case, one can pose

$$g_m(\boldsymbol{\theta}_i, \boldsymbol{\theta}_j) = \pi(i, \boldsymbol{\theta}_i|\mathbf{z})\pi_{ij}\varphi_{n-n_i}(\mathbf{u}_i)$$

$$g_m(\boldsymbol{\theta}_j, \boldsymbol{\theta}_i) = \pi(j, \boldsymbol{\theta}_j|\mathbf{z})\pi_{ji}\varphi_{n-n_j}(\mathbf{v}_j)\left|\frac{\partial f_{i,j}(\boldsymbol{\theta}_i, \mathbf{u}_i)}{\partial(\boldsymbol{\theta}_i, \mathbf{u}_i)}\right|$$

The probability of acceptance for the move from $\mathcal{M}_i$ to $\mathcal{M}_j$ is

$$\rho_m(\boldsymbol{\theta}_i, \boldsymbol{\theta}_j) = \min\left(1, (\text{likelihood ratio})\frac{\pi(j, \boldsymbol{\theta}_j)}{\pi(i, \boldsymbol{\theta}_i)}\frac{\pi_{ji}\varphi_{n-n_j}(\mathbf{v}_j)}{\pi_{ij}\varphi_{n-n_i}(\mathbf{u}_i)}\left|\frac{\partial f_{i,j}(\boldsymbol{\theta}_i, \mathbf{u}_i)}{\partial(\boldsymbol{\theta}_i, \mathbf{u}_i)}\right|\right)$$

4. Design of the jumps in the non-diagonal case is much more difficult as the moment condition (12) is expressed only as a condition on the trace of the matrix $\mathbf{Q}$.

5. Notice that it is always possible to reorder the factors at each step so that this is true.

6. Available upon request from the author.

REFERENCES

Amenc, N., Géhin, W., Martellini, L., & Meyfredi., J.-C. (2007, June). *The myths and limits of passive hedge fund replication*. Working Paper. EDHEC Risk and Asset Management Research Centre. Retrieved from http://faculty-research.edhec.com/servlet/com.univ.collaboratif.utils.LectureFichiergw?ID_FICHIER=1328885972635

Amenc, N., Martellini, L., Meyfredi, J.-C., & Ziemann, V. (2010). Passive hedge fund replication – Beyond the linear case. *European Financial Management*, *16*(2), 191–210.

Bai, J. (2003). Inferential theory for factor models of large dimensions. *Econometrica*, *71*(1), 135–171.

Bai, J., & Ng, S. (2002). Determining the number of factors in approximate factor models. *Econometrica*, *70*(1), 191–221.

Bai, J., & Ng, S. (2006). Evaluating latent and observed factors in macroeconomics and finance. *Journal of Econometrics*, *131*(1-2), 507–537.

Brookes, M. (2005). *The matrix reference manual*. Retrieved from http://www.ee.ic.ac.uk/hp/staff/dmb/matrix/intro.html

Brown, S. J. (1989). The number of factors in security returns. *The Journal of Finance*, *44*(5), 1247–1262.

Corielli, F., & Marcellino, M. (2006). Factor based index tracking. *Journal of Banking & Finance*, *30*(8), 2215–2233.

Darolles, S., & Mero, G. (2011). *Hedge fund returns and factor models: A cross-section approach*. Retrieved from http://ssrn.com/paper=1346319

Durbin, J., & Koopman, S. J. (2001). *Time series analysis by state space methods*. Oxford Statistical Science Series. New York, NY: Oxford University Press.

Forni, M., Hallin, M., Lippi, M., & Reichlin, L. (2000). The generalized dynamic-factor model: Identification and estimation. *Review of Economics and Statistics*, *82*(4), 540–554. Published by MIT Press Journals. doi:10.1162/003465300559037

Forni, M., Hallin, M., Lippi, M., & Reichlin., L. (2004). The generalized dynamic factor model: Consistency and rates. *Journal of Econometrics*, *119*(2), 231–255.

Forni, M., Hallin, M., Lippi, M., & Reichlin, L. (2005). The generalized dynamic factor model: Onesided estimation and forecasting. *Journal of the American Statistical Association*, *100*(471), 830–840.

Forni, M., & Lippi, M. (2001). The generalised factor model: Representation theory. *Econometric Theory*, *17*(6), 1113–1141. Published by Cambridge University Press. Retrieved from http://www.jstor.org/stable/3533246

Fung, W., & Hsieh, D. A. (1997). Empirical characteristics of dynamic trading strategies: The case of hedge funds. *Review of Financial Studies*, *10*(2), 275–302.

Fung, W., & Hsieh, D. A. (2007). Hedge fund replication strategies: Implications for investors and regulators. *Financial Stability Review*, *10*, 55–66.

Glabadanidis, P., & Zolotoy, L. (2013). Benchmark replication portfolio strategies. *Journal of Asset Management*, *14*(2), 95–110.

Green, P. J. (1995). Reversible jump Markov chain Monte Carlo computation and Bayesian model determination. *Biometrika*, *82*(4), 711–732.

Hallin, M., & Liška, R. (2007). Determining the number of factors in the general dynamic factor model. *Journal of the American Statistical Association*, *102*(478), 603.

Harding, M. C. (2007, May). *Essays in econometrics and random matrix theory*. PhD dissertation. MIT, Cambridge, MA.

Harding, M. C. (2008). Explaining the single factor bias of arbitrage pricing models in finite samples. *Economics Letters*, *99*(1), 85–88.

Kass, R. E., & Raftery, A. E. (1995). Bayes factors. *Journal of the American Statistical Association*, *90*(430), 773–795.

Lewis, S. M., & Raftery, A. E. (1997). Estimating Bayes factors via posterior simulation with the Laplace-Metropolis estimator. *Journal of the American Statistical Association*, *92*(438), 648–655.

Neal, R. M. (2011). MCMC using Hamiltonian dynamics. In *Handbook of Markov chain Monte Carlo*. CRC Press. Retrieved from http://www.mcmchandbook.net/HandbookChapter5.pdf

Nott, D. J., & Kohn, R. (2005). Adaptive sampling for Bayesian variable selection. *Biometrika*, *92*(4), 747–763.

Onatski, A. (2009). Testing hypotheses about the number of factors in large factor models. *Econometrica*, *77*(5), 1447–1479.

Onatski, A. (2010). Determining the number of factors from empirical distribution of eigenvalues. *The Review of Economics and Statistics*, *92*(4), 1004–1016.

Onatski, A. (2012). Asymptotics of the principal components estimator of large factor models with weakly influential factors. *Journal of Econometrics*, *168*(2), 244–258.

Petersen, K. B., & Pedersen, M. S. (2012). *The matrix cookbook*. Denmark: Technical University of Denmark. Retrieved from http://www2.imm.dtu.dk/pubdb/p.php?3274

Richardson, S., & Green, P. J. (1997). On Bayesian analysis of mixtures with an unknown number of components. *Journal of the Royal Statistical Society. Series B (Methodological)*, *59*(4), 731–792.

Robert, C. P., & Casella, G. (2004). *Monte Carlo statistical methods* (2nd ed.). Springer Texts in Statistics. New York, NY: Springer.

Roncalli, T., & Teiletche, J. (2008). An alternative approach to alternative beta. *Journal of Financial Transformation, Cass-Capco Institute Paper Series on Risk*, *24*, 43–52.

Roncalli, T., & Weisang, G. (2011). Tracking problems, hedge fund replication and alternative beta. *Journal of Financial Transformation*, *31*, 19–29.

Ross, S. A. (1976). The arbitrage theory of capital asset pricing. *Journal of Economic Theory*, *13*(3), 341–360.

Schäfer, C., & Chopin, N. (2010, July). *Adaptive monte carlo on multivariate binary sampling spaces*. Retrieved from http://arxiv.org/abs/1008.0055v1

Schäfer, C., & Chopin, N. (2011, February). *Sequential monte carlo on large binary sampling spaces*. Retrieved from http://arxiv.org/abs/1101.6037v2

Sharpe, W. F. (1992). Asset allocation: Management style and performance measurement. *Journal of Portfolio Management*, *18*(2), 7–19.

Whittle, P. (July 1996). *Optimal control: Basics and beyond* (1st ed.). Series in Systems and Optimization. Chichester, UK: Wiley-Interscience.

APPENDIX A: DERIVATION OF THE MOMENT CONDITION

I will denote $\mathbf{F} = \mathbb{E}(\mathbf{F}'_\tau)$ and $\mathbf{S}_F = \mathrm{Var}(\mathbf{F}'_\tau)$.

Using the measurement in Eq. (4), we find that

$$\begin{aligned} r_\tau^{\mathrm{HF}} &= \mathbf{F}'_\tau \left(\prod_{\ell=0}^{\tau-1} \mathbf{T}_{\tau-\ell} \right) \mathbf{w}_0 + \mathbf{F}'_\tau \sum_{\ell=0}^{\tau-1} \left(\prod_{i=0}^{\ell-1} \mathbf{T}_{\tau-i} \right) \nu_{\tau-\ell} + \eta_\tau \\ &= (\mathbf{F}'_\tau - \mathbf{F}') \left(\prod_{\ell=0}^{\tau-1} \mathbf{T}_{\tau-\ell} \right) \mathbf{w}_0 + \mathbf{F}' \left(\prod_{\ell=0}^{\tau-1} \mathbf{T}_{\tau-\ell} \right) \mathbf{w}_0 \\ &\quad + (\mathbf{F}'_\tau - \mathbf{F}') \sum_{\ell=0}^{\tau-1} \left(\prod_{i=0}^{\ell-1} \mathbf{T}_{\tau-i} \right) \nu_{\tau-\ell} + \mathbf{F}' \sum_{\ell=0}^{\tau-1} \left(\prod_{i=0}^{\ell-1} \mathbf{T}_{\tau-i} \right) \nu_{\tau-\ell} + \eta_\tau \end{aligned}$$

One can now use the following identity

$$\mathrm{Var}(r_\tau^{\mathrm{HF}}) = \mathrm{Var}[\mathbb{E}(r_\tau^{\mathrm{HF}} | \mathcal{F}_{\tau-1})] + \mathbb{E}[\mathrm{Var}(r_\tau^{\mathrm{HF}} | \mathcal{F}_{\tau-1})]$$

where $\mathcal{F}_{\tau-1} = \{\mathbf{F}_\ell, \mathbf{T}_\ell, \nu_\ell, \eta_\ell : \; 0 \leq \ell \leq \tau - 1\}$.

Denoting $\mathbb{E}(\cdot | \mathcal{F}_{\tau-1})$ by $\mathbb{E}_{\tau-1}(\cdot)$ and $\mathrm{Var}(\cdot | \mathcal{F}_{\tau-1})$ by $\mathrm{Var}_{\tau-1}(\cdot)$, we have

$$\mathbb{E}_{\tau-1}(r_\tau^{\mathrm{HF}}) = \mathbf{F}' \left(\prod_{\ell=0}^{\tau-1} \mathbf{T}_{\tau-\ell} \right) \mathbf{w}_0 + \mathbf{F}' \sum_{\ell=0}^{\tau-1} \left(\prod_{i=0}^{\ell-1} \mathbf{T}_{\tau-i} \right) \nu_{\tau-\ell} \tag{A.1}$$

and

$$\begin{aligned} \mathrm{Var}_{\tau\text{-}1}(r_{\tau-1}^{\mathrm{HF}}) &= \mathrm{Var}\left((\mathbf{F}'_\tau - \mathbf{F}') \left(\prod_{\ell=0}^{\tau-1} \mathbf{T}_{\tau-\ell} \right) \mathbf{w}_0 \right) \\ &\quad + \mathrm{Var}_{\tau\text{-}1} \left((\mathbf{F}'_\tau - \mathbf{F}') \sum_{\ell=0}^{\tau-1} \left(\prod_{i=0}^{\ell-1} \mathbf{T}_{\tau-i} \right) \nu_{\tau-\ell} \right) + S \\ &= \left(\prod_{\ell=0}^{\tau-1} \mathbf{T}_{\tau-\ell} \mathbf{w}_0 \right)' \Sigma_F \left(\prod_{\ell=0}^{\tau-1} \mathbf{T}_{\tau-\ell} \mathbf{w}_0 \right) \\ &\quad + \left(\sum_{\ell=0}^{\tau-1} \left(\prod_{i=0}^{\ell-1} \mathbf{T}_{\tau-i} \right) \nu_{\tau-\ell} \right)' \Sigma_F \left(\sum_{\ell=0}^{\tau-1} \left(\prod_{i=0}^{\ell-1} \mathbf{T}_{\tau-i} \right) \nu_{\tau-\ell} \right) + S \end{aligned}$$

Using $\mathbf{T}_\tau = \mathbf{T} = \lambda I$,

$$\begin{aligned}\mathrm{Var}[\mathbb{E}_{\tau-1}(r_\tau^{\mathrm{HF}})] &= \lambda^{2\tau}\mathbf{F}'\mathbf{P}_0\mathbf{F} + \mathbf{F}'\left(\sum_{\ell=0}^{\tau-1}\lambda^{2\ell}\mathbf{Q}\right)\mathbf{F} \\ &= \lambda^{2\tau}\mathbf{F}'\mathbf{P}_0\mathbf{F} + \frac{1-\lambda^{2\tau}}{1-\lambda^2}\mathbf{F}'\mathbf{Q}\mathbf{F}\end{aligned} \tag{A.2}$$

and

$$\begin{aligned}\mathbb{E}[\mathrm{Var}_{\tau-1}(r_{\tau-1}^{\mathrm{HF}})] &= \lambda^{2\tau}\mathbb{E}(\mathbf{w}'_0\Sigma_F\mathbf{w}_0) + \mathbb{E}\left[\left(\sum_{\ell=0}^{\tau-1}\lambda^\ell\nu_{\tau-\ell}\right)'\Sigma_F\left(\sum_{\ell=0}^{\tau-1}\lambda^\ell n_{\tau-\ell}\right)\right] \\ &= \lambda^{2\tau}[\mathrm{Tr}(\Sigma_F\mathbf{P}_0) + \mathbf{a}'_0\Sigma_F\mathbf{a}_0] + \sum_{\ell=0}^{\tau-1}\sum_{\ell'=0}^{\tau-1}\lambda^{\ell+\ell'}\mathbb{E}(\nu'_{\tau-\ell}\Sigma_F\nu_{\tau-\ell'}) + S\end{aligned}$$

where I used the fact that for any random vector ϵ, with mean $\mathbf{m}$ and covariance matrix $\mathbf{S}$, and for any symmetric matrix $\mathbf{A}$, $\mathbb{E}(\epsilon'\mathbf{A}\epsilon) = \mathrm{Tr}(\mathbf{AS}) + \mathbf{m}'\mathbf{Am}$ and $\mathrm{Tr}(\cdot)$ is the trace operator (Brookes, 2005; Petersen & Pedersen, 2012).

Since $\nu_\ell \sim NIID(0,\mathbf{Q})$, we have

$$\mathbb{E}(\nu'_\ell\Sigma_F\nu_{\ell'}) = \begin{cases}\mathrm{Tr}(\Sigma_F\mathbf{Q}) & \text{if } \ell = \ell' \\ 0 & \text{otherwise}\end{cases}$$

Hence,

$$\mathbb{E}\left[\mathrm{Var}_{\tau-1}(r_{\tau-1}^{\mathrm{HF}})\right] = \lambda^{2\tau}[\mathrm{Tr}(\Sigma_F\mathbf{P}_0) + \mathbf{a}'_0\Sigma_F\mathbf{a}_0] + \frac{1-\lambda^{2\tau}}{1-\lambda^2}\mathrm{Tr}(\Sigma_F\mathbf{Q}) + S \tag{A.3}$$

Therefore,

$$\begin{aligned}\mathrm{Var}(r_\tau^{\mathrm{HF}}) &= \lambda^{2\tau}[\mathbf{F}'\mathbf{P}_0\mathbf{F} + \mathrm{Tr}(\Sigma_F\mathbf{P}_0) + \mathbf{a}'_0\Sigma_F\mathbf{a}_0] \\ &\quad + \frac{1-\lambda^{2\tau}}{1-\lambda^2}[\mathrm{Tr}(\Sigma_F\mathbf{Q}) + \mathbf{F}'\mathbf{Q}\mathbf{F}] + S\end{aligned} \tag{A.4}$$

Since, $\mathrm{Var}(r_\tau^{\mathrm{HF}})$ is time-invariant, we deduce the result given by Eq. (12).

APPENDIX B: ACCEPTANCE PROBABILITIES

B.1. Split *Move*

Using Eqs. (8), (30), and (31), I find that the acceptance probability for the *Split* move is given by

$$\rho_m(\boldsymbol{\theta}_{m(s)}, \boldsymbol{\theta}^\dagger) = 1 \wedge \frac{\pi(m^\dagger, \boldsymbol{\theta}^\dagger_{\mathcal{F}}, \boldsymbol{\theta}^\dagger)}{\pi(m(s), \boldsymbol{\theta}_{\mathcal{F}}(s), \boldsymbol{\theta}(s))} \frac{\ell_{m^\dagger}(\lambda^\dagger, \mathbf{w}^\dagger_0, Q^\dagger | \mathbf{z}, m^\dagger, \boldsymbol{\theta}^\dagger_{\mathcal{F}})}{\ell_{m(s)}(\lambda(s), \mathbf{w}_0(s), Q(s) | \mathbf{z}, m(s), \boldsymbol{\theta}_{\mathcal{F}}(s))} \times \frac{\pi_{\text{Combine}}}{\pi_{\text{Split}} \varphi_{\text{Split}}(\mathbf{u})} \left| \frac{\partial f_{\text{Split}}(\boldsymbol{\theta}(s), \mathbf{u})}{\partial(\boldsymbol{\theta}(s), \mathbf{u})} \right| \tag{B.1}$$

$$= 1 \wedge \frac{\rho(m^\dagger, \boldsymbol{\theta}^\dagger_{\mathcal{F}})}{\rho(m(s), \boldsymbol{\theta}_{\mathcal{F}}(s))} \frac{\pi_{m^\dagger}(\boldsymbol{\theta}^\dagger | m^\dagger, \boldsymbol{\theta}^\dagger_{\mathcal{F}})}{\pi_{m(s)}(\boldsymbol{\theta}(s) | m(s), \boldsymbol{\theta}_{\mathcal{F}}(s))} \times \frac{\ell_{m^\dagger}(\lambda^\dagger, \mathbf{w}^\dagger_0, Q^\dagger | \mathbf{z}, m^\dagger, \boldsymbol{\theta}^\dagger_{\mathcal{F}})}{\ell_{m(s)}(\lambda(s), \mathbf{w}_0(s), Q(s) | \mathbf{z}, m(s), \boldsymbol{\theta}_{\mathcal{F}}(s))} \times \frac{\pi_{\text{Combine}}}{\pi_{\text{Split}} \varphi_{\text{Split}}(\mathbf{u})} \left| \frac{\partial f_{\text{Split}}(\boldsymbol{\theta}(s), \mathbf{u})}{\partial(\boldsymbol{\theta}(s), \mathbf{u})} \right| \tag{B.2}$$

However, by construction (cf. Section 2.2), $\varrho(m, \boldsymbol{\theta}_{\mathcal{F}}) = \prod_{i=1}^{m^{star}} \mathscr{B}_{p_i(\boldsymbol{\theta}_{\mathcal{F}})}(\gamma)$. Therefore

$$\frac{\varrho(m^\dagger, \boldsymbol{\theta}^\dagger_{\mathcal{F}})}{\varrho(m(s), \boldsymbol{\theta}_{\mathcal{F}}(s))} = \frac{p_i(\boldsymbol{\theta}_{\mathcal{F}})}{1 - p_i(\boldsymbol{\theta}_{\mathcal{F}})} \tag{B.3}$$

Now, considering π_{Combine} and π_{Split} the probabilities of choosing the particular models described by $\boldsymbol{\theta}_{\mathcal{F}}(s)$ and $\boldsymbol{\theta}^\dagger_{\mathcal{F}}$, we have

$$\pi_{\text{Split}} = \frac{1}{m(s)(m(s))!}$$

$$\pi_{\text{Combine}} = \frac{1}{m(s)(m(s))!}$$

where the factorials account for the different possible arrangements of the factors. Hence,

$$\frac{\pi_{\text{Combine}}}{\pi_{\text{Split}}} = 1 \tag{B.4}$$

The Jacobian of the *Split* transformation between the two vectors $(\boldsymbol{\theta}(s), u)$ and $\boldsymbol{\theta}^{\dagger}$ defined by Eq. (30) is given by

$$\left|\frac{\partial f_{\text{Split}}(\boldsymbol{\theta}(s), \mathbf{u})}{\partial(\boldsymbol{\theta}(s), \mathbf{u})}\right| = \begin{vmatrix} u_3 & w_0^{m(s)} & 0 & 0 \\ 1-u_3 & -w_0^{m(s)} & 0 & 0 \\ 0 & 0 & u_4 & q_{m(s)} \\ 0 & 0 & (1-u_4)A & -q_{m(s)}A \end{vmatrix} = w_0^{m(s)} q_{m(s)} A \tag{B.5}$$

where $A = \dfrac{\mu_{m(s)}^2 + \sigma_{m(s)}^2}{\mu_{m^{\dagger}}^2 + \sigma_{m^{\dagger} m^{\dagger}}}$.

B.2. Acceptance Probability of a Combine *Move*

One can derive in a similar way the acceptance probability of a *Combine* move. Indeed, using a similar expression as (B.1), and noting that in the case of *Combine* move, we have

$$\frac{\varrho(m^{\dagger}, \boldsymbol{\theta}_{\mathcal{F}}^{\dagger})}{\varrho(m(s), \boldsymbol{\theta}_{\mathcal{F}}(s))} = \frac{1 - p_i(\boldsymbol{\theta}_{\mathcal{F}})}{p_i(\boldsymbol{\theta}_{\mathcal{F}})}$$

Furthermore, we have

$$\left|\frac{\partial f_{\text{Combine}}(\boldsymbol{\theta}(s))}{\partial(\boldsymbol{\theta}(s))}\right|$$

$$= \begin{vmatrix} \dfrac{\frac{1}{w_0^{(m(s))}(s)}}{(w_0^{(m(s)-1)}(s) + w_0^{(m(s))}(s))^2} & -\dfrac{\frac{1}{w_0^{(m(s)-1)}(s)}}{(w_0^{(m(s)-1)}(s) + w_0^{(m(s))}(s))^2} & 0 & 0 \\ & & 0 & 0 \\ 0 & 0 & \dfrac{\frac{1}{A^{-1}q(s)_{m(s)}}}{(q(s)_{m(s)-1} + A^{-1}q(s)_{m(s)})^2} & -\dfrac{\frac{A^{-1}}{A^{-1}q(s)_{m(s)-1}}}{(q(s)_{m(s)-1} + A^{-1}q(s)_{m(s)})^2} \\ 0 & 0 & & \end{vmatrix}$$

$$= \frac{1}{w_0^{(m(s)-1)}(s) + w_0^{(m(s))}(s)} \frac{1}{Aq(s)_{m(s)-1} + q(s)_{m(s)}} \tag{B.6}$$

where $A = \dfrac{\mu_{m(s)}^2 + \sigma_{m(s)}^2}{\mu_{m^\dagger}^2 + \sigma_{m^\dagger m^\dagger}}$.

B.3. Acceptance Probability of a Birth *Move*

In a similar way to the derivations above, the acceptance probability of a *Birth* move is given by

$$\rho_m(\boldsymbol{\theta}_{m(s)}, \boldsymbol{\theta}^\dagger) = 1 \wedge \frac{\varrho(m^\dagger, \boldsymbol{\theta}_{\mathcal{F}}^\dagger)}{\varrho(m(s), \boldsymbol{\theta}_{\mathcal{F}}(s))} \frac{\pi_{m^\dagger}(\boldsymbol{\theta}^\dagger | m^\dagger, \boldsymbol{\theta}_{\mathcal{F}}^\dagger)}{\pi_{m(s)}(\boldsymbol{\theta}(s) | m(s), \boldsymbol{\theta}_{\mathcal{F}}(s))}$$

$$\times \frac{\ell_{m^\dagger}(\lambda^\dagger, \mathbf{w}_0^\dagger, Q^\dagger | \mathbf{z}, m^\dagger, \boldsymbol{\theta}_{\mathcal{F}}^\dagger)}{\ell_{m(s)}(\lambda(s), \mathbf{w}_0(s), Q(s) | \mathbf{z}, m(s), \boldsymbol{\theta}_{\mathcal{F}}(s))}$$

$$\times \frac{\pi_{\text{Death}}}{\pi_{\text{Birth}} \varphi_{\text{Birth}}(\mathbf{u})} \left| \frac{\partial f_{\text{Birth}}(\boldsymbol{\theta}(s), \mathbf{u})}{\partial(\boldsymbol{\theta}(s), \mathbf{u})} \right| \tag{B.7}$$

Considering π_{Birth} and π_{Death} the probabilities of choosing the particular models described by $\mathbf{w}_{\mathcal{F}}(s)$ and $\mathbf{w}_{\mathcal{F}}^\dagger$, we have

$$\frac{\pi_{\text{Death}}}{\pi_{\text{Birth}}} = 1 \tag{B.8}$$

Also, we have

$$\varphi_{\text{Birth}}(u_1, u_2) = \frac{1}{\sqrt{2\pi}} e^{-\frac{1}{2}\left(u_1 - \frac{1}{m^\dagger}\right)^2} \frac{\theta_2^{\theta_1}}{\Gamma(\theta_1)} e^{-\frac{\theta_2}{u_2}} \tag{B.9}$$

with $\mathbf{u} = (u_1, u_2)$. Considering the Jacobian matrix of the transformation f_{Birth}, it is easy to see that it is upper triangular, and therefore to calculate the Jacobian of the transformation as

$$
\left|\frac{\partial f_{\text{Birth}}\left(\boldsymbol{\theta}\left(s\right),\mathbf{u}\right)}{\partial\left(\boldsymbol{\theta}\left(s\right),\mathbf{u}\right)}\right| = \left|\begin{array}{ccc|ccccc}
1-w_0^{(m^\dagger)} & \cdots\cdots & \cdots\cdots & & & & & \\
 & \ddots & \cdots\cdots & & & & & \\
 & & 1-w_0^{(m^\dagger)} & & & & & \\
\hline
 & & & 1 & \cdots\cdots & & & \\
 & & & & 1 & \cdots\cdots & & \\
 & & & & & \ddots & \cdots & \\
 & & & & & & 1 & \cdots \\
 & & & & & & & 1
\end{array}\right| \tag{B.10}
$$

$$
= \left(1-w_0^{(m^\dagger)}\right)^{m(s)}
$$

B.4. Acceptance Probability of a Death *Move*

Finally, the probability of acceptance of a *Death* move can be derived similarly as for the *Birth* move. Hence, it is easy to find that

$$
\left|\frac{\partial f_{\text{Death}}(\boldsymbol{\theta}(s),\mathbf{u})}{\partial(\boldsymbol{\theta}(s),\mathbf{u})}\right| = (1-w_0^{(m(s))})^{-m^\dagger} \tag{B.11}
$$

DETERMINING THE PROPER SPECIFICATION FOR ENDOGENOUS COVARIATES IN DISCRETE DATA SETTINGS

Angela Vossmeyer

Department of Economics, University of California, Irvine, CA, USA

ABSTRACT

An important but often overlooked obstacle in multivariate discrete data models is the specification of endogenous covariates. Endogeneity can be modeled as latent or observed, representing competing hypotheses about the outcomes being considered. However, little attention has been applied to deciphering which specification is best supported by the data. This paper highlights the use of existing Bayesian model comparison techniques to investigate the proper specification for endogenous covariates and to understand the nature of endogeneity. Consideration of both observed and latent modeling approaches is emphasized in two empirical applications. The first application examines linkages for banking contagion and the second application evaluates the impact of education on socioeconomic outcomes.

Bayesian Model Comparison
Advances in Econometrics, Volume 34, 223–247

ISSN: 0731-9053/doi:10.1108/S0731-905320140000034010

Keywords: Bayesian estimation; data augmentation; educational attainment; financial contagion; latent data; marginal likelihood

JEL classifications: C35; C52; G01; I00

1. INTRODUCTION

In empirical applications, endogenous regressors are generally the key variables of interest. Treatment models, triangular systems with recursive endogeneity, and sequential decision-making all feature endogenous covariates that often represent the main components of the study. In continuous data settings, modeling endogeneity is simple and interpretation is straightforward. In discrete data settings, modeling endogeneity is complicated because it can take several forms based on latent or observed data. This is not a limitation of the system. Instead, this feature increases the flexibility of such models because unobservables can be captured as explanatory variables separately from variables that are observable to the econometrician. This is only beneficial if a formal model comparison can be performed to decipher which depicted relationship is best supported by the data and to resolve competing hypotheses about the type of endogeneity. Investigating both approaches strengthens the eventual results by increasing a researcher's understanding of the relationships and the dependence structure being modeled. Model testing is easily afforded by existing Bayesian model comparison techniques.

Initial latent data modeling innovations occurred in psychometrics, where the traditional usage of latent variables focused on measurement error and hypothetical constructs (Muthen, 2002). In econometrics, latent data analysis advances discrete choice methods in which choice outcomes are linked to latent utility. For a review, see Jeliazkov and Rahman (2012). Bayesian econometrics further benefited by the association between data augmentation and latent variables (Tanner & Wong, 1987). Although latent modeling approaches have captured a variety of statistical concepts, including random coefficients, missing data, discrete choice, and finite mixture modeling, latent variables are rarely employed as regressors since an investigator has not or can not measure or observe them. Macroeconomics has moved in this direction with factor models; however, latent covariates remain unexplored in applied microeconomic research. Furthermore, across all fields, little attention has been applied to formally compare such models. In a recent marketing paper, Mintz, Currim, and Jeliazkov (2013) look at both specifications and find that a latent measure of information processing pattern better explains an individual's propensity to buy. Overlooking the consideration of both approaches can lead to specification errors and misrepresent the relationships being examined.

This paper employs Bayesian model selection methods for comparing latent and observed endogeneity models in two empirical applications. Each application features competing hypotheses discussed in the literature and a formal motivation for using observed or latent endogeneity. The first application examines banking contagion and the relative influence or spread of contagion from both regional and network linkages. The second application considers the impact of education on adult socioeconomic status. These applications highlight a key aspect of this research topic. While an applied researcher may have *a priori* expectations of the "correct" model, in most cases and especially in these examples, arguments for both approaches are easily formed, making it difficult for a researcher to completely rule out a specification without performing model comparison.

The rest of this paper is organized as follows: Section 2.1 reviews each specification in a simple bivariate system of equations and Section 2.2 discusses existing techniques for model selection. Section 3 considers the application to financial contagion and Section 4 considers the application to education. Finally, Section 5 offers concluding remarks.

2. METHODOLOGY

2.1. Model

To exemplify the approaches discussed in this paper, consider a bivariate model with recursive endogeneity, where latent data are referred to as $\mathbf{y}_i^*$ and observed data are referred to as $\mathbf{y}_i$. The two different modeling techniques, latent and observed, are shown as:

Observed Endogeneity

$$\begin{aligned} y_{i1}^* &= \mathbf{x}'_{i1}\boldsymbol{\beta}_1 + \varepsilon_{i1} \\ y_{i2}^* &= \mathbf{x}'_{i2}\boldsymbol{\beta}_2 + y_{i1}\gamma_2 + \varepsilon_{i2} \end{aligned} \tag{1}$$

Latent Endogeneity

$$\begin{aligned} y_{i1}^* &= \mathbf{x}'_{i1}\boldsymbol{\beta}_1 + \varepsilon_{i1} \\ y_{i2}^* &= \mathbf{x}'_{i2}\boldsymbol{\beta}_2 + y_{i1}^*\gamma_2 + \varepsilon_{i2}. \end{aligned} \tag{2}$$

The latent data are related to the observed outcomes by a link function depending on the values y_{ik} can take, for equations $k = 1, 2$. The binary setting occurs when $y_{ik} = 1\{y_{ik}^* > 0\}$, and the censored setting occurs when $y_{ik} = y_{ik}^* \cdot 1\{y_{ik}^* > 0\}$. The link function for ordered data is $y_{ik} = \sum_{j=1}^{J} 1\{y_{ik}^* > \alpha_{k,j-1}\}$ for J ordered alternatives, where α_{kj} is a cut-point between the categories. In this context, the observed endogeneity system differs from (1) and instead is:

$$\begin{aligned} y_{i1}^* &= \mathbf{x}'_{i1}\boldsymbol{\beta}_1 + \varepsilon_{i1} \\ y_{i2}^* &= \mathbf{x}'_{i2}\boldsymbol{\beta}_2 + 1\{y_{i1} = 2\}\gamma_{22} + 1\{y_{i1} = 3\}\gamma_{23} + \cdots + 1\{y_{i1} = J\}\gamma_{2J} + \varepsilon_{i2}, \end{aligned} \tag{3}$$

where there is a set of endogenous indicator variables for $J - 1$ categories, as opposed to a single endogenous regressor as in (1). This case is explored in the second application to education in Section 4. For simplicity, assume $\varepsilon_i \equiv (\varepsilon_{i1}, \varepsilon_{i2})' \sim N_2(0,\Omega)$ and $\Omega = \begin{pmatrix} \omega_{11} & \omega_{12} \\ \omega_{21} & \omega_{22} \end{pmatrix}$ in Eqs. (1), (2), and (3). Models with endogeneity have been difficult to estimate when the response variables of interest are not continuous because standard two-stage estimators are inapplicable in this context. Therefore, estimation in this paper relies on Markov chain Monte Carlo (MCMC) methods, which are discussed in detail in each application.

The latent data have the customary random utility interpretation underlying the theory on discrete choice analysis in econometrics. Therefore, even though the observed data can only take certain values, the latent variables that determine those outcomes are unrestricted. The fact that the latent utilities can be changed without necessarily inducing a corresponding change in the observed variable is a key distinction between these models (Mintz et al., 2013). Furthermore, the observed and latent specifications pose different relationships between the variables of interest because latent data measure intentions and observed data measure actual actions or outcomes. In (2), latent endogeneity says that intentions about y_{i1} determine intentions about y_{i2}. In (1), observed endogeneity says that actions about y_{i1} determine intentions about y_{i2} (Maddala, 1983). Despite the clear interpretation of each model, in most cases, it is difficult for a researcher to decipher which specification is correct. Generally, convincing hypotheses or arguments can be made in support of either modeling approach, hence motivating the need for model comparison.

It is important to note that these considerations extend to larger systems of equations, models for sample selection, potential outcomes, simultaneous equations, and more. For a review of some of these models, see van Hasselt (2014) and Li and Tobias (2014). Any multivariate discrete outcome model should not overlook this problem. The bivariate system is considered here to stress the importance of the issue, highlight the ease of considering both models, and offer a more complete understanding of the relationships in each application.

2.2. Model Comparison

Model comparison techniques are often employed to deal with issues of model uncertainty and variable selection. This paper utilizes these same approaches to determine the nature of endogeneity. The methods used in this paper are from Chib (1995) and Chib and Jeliazkov (2001), which are computationally convenient and do not require much additional coding. The applications in this paper span a number of discrete choice models, including ordered probit, Tobit, and binary probit. Therefore, the model comparison methods discussed here are general across these classes of models.

Given the data $\mathbf{y}$, interest centers upon the models $\{\mathcal{M}_l, \mathcal{M}_o\}$ where $\mathcal{M}_l$ represents the latent endogeneity model and $\mathcal{M}_o$ represents the observed endogeneity model. Each model is characterized by a sampling density $\{f(\mathbf{y}|\mathcal{M}_l, \boldsymbol{\theta}_l), f(\mathbf{y}|\mathcal{M}_o, \boldsymbol{\theta}_o)\}$ where $\{\boldsymbol{\theta}_l, \boldsymbol{\theta}_o\}$ are model-specific parameter vectors. Bayesian model selection proceeds by comparing the models through their posterior odds ratio

$$\frac{\Pr(\mathcal{M}_l|\mathbf{y})}{\Pr(\mathcal{M}_o|\mathbf{y})} = \frac{\Pr(\mathcal{M}_l)}{\Pr(\mathcal{M}_o)} \times \frac{m(\mathbf{y}|\mathcal{M}_l)}{m(\mathbf{y}|\mathcal{M}_o)}. \tag{4}$$

Chib (1995) recognized the *basic marginal likelihood identity* in which the marginal likelihood for model $\mathcal{M}_l$ can be expressed as

$$m(\mathbf{y}|\mathcal{M}_l) = \frac{f(\mathbf{y}|\mathcal{M}_l, \boldsymbol{\theta}_l)\pi(\boldsymbol{\theta}_l|\mathcal{M}_l)}{\pi(\boldsymbol{\theta}_l|\mathbf{y}, \mathcal{M}_l)}. \tag{5}$$

Calculation of the marginal likelihood is then reduced to finding an estimate of the posterior ordinate $\pi(\boldsymbol{\theta}_l^*|\mathbf{y}, \mathcal{M}_l)$ at a single point $\boldsymbol{\theta}_l^*$, which is often taken as the posterior mean or mode. Since the topics in this paper involve multivariate discrete data, sampling densities are often analytically intractable. A straightforward approach for evaluating the likelihood function employed in this paper is the Chib–Ritter–Tanner (CRT) method, which was developed in Jeliazkov and Lee (2010).

Decomposition of the posterior ordinate varies across the examples in this paper, with the most difficult being the multivariate ordered probit model used in the education application. This case is a bit more complex due to the additional cut-point parameters $\boldsymbol{\delta}$.[1] This section outlines the decomposition used for the ordered probit model, which follows from Jeliazkov, Graves, and Kutzbach (2008). It should be noted that the decomposition for the binary probit and Tobit models are simplified versions of the ordered probit without computations for the $\boldsymbol{\delta}$ parameter vector. Let $\boldsymbol{\lambda} = (\boldsymbol{\beta}', \boldsymbol{\gamma}')'$ be a parameter vector for all endogenous and exogenous covariates. Estimation of the posterior ordinate can be facilitated using the decomposition

$$\pi(\boldsymbol{\lambda}^*, \Omega^*, \boldsymbol{\delta}^*|\mathbf{y}) = \pi(\boldsymbol{\lambda}^*|\mathbf{y})\pi(\Omega^*|\mathbf{y}, \boldsymbol{\lambda}^*)\pi(\boldsymbol{\delta}^*|\mathbf{y}, \Omega^*, \boldsymbol{\lambda}^*).$$

Estimation of $\pi(\boldsymbol{\lambda}^*|\mathbf{y})$ is done by averaging the full conditional density with draws $\{\mathbf{y}^{*(g)}, \Omega^{(g)}\} \sim \pi(\mathbf{y}^*, \Omega|\mathbf{y})$ from the main MCMC run for $g = 1, \ldots, G$,

$$\pi(\boldsymbol{\lambda}^*|\mathbf{y}) \approx G^{-1} \sum_{g=1}^{G} \pi(\boldsymbol{\lambda}^*|\mathbf{y}, \mathbf{y}^{*(g)}, \Omega^{(g)}).$$

The next ordinate, $\pi(\Omega^*|\mathbf{y}, \boldsymbol{\lambda}^*)$, can be estimated using a reduced run to obtain

$$\pi(\Omega^*|\mathbf{y}, \boldsymbol{\lambda}^*) \approx G^{-1} \sum_{g=1}^{G} \pi(\Omega^*|\mathbf{y}, \boldsymbol{\lambda}^*, \mathbf{y}^{*(g)}).$$

The last ordinate, $\pi(\boldsymbol{\delta}^*|\mathbf{y}, \Omega^*, \boldsymbol{\lambda}^*)$, which is unique to the ordered probit setting, is estimated using the methods in Chib and Jeliazkov (2001).

The benefits of Bayesian model comparison go beyond dealing with issues of variable selection and model uncertainty. In this context, additional benefits include understanding the type of endogeneity and the dependence structure between a set of outcome variables. This allows a researcher to distinguish between several competing specifications and further investigate the relationships of interest. Emphasis is placed on Bayesian techniques because considering both specifications is a straightforward extension of the methodology. Gibbs sampling methods employed in discrete data models already generate the latent $\mathbf{y}^*$s in the data augmentation part of a sampler. Employing these draws as data involves minimal additional coding as discussed in the next section.

3. BANK CONTAGION

The first application addresses financial contagion in two ways. First, this application examines both latent and observed measures of a bank failure and how these determine nearby bank performance. The impact of a regional bank failure on bank health remains unclear in the existing literature. Calomiris and Mason (2003b) find that a nearby failure decreases the probability of survival for the remaining banks. However, there is also research on efficient bank runs (Freixas & Rochet, 2008), which finds positive effects stemming from nearby bank failures due to market competition. If an inefficient bank fails, its customers can go to the remaining banks in the market for deposits and lending, thereby benefiting the existing depository institutions. Second, the application evaluates different linkages for the spread of contagion. Channels for contagion have been found in both regional and correspondent[2] networks (Aharony & Swary, 1996; Kaufman, 1994; Richardson, 2007; Richardson & Troost, 2009). While both linkages may be present, financial regulators need to determine which channel is stronger in order to implement policies for restricting the spread of contagion and preventing bank runs. This paper addresses these issues and examines the consequences of bank failures during the 1930s by looking at town-wide failure rates and changes to correspondent networks.

The models for banks $i = 1,\ldots,n$, are those in Eqs. (1) and (2) where $\varepsilon_i \equiv (\varepsilon_{i1}, \varepsilon_{i2})' \sim N_2(0,\Omega)$ and $\Omega = \begin{pmatrix} 1 & \omega_{12} \\ \omega_{21} & \omega_{22} \end{pmatrix}$. The models are characterized with two dependent variables in which $\mathbf{y}_i^* \equiv (y_{i1}^*, y_{i2}^*)'$ are the continuous latent data and $\mathbf{y}_i \equiv (y_{i1}, y_{i2})'$ are the corresponding discrete observed data. For the first outcome, the latent variables relate to the observed binary outcomes by $y_{i1} = 1\{y_{i1}^* > 0\}$ and

$$y_{i1} = \begin{cases} 0 & \text{No bank failure occurred nearby between 1929 and 1932} \\ 1 & \text{Bank failure occurred nearby between 1929 and 1932.} \end{cases}$$

The first outcome y_{i1} indicates whether or not a bank failure occurred between 1929 and 1932 in the town the subject bank does business. Ω incorporates the usual unit variance restriction in probit models, which is a normalization for identification. The second outcome y_{i2} measures a bank's performance in 1933, where there is point mass at 0 for banks that were suspended since 1932 and a continuous distribution with "loans and discounts" (hereafter referred to as LD) representing bank health. LD is chosen to measure a bank's performance following the literature on the credit crunch and its relation to economic activity (Bernanke, 1983; Calomiris & Mason, 2003a). The latent variables $\{y^*_{i2}\}$ relate to the observed censored outcomes by $y_{i2} = y^*_{i2} \cdot 1\{y^*_{i2} > 0\}$.

The endogenous covariate, y_{i1} in (1) or y^*_{i1} in (2), displays the impact of a nearby or regional bank failure on a bank's health and lending. Both the latent and observed specifications are easily motivated by hypotheses discussed in the existing literature. The latent counterpart of a regional bank failure can reveal unobserved factors that affect bank distress and profitability, such as corporate governance, risk behavior, and loss of confidence. Although an econometrician can observe whether a bank fails, additional information bankers have on the performance of their bank and business environment remains unobservable. In addition, existing evidence suggests that bank runs, which were the main mechanisms that caused bank failure (Bernanke, 1983), were often facilitated through "word-of-mouth" or information-based contagion (Park, 1991). For instance, if an individual's neighbor speculated about a pending bank failure, the individual is likely to withdraw deposits from his bank to avoid undue losses. Depositors lack financial information, resulting in withdrawal decisions based on the condition of the banking system as a whole (Park, 1991). Although researchers cannot measure the speculative nature of banking panics, the latent specification can act as a proxy for these factors. These arguments formally motivate the latent endogeneity model.

On the other hand, the literature notes that bank failures trigger panic (Chen, 1999). Despite speculation about bank health, depositors do not react until an indicator for failure is triggered. The literature also notes that publicizing the names of failed banks worsened remaining bank health. Additionally, the publication of the names of banks receiving financial assistance mitigated lender of last resort relief efforts (Butkiewicz, 1995). These observed outcomes, or triggers, support the observed endogeneity model. As mentioned previously, although a researcher may have an *a priori* expectation of the correct specification, it is hard to completely rule out the opposing approach. Therefore, a formal model comparison is necessary to better understand how a nearby failure affects bank performance and to ensure the employed specification accurately captures the interactions and decisions of banks during financial crises.

The data collected for this application are from the *Rand McNally Bankers' Directory*. This directory details balance sheets, correspondent relationships, and characteristics for all banks in a given state. Additional data are gathered from the 1930 U.S. census of agriculture, manufacturing and population, which describe the characteristics of the county and banks' business environment. The sample includes all banks operating in 1932 in Alabama, Arkansas, Michigan, Mississippi, and Tennessee for a total of 1,794 banks. These five states are considered because they provide variation across bank characteristics, size, Federal Reserve districts, and county characteristics. Table 1 presents descriptive statistics on the banks and average county characteristics for each state. For further information on the data set, see Vossmeyer (2014).

A key covariate of interest listed in Table 1 is Δ *Correspondents* – an indicator variable that takes the value 1 if a correspondent was removed from a bank's network and 0 otherwise. Recall that there are two linkages for contagion, regional which the nearby bank failure variable (y_{i1}) captures, and correspondent networks which this variable captures. Correspondent banks are usually designated in reserve cities of the Federal Reserve system and often provided smaller, local banks with liquidity (Richardson & Troost, 2009). Correspondent relationships between bigger and smaller banks built a structure for the Federal Reserve to influence nonmember institutions. However, this structure created pathways for contagion to spread. Therefore, controlling for it in the model is important and of interest to policy-makers in order to mitigate contagion through its many channels.

3.1. Estimation

The model is completed by specifying the prior distributions. For $\boldsymbol{\lambda} = (\boldsymbol{\beta}', \boldsymbol{\gamma}')'$, $\pi(\boldsymbol{\lambda}) = \mathcal{N}(\boldsymbol{\lambda}|\mathbf{d}_0, \mathbf{D}_0)$, and $\pi(\Omega) \propto \mathcal{IW}(\nu_0, \mathbf{R}_0)1\{\omega_{11} = 1\}$, where the prior on Ω (inverse Wishart) is on the derived quantities that appear in Algorithm 1. The hyperparameters for the priors are selected using a training sample of 100 banks. A thorough sensitivity analysis is provided in Section 3.2. Algorithm 1 presents the Gibbs sampling and data augmentation methods to simulate the posterior distribution for the observed endogeneity specification.

Table 1. Financial Characteristics of the Banks in Each State in 1932 and County Characteristics.

Variable	Alabama	Arkansas	Michigan	Mississippi	Tennessee
Number of banks	250	278	638	235	393
Average age	24	22	30	25	25
Federal Reserve district	6	8	7, 9	6, 8	6, 8
Financial characteristics (avg. $1,000)					
Deposits	716	437	1750	528	765
Deposits/Liab. (*ratio*)	0.612	0.750	0.751	0.730	0.675
Loans & discounts (LD)	582	273	1203	356	649
Charters and memberships (counts)					
State bank	166	222	438	208	308
National bank	82	44	102	26	83
Amer Bk Ass'n (ABA)	160	188	370	171	185
Correspondents (averages)					
Total correspondents	2.6	2.4	2.8	2.9	2.4
Out of state corres.	1.5	1.4	1.5	2.5	1
Δ Correspondents	0.19	0.31	0.22	0.28	0.40
Market shares (averages)					
Liab./County Liab.	0.27	0.26	0.13	0.33	0.24
Liab./Town Liab.	0.68	0.75	0.17	0.74	0.69
Herfindahl index (HHI)	0.66	0.60	0.29	0.70	0.54
County characteristics (averages)					
No. of wholesale retailers	31.3	22.4	45.3	15.4	25.3
Cropland (×1,000 acres)	115.6	96.6	122.9	81.9	78.1
Town pop. (×1,000)	14.6	4.7	49.6	4.5	14.2

Algorithm 1. *MCMC Estimation Algorithm – Observed Specification.*

1. *Sample* $[\lambda|\mathbf{y}^*, \Omega] \sim N(\hat{\boldsymbol{d}}, \hat{\boldsymbol{D}})$, *where* $\hat{\boldsymbol{d}}$ *and* $\hat{\boldsymbol{D}}$ *are given by* $\hat{\boldsymbol{d}} = \hat{\boldsymbol{D}}\left(\boldsymbol{D}_0^{-1}\boldsymbol{d}_0 + \sum_{i=1}^{n} \mathbf{W}_i'\Omega^{-1}\mathbf{y}_i^*\right)$ *and* $\hat{\boldsymbol{D}} = \left(\boldsymbol{D}_0^{-1} + \sum_{i=1}^{n} \mathbf{W}_i'\Omega^{-1}\mathbf{W}_i\right)^{-1}$ *where*

$$\mathbf{W}_i = \begin{pmatrix} \mathbf{x}'_{i1} & 0 & 0 \\ 0 & \mathbf{x}'_{i2} & y_{i1} \end{pmatrix}.$$

2. *Sample* Ω *in a one-block, two-step procedure by drawing* $\omega_{22\cdot 1} \equiv \omega_{22} - \omega_{21}\omega_{11}^{-1}\omega_{12}$ *and* ω_{21}, *then reconstructing* Ω *from these quantities,*
   (a) $\omega_{22\cdot 1} \sim \mathcal{IW}(\nu_0 + n, Q_{22})$
   (b) $\omega_{21} \sim \mathcal{N}(Q_{11}^{-1}Q_{12}, \omega_{22\cdot 1}Q_{11}^{-1})$, *where*

$$\mathbf{Q} = \mathbf{R}_0 + \sum_{i=1}^{n} (\mathbf{y}_i^* - \mathbf{W}_i\boldsymbol{\lambda})(\mathbf{y}_i^* - \mathbf{W}_i\boldsymbol{\lambda})'.$$

and $\mathbf{Q}$ is partitioned conformably with Ω, i.e.,

$$\mathbf{Q} = \begin{pmatrix} Q_{11} & Q_{12} \\ Q_{21} & Q_{22} \end{pmatrix}.$$

3. *For i=1,..., n, sample $y_{i1}^* | y_{i2}^*, y_{i1}, \boldsymbol{\lambda}, \Omega \sim \mathcal{TN}_{\mathcal{A}_i}(\mu_{1|2}, V_{1|2})$, where $\mu_{1|2}$ and $V_{1|2}$ are the usual conditional mean and conditional variance, respectively. If $y_{i1} = 0$, $\mathcal{A}_i$* is $(-\infty, 0)$, *and if $y_{i1} = 1$, $\mathcal{A}_i$ is $(0,\infty)$.*
4. *For $i{:}y_{i2} = 0$, sample $y_{i2}^* | y_{i1}^*, y_{i2}, \boldsymbol{\lambda}, \Omega \sim \mathcal{TN}_{\mathcal{A}_i}(\mu_{2|1}, V_{2|1})$, where the region $\mathcal{A}_i$ is $(-\infty, 0)$ implied by the censoring of y_{i2}* in the truncated normal distribution.

This sampler can be easily adapted to handle the latent specification. The most convenient approach is to move to the reduced-form where the system in (2) can be rewritten as,

$$\begin{aligned} \begin{pmatrix} 1 & 0 \\ -\gamma_2 & 1 \end{pmatrix} \begin{pmatrix} y_{i1}^* \\ y_{i2}^* \end{pmatrix} &= \begin{pmatrix} \mathbf{x}'_{i1} & 0 \\ 0 & \mathbf{x}'_{i2} \end{pmatrix} \begin{pmatrix} \boldsymbol{\beta}_1 \\ \boldsymbol{\beta}_2 \end{pmatrix} + \begin{pmatrix} \varepsilon_{i1} \\ \varepsilon_{i2} \end{pmatrix} \\ \mathbf{A}\mathbf{y}_i^* &= \mathbf{X}_i\boldsymbol{\beta} + \varepsilon_i \\ \Leftrightarrow \mathbf{y}_i^* &= \mathbf{A}^{-1} X_i\boldsymbol{\beta} + \mathbf{A}_i^{-1}\varepsilon_i \\ \Leftrightarrow \mathbf{y}_i^* &= \mu_i + \nu_i \\ \nu_i &\sim \mathcal{N}(0, \mathbf{A}^{-1}\Omega\mathbf{A}^{-1'}). \end{aligned} \tag{6}$$

Simply change the data augmentation steps of the sampler (steps 3–4) to use the conditional mean and conditional variance from the reduced form, avoiding any additional computational burden brought on by considering both specifications. This is a straightforward adjustment because the $\mathbf{y}^*$s are already being generated. In the latent specification, they are generated and passed through the sampler as data, so $\mathbf{W}_i = \begin{pmatrix} \mathbf{x}'_{i1} & 0 & 0 \\ 0 & \mathbf{x}'_{i2} & y_{i1}^* \end{pmatrix}$ in steps 1 and 2 of Algorithm 1.

3. 2. *Results*

The results for the application are based on 11,000 MCMC draws with a burn-in of 1,000. The inefficiency factors for the parameters remain low with slightly higher values occurring for the parameters on the endogenous covariates and variables common to both equations. The point estimates for the exogenous covariates are similar across both specifications except in cases where the variable has a 95% credibility interval that includes 0. The following discussion covers the basic results for each equation, followed by the model comparison and sensitivity analysis.

Table 2 presents the posterior means and standard deviations for both specifications. The results indicate that a nearby bank failure (y_{i1}) positively affects lending for existing banks, which disagrees with some of the literature. However, unlike other studies, this paper is looking at a longer window of impact. Previous papers examine the immediate impact of a nearby failure, which is distress. Whereas, this paper finds that the long-run impact is positive, corroborating the research on efficient bank runs (Freixas & Rochet, 2008). This result supports the market competition hypothesis where failing banks leave additional depositors in the market as customers for the remaining banks. As a result, the failure of a nearby bank

Table 2. Banking Application – Posterior Means and Standard Deviations for the Bivariate System of Equations.

	Observed Endogeneity		Latent Endogeneity	
	Nearby Fail	Lending	Nearby Fail	Lending
Intercept	−5.023 (0.250)	−0.480 (0.137)	−6.514 (0.349)	−0.122 (0.138)
Financial characteristics				
Bank age		0.009 (0.002)		0.013 (0.002)
Lagged LD		0.134 (0.016)		0.107 (0.017)
National bank		0.367 (0.093)		0.488 (0.102)
Deposits/Liabilities		0.115 (0.094)		0.143 (0.111)
Δ Correspondents		−0.176 (0.081)		−0.138 (0.087)
Total correspondents	0.041 (0.029)	−0.014 (0.029)	−0.021 (0.039)	−0.032 (0.031)
County characteristics				
Town population	1.262 (1.008)	0.193 (0.327)	−0.658 (1.046)	−2.627 (0.540)
Wholesale retailers	0.001 (0.001)		0.001 (0.001)	
Acres of cropland	0.218 (0.481)		−0.025 (0.570)	
Liab./Town Liab.	2.273 (0.180)	−0.025 (0.119)	3.106 (0.238)	−0.152 (0.119)
Herfindahl index (HHI)	0.441 (0.111)		0.548 (0.131)	
No. of banks in town	1.000 (0.051)		1.426 (0.074)	
Fed. Dist 7	0.123 (0.110)	−0.619 (0.097)	0.181 (0.132)	−0.715 (0.097)
Fed. Dist 8	0.265 (0.101)	−0.027 (0.093)	0.280 (0.113)	−0.012 (0.092)
Fed. Dist 9	0.069 (0.207)	−0.239 (0.206)	0.021 (0.238)	−0.394 (0.213)
y_1		1.314 (0.090)		
y_1^*				0.122 (0.017)

strengthens the balance sheets for surviving depository institutions. The results also illustrate that the negative impacts of regional contagion diminish over time. Although the initial panic is not captured here, the contagion channel disappears and presents market benefits. The increased lending by the remaining banks in the town increases economic activity and restores confidence in the financial system.

The second channel for the spread of contagion is represented by the variable Δ *Correspondents*. The result for this covariate demonstrates that a reduction in a bank's correspondent network has a negative impact on bank lending. This result aligns with intuition because correspondent banks often provided short-term commercial paper to smaller banks and urged them to extend credit (Richardson & Troost, 2009). When a bank was removed or a failure occurred in these networks, it constrained the liquid assets available and, as a result, banks extended less credit. The marginal effect of a correspondent removal, averaged over both observations and MCMC draws, is approximately −0.037 and −0.051 for the latent and observed specifications, respectively. In other words, a reduction in a bank's correspondent network decreases lending by about $370–$510 for every $10,000. This contagion channel presents moderate long-term negative effects. Policy-makers providing *ex post* liquidity following a downturn may want to focus relief efforts on banks experiencing failures across their correspondent networks as the negative effects may still be lingering.

Other results for the first equation display that the higher share of liability held at an individual bank (*Liab./Town Liab.*) and the higher the market concentration in a town (*HHI*), the more likely that town was to experience a bank failure between 1929–1932. In addition, relative to the sixth Federal Reserve district, the eighth district is more likely to experience bank failures, which accords well with Richardson and Troost's (2009) paper. The results for the second equation show that older banks and national banks have higher lending in 1933 relative to younger and nonmember institutions.

Model comparison results are presented in Table 3 and reveal that the data strongly support the latent endogeneity specification, where the full magnitude of y_{i1}^*, even values whose extent is driven by observed covariates and unobserved factors outside {0,1}, is relevant for bank lending. The marginal likelihood is nearly 50 points higher on the log scale, giving the observed specification a posterior model probability of approximately 0. The latent measure of a nearby bank failure captures unobserved aspects of bank profitability and the loss of confidence cultivating through these struggling local economies, which better explain bank lending. This result highlights an interesting point from Calomiris and Mason's (2003b) paper where they state, "Indicator variables are uninformative about the particular mechanism through which illiquidity and contagion produces a bank failure." Calomiris and Mason (2003b) further urge researchers to interpret indicators with caution when examining contagion as there may be evidence of missing fundamentals and loss of financial confidence. The issues these authors refer to can be mitigated by employing a latent measure for a nearby bank failure. The authors' analysis explains the strong support from the data for the latent specification and further corroborates the underlying hypothesis.

Without considering both modeling approaches, a researcher using observed endogeneity could misinterpret the relationship between regional failures and bank lending, and fundamentally misspecify financial panics. The nonlinear dichotomizing mechanism in which regional failures determine bank performance is inadequate. The latent variable approach encompasses the speculative nature of bank runs. One of the most documented and well understood features of banking crises is asymmetric information, which dates back to the original research by Bagehot (1873). Depositors, bankers, and central bankers lack credible information (Diamond & Dybvig, 1983; Gorton, 1985), resulting in speculative and fundamental bank runs. When an econometrician is examining the interactions involved in banking panics, asymmetric information should be apparent and the models being considered should reflect this feature.

Table 3 also displays the results for the variance–covariance matrix Ω. There is a negative correlation between the equations for a nearby bank failure and bank performance. After controlling for a number of balance sheet and county characteristics, the errors present a negative relationship, which implies there are harmful effects of a nearby bank failure not controlled for in the model. This result holds for both model specifications but there is a clear magnitude difference between the two. The observed

Table 3. Banking Application – Results for the Variance–Covariance Matrix and Marginal Likelihood Estimates.

	Observed Endogeneity	Latent Endogeneity
ω_{12}	−1.078 (0.053)	−0.305 (0.130)
ω_{22}	2.125 (0.087)	2.044 (0.088)
Ω_{corr}	$\begin{pmatrix} 1 & -0.740 \\ -0.740 & 1 \end{pmatrix}$	$\begin{pmatrix} 1 & -0.214 \\ -0.214 & 1 \end{pmatrix}$
Log-marginal lik. (numerical S.E.)	−1713.1 (0.423)	−1666.3 (0.245)

Table 4. Banking Application – Sensitivity Analysis for Different Training Sample Sizes.

Training Sample Size	Observed Log-Marg. Lik.	Latent Log-Marg. Lik.
No sample (priors centered at zero)	−1837.2	−1778.9
50	−1769.5	−1736.1
100	−1713.1	−1666.3
150	−1666.1	−1621.1
200	−1627.5	−1556.8

specification may overstate the relationship and the "true incidence of panic, since relevant fundamentals are likely omitted" from the model (Calomiris & Mason, 2003b).

The results discussed thus far employ a training sample prior of 100 banks. The sensitivity of the results to the hyperparameters is displayed in Table 4. The model rankings do not change across different training sample sizes. This indicates that the data speak loudly for the results and support the latent endogeneity model. Researchers interested in modeling and understanding decisions and relationships of banks in adverse macroeconomic conditions can employ latent variables to accommodate asymmetric information and to better capture the interactions between the outcomes of interest.

4. EDUCATION

The second application considers the impact of education on adult socioeconomic status. The return on schooling has been an ongoing area of research for empirical economists. Despite the great deal of attention focused on accommodating issues with survey responses, latent endogenous variables are lacking in this literature.

Education's impact on adult socioeconomic status is of particular interest because there are convincing arguments for both observed and latent endogeneity. Observed endogeneity is motivated by societal evaluations of education, which are generally looked at by crossing particular achievement thresholds or cut-points, e.g., high school degree and college degree. Without a college degree, individuals cannot apply for many jobs notwithstanding 15 years of schooling. Therefore, labor market outcomes are determined by these observed degree or threshold crossing indicators. Alternatively, a common issue in the analysis of returns to schooling is the inadequacy of measures for motivation and ability, which also explain socioeconomic status. Observed measures for ability have been proposed, such as standardized test scores. However, these data are often not available. Therefore, the latent representation of education can act as a proxy for unobservable characteristics linking education and socioeconomic outcomes. While the hypotheses for each approach are convincing, there have been no attempts to compare these models, which is necessary to ensure the most accurate specification is employed. Furthermore, knowing which approach is best supported by the data advances our understanding of how education determines socioeconomic status.

Studying education provides a unique opportunity for model comparison. Most education and socioeconomic outcomes are discretized by ordered categories. For instance, education categories can be defined by degree level and socioeconomic status can be categorized by income brackets. This ordinal data setting makes the observed specification in (1) invalid, and instead, the system in (3) can be employed. The specification for observed endogeneity for five education categories (defined shortly) and individuals $i = 1,\ldots,n$ is:

$$\begin{aligned} y_{i1}^* &= \mathbf{x}'_{i1}\boldsymbol{\beta}_1 + \varepsilon_{i1} \\ y_{i2}^* &= \mathbf{x}'_{i2}\boldsymbol{\beta}_2 + 1\{y_{i1} = 2\}\gamma_{22} + 1\{y_{i1} = 3\}\gamma_{23} + 1\{y_{i1} = 4\}\gamma_{24} + 1\{y_{i1} = 5\}\gamma_{25} + \varepsilon_{i2}, \end{aligned} \tag{7}$$

where $\varepsilon_i \equiv (\varepsilon_{i1}, \varepsilon_{i2})' \sim N_2(0,\Omega)$ and $\Omega = \begin{pmatrix} \omega_{11} & \omega_{12} \\ \omega_{21} & \omega_{22} \end{pmatrix}$. This specification differs from (1) because the endogenous covariate enters as a set of dummy variables for each category, whereas previously it entered as a single endogenous regressor.[3] The latent specification remains identical to the system in (2), which is now a more parsimonious model relative to the observed approach. It is important to note that the elements in Ω are left free. Location and scaling restrictions are accommodated by fixing two cutpoints. The different approaches for identification in multivariate ordered probit models are discussed in Jeliazkov et al. (2008).

The model is characterized by two dependent variables, where $\mathbf{y}_i^* \equiv (y_{i1}^*, y_{i2}^*)'$ are the continuous latent data and $\mathbf{y}_i \equiv (y_{i1}, y_{i2})'$ are the corresponding discrete observed data. For equations $k = 1, 2$, the latent variables relate to the observed ordered outcomes by $y_{ik} = \sum_{j=1}^{J} 1\{y_{ik}^* > \alpha_{k,j-1}\}$ for J ordered alternatives where α_{kj} is a cut-point between the categories given by

$$y_{i1} = \begin{cases} 1 & \text{Less than high school} \\ 2 & \text{High school degree} \\ 3 & \text{Some college} \\ 4 & \text{College degree} \\ 5 & \text{Graduate education} \end{cases}$$

$$y_{i2} = \begin{cases} 1 & \text{Poverty line and below} \\ 2 & \text{Lower-middle class} \\ 3 & \text{Middle class and up.} \end{cases}$$

The outcome y_{i1} represents the amount of education an individual completes and y_{i2} represents an individual's socioeconomic status. The second outcome is measured by an income-to-needs ratio. Income is measured using the actual amount of total income, which is the sum of taxable income and transfer income. Needs is measured as the poverty threshold taken from the Census Bureau. These thresholds are based on family size and age of the household. An income-to-needs ratio below 1.3 indicates the poverty line and below, between 1.3 and 3 indicates lower-middle class, and above 3 represents the middle class and up.[4] The endogenous covariate y_{i1} represents the impact of education on adult socioeconomic status.

The data collected for this application are from the Panel Study of Income Dynamics (PSID). The sample includes 2,779 respondents from the 1999 survey. The data set contains information on childhood health, parental socioeconomic status, parental education, adult socioeconomic status, adult health, and educational attainment for individuals between the ages of 30 and 50. The year 1999 is selected because it features retrospective reports on childhood health. Table 5 offers descriptive statistics on the data and details the discretization for a number of variables.

4.1. Estimation

The model is completed by specifying the prior distributions,

$$\boldsymbol{\lambda} \sim \mathcal{N}(\boldsymbol{d}_0, \mathbf{D}_0)$$

Table 5. Descriptive Statistics for the Sample of 2,779 Respondents from the PSID.

Variable	Sample Proportion	Variable	Sample Proportion
Respondent education (Educ)		**Mother's education** (Meduc)	
<High school degree	0.17	<High school degree	0.33
High school degree	0.32	High school degree	0.47
Some college	0.25	Some college	0.09
College degree	0.17	College degree	0.08
Graduate school	0.09	Graduate school	0.03
Father's education (Feduc)		**Childhood health**	
<High school degree	0.39	Poor	0.16
High school degree	0.38	Average	0.38
Some college	0.07	Excellent	0.46
College degree	0.10	**Adult health**	
Graduate school	0.06	Poor	0.10
Adult socioeconomic status		Average	0.62
Low	0.13	Excellent	0.28
Medium	0.27	**Marital status**	
High	0.60	Single	0.16
Parental socioeconomic status (pSES)		Divorced	0.24
Low	0.25	Married	0.60
Medium	0.45	**Race**	
High	0.30	White/Asian	0.63
Debt		Non-white	0.37
Debt	0.55	**Employment**	
No debt	0.45	Employed	0.88
Sex		Unemployed	0.12
Male	0.76	**Age (average)**	40
Female	0.24		

$$\boldsymbol{\Omega} \sim \mathcal{IW}(\nu_0, \mathbf{R}_0).$$

The hyperparameters are selected using a training sample of 200 individuals. Algorithm 2 presents the sampling methods to simulate the posterior distribution for the observed endogeneity specification, which follow from Jeliazkov et al. (2008). Note, the cut-point parameters α_{kj} are transformed to ensure the ordering constraints, so

$$\delta_{kj} = \ln(\alpha_{kj} - \alpha_{k,j-1}),$$

where $2 \leq j \leq J - 1$ for equations $k = 1, 2$.

Algorithm 2 *MCMC Estimation Algorithm – Observed Specification*

1. *For each equation k, sample* $\boldsymbol{\delta}_k, \mathbf{y}_k^* | \mathbf{y}, \boldsymbol{\lambda}, \mathbf{y}_{\backslash k}^*$ *as follows*
   (a) *Sample* $\boldsymbol{\delta}_k | \mathbf{y}, \boldsymbol{\lambda}, \Omega, \mathbf{y}_{\backslash k}^*$ *using the Metropolis–Hastings algorithm*
   (b) *Sample* $y_{ik}^* | \mathbf{y}, \boldsymbol{\lambda}, \Omega, \mathbf{y}_{\backslash k}^* \sim \mathcal{TN}_{(\alpha_{k,j-1}, \alpha_{kj})}(\mu_{k|\backslash k}, V_{k|\backslash k})$ *for* $i = 1, \ldots, n$.
2. *Sample* $[\boldsymbol{\lambda} | \mathbf{y}^*, \Omega] \sim N(\hat{\boldsymbol{d}}, \hat{\boldsymbol{D}})$, *where* $\hat{\boldsymbol{d}}$ *and* $\hat{\boldsymbol{D}}$ *are given by* $\hat{\boldsymbol{d}} = \hat{\boldsymbol{D}}\left(\boldsymbol{D}_0^{-1}\boldsymbol{d}_0 + \sum_{i=1}^n \mathbf{W}_i' \Omega^{-1} \mathbf{y}_i^*\right)$ *and* $\hat{\boldsymbol{D}} = \left(\boldsymbol{D}_0^{-1} + \sum_{i=1}^n \mathbf{W}_i' \Omega^{-1} \mathbf{W}_i\right)^{-1}$.
3. *Sample* $\Omega \sim \mathcal{IW}(\nu_0 + n, \mathbf{R}_0 + (\mathbf{y}^* - \mathbf{W}\boldsymbol{\lambda})'(\mathbf{y}^* - \mathbf{W}\boldsymbol{\lambda}))$.

As a matter of notation, "$\backslash k$" is used to represent all elements in a set except the kth one. Estimation of the latent specification follows these steps closely, however, employs the reduced-form trick discussed in (6) for the data augmentation step in 1(*b*).

4.2. Results

The results are based on 11,000 MCMC draws with a burn-in of 1,000. Analysis of the sensitivity of the results to the training sample size is conducted as in the first application. The results again show no sensitivity to the training sample size and model rankings do not change for different hyperparameters. The inefficiency factors for the parameters remain low with the highest values (≈20) occurring for the parameters on the endogenous covariates in both specifications. The following discussion reviews the basic results for each equation, then the model comparison results.

Table 6 presents the posterior means and standard deviations for both specifications. The results from the first equation accord well with the existing literature on the determinants of educational attainment (Haveman & Wolfe, 1995). The results show that parental education and parental socioeconomic status play a positive role in educational attainment. Parents with more income are able to invest in their children's education with schooling supplies, tutors, and financial assistance in college. Furthermore, parents who themselves achieve a higher level of education often motivate their children to do the same. The results also indicate that whites complete more education relative to non-whites, and males complete more schooling relative to females.

The results of the second equation coincide well with intuition and what is often found in the literature on socioeconomic achievement. Parental socioeconomic status positively affects adult socioeconomic status and individuals who are married have a higher income-to-needs ratio relative to divorced individuals. An interesting result is the positive coefficient on the

Table 6. Education Application – Posterior Means and Standard Deviations for the Bivariate System of Equations.

	Observed Endogeneity		Latent Endogeneity	
	Education	SES	Education	SES
Intercept	−0.683 (0.593)	−1.706 (0.193)	−0.696 (0.597)	−1.353 (0.199)
Child health – Exc	−0.269 (0.586)		−0.265 (0.586)	
Child health – Avg	−0.429 (0.587)		−0.425 (0.586)	
Feduc – High school	0.188 (0.058)		0.189 (0.059)	
Feduc – Some college	0.616 (0.097)		0.613 (0.095)	
Feduc – College	0.840 (0.094)		0.835 (0.093)	
Feduc – Graduate	1.033 (0.114)		1.028 (0.115)	
Meduc – High school	0.197 (0.060)		0.204 (0.060)	
Meduc – Some college	0.499 (0.093)		0.498 (0.092)	
Meduc – College	0.534 (0.103)		0.531 (0.102)	
Meduc – Graduate	0.780 (0.147)		0.775 (0.147)	
pSES – High	0.339 (0.067)	0.073 (0.065)	0.351 (0.064)	0.035 (0.067)
pSES – Med	0.344 (0.053)	0.079 (0.058)	0.351 (0.060)	0.049 (0.059)
Single		−0.003 (0.064)		0.005 (0.064)
Married		0.154 (0.062)		0.168 (0.062)
Employed		0.634 (0.067)		0.637 (0.067)
Debt		0.202 (0.042)		0.196 (0.042)
Adult health – Exc		0.359 (0.080)		0.352 (0.080)
Adult health – Avg		0.244 (0.069)		0.245 (0.070)
Educ – High school		0.553 (0.086)		
Educ – Some college		0.813 (0.127)		
Educ – College		1.312 (0.172)		
Educ – Graduate		1.702 (0.241)		
Latent Educ (y_1^*)				0.424 (0.055)
White	0.344 (0.053)	0.416 (0.056)	0.347 (0.053)	0.389 (0.056)
Age	0.030 (0.004)	0.015 (0.004)	0.030 (0.004)	0.014 (0.004)
Male	0.101 (0.055)	0.359 (0.064)	0.101 (0.054)	0.358 (0.063)

debt variable. The debt variable is measured by summing all debt excluding debt from the purchase of a house. However, debt does not necessarily indicate financial distress, as wealthier individuals have more access to debt and may finance other purchases, including vehicles, businesses, and school loans. The results also indicate that health has a positive relationship with wealth. Healthy individuals are less likely to miss work due to illness or disability and are likely to be more productive.

The covariate of interest, education, positively impacts adult socioeconomic status in both specifications. Relative to no high school degree, the coefficient on each discrete category for additional schooling gets

Table 7. Education Application – Results for the Variance–Covariance Matrix and Marginal Likelihood Estimates.

	Observed Endogeneity	Latent Endogeneity
ω_{11}	1.258 (0.043)	1.263 (0.043)
ω_{12}	−0.168 (0.074)	−0.195 (0.072)
ω_{22}	0.739 (0.049)	0.744 (0.057)
Ω_{corr}	$\begin{pmatrix} 1 & -0.174 \\ -0.174 & 1 \end{pmatrix}$	$\begin{pmatrix} 1 & -0.200 \\ -0.200 & 1 \end{pmatrix}$
Log-marginal lik. (numerical S.E.)	−7864.0 (0.053)	−7862.9 (0.044)
$\Pr(\mathcal{M}_k \mid y)$	0.333	0.667

incrementally larger in the observed specification. The results for Ω, presented in Table 7, demonstrate that after controlling for family background, health, and other demographics, there is a negative correlation between education and adult socioeconomic status. Although the direct effect of education on wealth is positive, there is a negative relationship between the errors. This result has been noted in the literature (Becker & Chiswich, 1966; Griliches, 1977), and a simple explanation for this is luck. The basic explanatory variables do not control for instances of good fortune.

Table 7 also presents the model comparison results. The marginal likelihoods and posterior model probabilities are very close for both specifications with a slight preference toward the latent endogeneity model. The latent specification is more parsimonious, with three less covariates than the observed endogeneity model. Further investigation into this relationship is necessary due to the lack of model preference. An approach for understanding these interactions is to separate the sample by age cohort. The intuition for this comes from recognizing that individuals' degree level may only be influential in obtaining their first few jobs. As the amount of time an individual is in the labor force increases, the individual accumulates work history, skills, and references, which mitigate the importance of degree level. Consider job postings for entry-level positions, the salary level is often listed as "competitive", whereas for more senior positions, "depending on experience" is a common listing. Following this intuition, both models are compared for two separate age cohorts. The results are displayed in Table 8.

The results align with the age group hypotheses. The data support the observed specification for the 30–35 age cohort. Alternatively, the data support the latent specification for the 40–50 age cohort. This is a major

Table 8. Education Application – Model Comparison and Marginal Likelihood Estimates for Different Age Cohorts.

	Ages 30–35		Ages 40–50	
	Observed	Latent	Observed	Latent
Observations	843	843	1489	1489
Log-marginal lik.	−2450.6	−2455.3	−4196.1	−4192.0
$\Pr(\mathcal{M}_k \vert y)$	0.991	0.009	0.017	0.983

result because it displays how the dependence structure between educational attainment and adult socioeconomic status changes with age. For younger individuals, the primary way to signal intelligence and ability is through degree-level. As a result, society evaluates the amount of education completed by observed threshold crossing indicators, thus resulting in labor market outcomes for these individuals. On the other hand, for older individuals, the data support the underlying latent specification, capturing ability, work history, and other factors unobserved by the econometrician. The importance of education diminishes as other elements become more prominent and, eventually, better explain labor market outcomes and wealth. Additionally, the latent approach offers a unique opportunity for studies examining outcomes of older cohorts. Surveys often do not contain a comprehensive work history of the respondents, therefore, if these data are unavailable, the latent measure of education can offer some insight into these characteristics.

These results truly stress the importance of considering and comparing both specifications and contribute to the literature on returns to schooling by demonstrating the evolution in the dependence structure between education and socioeconomic status. Pathway models employed in labor economics attempt to capture life-cycle interactions and intergenerational transmissions of education, health, and wealth. The complexity of these models increases because these outcomes are dynamic and change over time. Future research can employ both observed and latent measures of these outcomes to better explain these relationships pertaining to age cohorts.

5. CONCLUDING REMARKS

This paper addresses an important but often overlooked issue, which is the proper specification of endogenous covariates. In multivariate discrete data models, endogeneity can be based on latent or observed data. Bayesian model comparison techniques can be employed to determine which approach is best supported by the data, thus increasing the understanding of the nature of endogeneity and the dependence structure between the relationships being modeled.

While most applied researchers have an *a priori* expectation of the correct model, it is extremely difficult to rule out hypotheses in support of the alternative approach, which is apparent from the two empirical applications considered in this paper. Therefore, model selection provides important insights that resolve competing hypotheses about the interactions of interest. The results from the first application show that the latent representation of a bank failure better explains regional financial contagion, relative to conditioning on its observed counterpart. The latent measure captures the speculative nature of banking panics and accommodates asymmetric information issues discussed in the literature, thus providing a more accurate model of banking crises. The results for the second application show that the observed representation of education better explains socioeconomic outcomes for younger cohorts. This dependence structure changes as individuals age and the latent measure of education becomes more meaningful. This paper employs a bivariate system of equations in the applications, however, these approaches are generalizable to a number of methodologies.

These results stress the importance of employing model selection techniques to distinguish between competing specifications. The issues discussed here are present in any multivariate discrete data setting and should be addressed in a number of applied literatures. Ignoring these techniques can cause a researcher to misinterpret the depicted relationships and misunderstand the nature of endogeneity.

NOTES

1. The cut-points α_{kj}, which are defined in the link function for ordinal data, are transformed such that $\delta_{kj} = \ln(\alpha_{kj} - \alpha_{k,j-1})$. A discussion of this transformation is offered in Section 4.1.

2. Correspondents were banks with ongoing relationships facilitated by deposits of funds (Richardson, 2007). These networks linked banks across the country and indicated the extent to which a bank was important within the national network of banking (Calomiris, Mason, Weidenmier, & Bobroff, 2013).

3. If y_{i1} entered directly, this would lead to a cardinal interpretation of the categories, which is incorrect.

4. The cut-point 1.3 is selected because it is the threshold for food stamps.

ACKNOWLEDGMENT

Special thanks to Ivan Jeliazkov for his invaluable guidance. I am also grateful to David Brownstone, Sean Dowsing, Ben Gillen, Dale Poirier, Arshad Rahman, Gary Richardson, Michael Sacks, and participants at the Advances in Econometrics Conference on Bayesian Model Comparison for their helpful comments.

REFERENCES

Aharony, J., & Swary, I. (1996). Additional evidence on the information-based contagion effects of bank failures. *Journal of Banking and Finance*, *20*, 57–69.

Bagehot, W. (1873). *Lombard street: A description of the money market*. London, NY: H.S. King.

Becker, G., & Chiswich, B. (1966). Education and the distribution of earnings. *American Economic Review*, *56*, 358–369.

Bernanke, B. S. (1983). Nonmonetary effects of the financial crisis in propagation of the great depression. *American Economic Review*, *73*, 257–276.

Butkiewicz, J. (1995). The impact of lender of last resort during the great depression: The case of the reconstruction finance corporation. *Explorations in Economic History*, *32*, 197–216.

Calomiris, C., & Mason, J. (2003a). Consequences of bank distress during the great depression. *American Economic Review*, *93*, 937–947.

Calomiris, C., & Mason, J. (2003b). Fundamentals, panics, and bank distress during the depression. *American Economic Review*, *93*, 1615–1645.

Calomiris, C. W., Mason, J. R., Weidenmier, M., & Bobroff, K. (2013). The effects of the reconstruction finance corporation assistance on Michigan's banks' survival in the 1930s. *Explorations in Economic History*, *50*, 525–547.

Chen, Y. (1999). The role of the first-come, first-served rule and information externalities. *Journal of Political Economy*, *107*, 946–968.

Chib, S. (1995). Marginal likelihood from the Gibbs output. *Journal of the American Statistical Association*, *90*, 1313–1321.

Chib, S., & Jeliazkov, I. (2001). Marginal likelihood from the Metropolis–Hastings output. *Journal of the American Statistical Association*, *96*, 270–281.

Diamond, D., & Dybvig, P. (1983). Bank runs, deposit insurance and liquidity. *Journal of Political Economy*, *91*, 401–419.

Freixas, X., & Rochet, J. C. (2008). *Microeconomics of banking*. Cambridge, MA: MIT Press.

Gorton, G. (1985). Bank suspension of convertibility. *Journal of Monetary Economics*, *15*, 177–193.

Griliches, Z. (1977). Estimating the returns to schooling: Some econometric problems. *Econometrica*, *45*, 1–22.

Haveman, R., & Wolfe, B. (1995). The determinants of children's attainments: A review of methods and findings. *Journal of Economic Literature*, *33*, 1829–1878.

Jeliazkov, I., Graves, J., & Kutzbach, M. (2008). Fitting and comparison of models for multivariate ordinal outcomes. In *Bayesian econometrics* (Vol. 23, pp. 115–156). Advances in Econometrics. Bingley, UK: Emerald Group Publishing Limited.

Jeliazkov, I., & Lee, E. H. (2010). MCMC perspectives on simulated likelihood estimation. In *Maximum simulated likelihood* (Vol. 26, pp. 3–39). Advances in Econometrics. Bingley, UK: Emerald Group Publishing Limited.

Jeliazkov, I., & Rahman, M. A. (2012). Binary and ordinal data analysis in economics: Modeling and estimation. In X.-S. Yang (Ed.), *Mathematical modelling with multidisciplinary applications* (pp. 123–150). Hoboken, NJ: Wiley.

Kaufman, G. G. (1994). Banking contagion: A review of the theory and evidence. *Journal of Financial Services Research*, *8*, 123–150.

Li, M., & Tobias, J. (2014). Bayesian analysis of treatment effect models, In I. Jeliazkov & X.-S. Yang (Eds.), *Bayesian inference in the social sciences* (pp. 63–90). Hoboken, NJ: Wiley.

Maddala, G. (1983). *Limited dependent and qualitative variables in econometrics*. Cambridge: Cambridge University Press.

Mintz, O., Currim, I. S., & Jeliazkov, I. (2013). Information processing pattern and propensity to buy: An investigation of online point-of-purchase behavior. *Marketing Science*, *32*, 716–732.

Muthen, B. O. (2002). Beyond SEM: General latent variable modeling. *Behaviormetrika*, *29*, 81–117.

Park, S. (1991). Bank failure contagion in historical perspective. *Journal of Monetary Economics*, *28*, 271–286.

Richardson, G. (2007). Categories and causes of bank distress during the Great Depression, 1929–1933: The illiquidity versus insolvency debate revisited. *Explorations in Economic History*, *44*, 588–607.

Richardson, G., & Troost, W. (2009). Monetary intervention mitigated banking panics during the great depression: Quasi-experimental evidence from a federal reserve district boarder 1929–1933. *Journal of Political Economy*, *117*, 1031–1073.

Tanner, M. A., & Wong, W. H. (1987). The calculation of posterior distributions by data augmentation. *Journal of the American Statistical Association*, *82*, 528–549.

van Hasselt, M. (2014). Bayesian analysis of sample selection models. In I. Jeliazkov & X.-S. Yang (Eds.), *Bayesian inference in the social sciences* (pp. 91–113). Hoboken, NJ: Wiley.

Vossmeyer, A. (2014). Treatment effects and informative missingness with an application to bank recapitalization programs. *American Economic Review*, *104*, 212–217.

VARIABLE SELECTION IN BAYESIAN MODELS: USING PARAMETER ESTIMATION AND NON PARAMETER ESTIMATION METHODS

Gail Blattenberger[a], Richard Fowles[a] and Peter D. Loeb[b]

[a]*Department of Economics, University of Utah, Salt Lake City, UT, USA*
[b]*Department of Economics, Rutgers University, Newark, NJ, USA*

ABSTRACT

This paper examines variable selection among various factors related to motor vehicle fatality rates using a rich set of panel data. Four Bayesian methods are used. These include Extreme Bounds Analysis (EBA), Stochastic Search Variable Selection (SSVS), Bayesian Model Averaging (BMA), and Bayesian Additive Regression Trees (BART). The first three of these employ parameter estimation, the last, BART, involves no parameter estimation. Nonetheless, it also has implications for variable selection. The variables examined in the models include

Bayesian Model Comparison
Advances in Econometrics, Volume 34, 249–278

ISSN: 0731-9053/doi:10.1108/S0731-905320140000034011

traditional motor vehicle and socioeconomic factors along with important policy-related variables. Policy recommendations are suggested with respect to cell phone use, modernization of the fleet, alcohol use, and diminishing suicidal behavior.

Keywords: Bayesian variable selection; Extreme Bounds Analysis; stochastic search model selection; Bayesian tree models; motor vehicle fatality rates; Bayesian Model Averaging

JEL classifications: C11; C14; L9

1. INTRODUCTION

Standard econometrics largely focuses on parameter uncertainty. A model is estimated conditional on the assumption that the model estimated and reported is the "true" model. Often, however, tests are then made on a multitude of alternative models; each of these sequentially assumed to be "true." Model uncertainty is manifest in this procedure, but it is generally ignored in practice. Bayesian statistics speaks directly to this problem and much work has been done in recent years on model choice and variable selection.[1] This paper illustrates several developments in this area, highlighting linear model selection relating to automobile fatalities. Within this context, the results of four methods that have been developed are compared. In the four procedures we examine a basic theme of variable selection. Given k potential explanatory variables considered in a particular model, there are 2^k possible models to estimate. The value of 2^k gets large quickly and how to choose among candidate models is an important research question. This is the enquiry we will address in our model selection exercise.

Section 2 describes the nature of automobile fatalities in the United States. Section 3 describes the set of data we utilize comprised of annual observations by US states from 1980 to 2010. Section 4 provides an overview of the econometric methods used and the results for the four procedures we examine. These are Extreme Bounds Analysis (EBA), Stochastic Search Variable Selection (SSVS), Bayesian Model Averaging (BMA), and Bayesian Additive Regression Trees (BART). The first three procedures are parametric in nature, the last it not. In Section 5 we compare the four procedures and highlight issues related to model choice in terms of variable importance.

2. BACKGROUND

Understanding the determinants of fatalities associated with motor vehicle crashes continues to be of major interest to public health officials, economists, statisticians, and policy makers.[2] Motor vehicle fatalities in the United States were at a high in 1972 with 54,589 deaths. Between that year and 2007, deaths due to crashes were generally above 40,000 per year. More recently the numbers have diminished. Between 2007 and 2010, deaths have fallen from 41,059 to 32,885.[3] Explaining this change is of interest, aside from a general time trend. For example, some of this decline may be due to economic events such as the Great Recession. Tastes have also changed. There seems to be a preference, especially among the youth, to move from the suburbs to the cities where there is a greater reliance on public transportation and walking. Furthermore, while the baby boomers had a strong desire to obtain powerful cars while in their teens, the current population of youths may be more inclined to desire powerful cell phones and electronic equipment. Although deaths have diminished, there remains a significant number of crash-related fatalities and these are the leading cause of death among young persons in the United States.

Many factors thought to contribute to motor vehicle accidents and crash fatalities have been examined over the last two decades.[4] These include motor vehicle speed, speed variance, alcohol use and alcohol-related legislation, speed limits, vehicle miles traveled, measures of income and wealth, unemployment rates, advances in technology, the age of the motor vehicle fleet, population characteristics, insurance effects, seat belts and seat belt legislation, and the deregulatory climate of the 1980s.

In general, these factors can be classified into three main groups which have guided our data collection. The categories are those associated with vehicles such as technology and design characteristics; those associated with roadways such as speed limits; and those associated with drivers themselves such as alcohol consumption and seat belt usage. Recently there has been keen interest in the effects of cell phone usage on motor vehicle accidents and fatalities along with the effects of other socioeconomic and technology factors such as the age of the fleet, education, and suicidal propensities.[5]

With even a relatively small set of potential explanatory variables that are available to include in a regression model, the number of potential models is large. The data we examine are described in the following section.

3. DATA

In order to better understand the effects of socioeconomic and policy-related variables on traffic fatality rates we utilize a rich set of data that were collected on 50 states and Washington, DC over the period from 1980 to 2010. Throughout our analysis, 1980–2005 is used as a training sample and 2006–2010 as a test sample for forecasting as explained in this section.

The choice of the measure of the dependent variable was of prime importance. Data are available on the number of fatalities and on four different fatality rates. Here we examine the most commonly reported dependent variable, fatalities per 100 million vehicle miles traveled. Our choice of explanatory variables is based on a rich literature highlighting the importance of policy, safety, demographic, and economic determinants of fatality rates. Issues related to the choice of these variables are well described in Blattenberger et al. (2012, 2013). During our coverage period there were significant changes in a host of variables. These include changes in states' speed limits, alcohol and seat belt laws, and major improvements in automobile safety such as new types of passive restraint and braking systems. Importantly, our data cover the time of the explosive growth in cell phone subscriptions from effectively zero to over 270 million. Annual cell phone subscription data at the state level were only available beginning in year 2000. For the earlier years we used national level data and imputed state level subscriptions to be proportional to state population proportions for the prior years.[6] Another major change observed in the data allowed states to modify the 55 mile per hour speed limit on Interstate highways. Our data records the highest posted urban Interstate speed limit that was in effect during the year for each state. Within the data, per se blood alcohol concentration (BAC) laws vary widely, even though by 2005 all states and the District of Columbia had mandated a .08 BAC illegal per se law.[7] Seat belt legislation varies widely across states. Our data records the years in which a state mandatory primary or secondary seat belt law came into effect. Regional effects are measured by state locations and climate data. The variables are defined and described in Table 1 along with their expected effects on fatality rates. In all four procedures, we have standardized these data to have mean zero and range one.

In previous versions on this work we used regional dummies to indicate regional differences in accident likelihood. This was unsatisfactory and here we have attempted to incorporate factors underlying these regional differences. LATITUDE, LONGITUDE, and the interaction term LATXLONG

Table 1. Explanatory Variables and Sources.

Name	Description	Source	Expected Sign
BACLAW	Dummy variable indicating the existence of a law defining intoxication of a driver in terms of blood alcohol concentration (BAC). BACLAW = 1 indicates the existence of such a law and BACLAW = 0 indicates the absence of such a law. (More precisely, BACLAW = 1 when the BAC indicating driving under the influence is 0.1 or lower.)	*Digest of State Alcohol-Highway Safety Related Legislation (various years), Traffic Laws Annotated 1979, Alcohol and Highway Safety Laws: A National Overview 1980, National Highway Traffic Safety Administration*	–
BEER	Per capita beer consumption (in gal) per year	*U.S. Census Bureau, National Institute on Alcohol Abuse and Alcoholism*	+
CELLPOP	Number of cell phone subscribers per capita	*Cellular Telecommunication and Internet Association Wireless Industry Survey, International Association for the Wireless Telecommunications Industry*	+
CRIME	Crime rate (reported crimes per 100,000 population)	*Statistical Abstract of the United States (various years), U.S. Census Bureau*	?
ED_COL	Percent of persons with a college degree	*Digest of Education Statistics (various years), National Center for Education Statistics, Educational Attainment in the United States (various years), U.S. Census Bureau*	–
ED_HS	Percent of persons with a high school diploma	*Digest of Education Statistics (various years), National Center for Education Statistics, Educational Attainment in the United States (various years), U.S. Census Bureau*	–
GINI	The Gini coefficient measuring income inequality	*US Census Bureau: Historical Income Tables and http://utip.gov.utexas.edu/data.html*	+

Table 1. *(Continued)*

Name	Description	Source	Expected Sign
INCOME	Real per household income in 2,000 dollars	*State Personal Income (various years), Bureau of Economic Analysis website http://www.bea.doc.gov/bea/regional/spi/dpcpi.htm*	?
INSPECT	Indicator for annual safety inspection	*Highway Statistics (various years), Federal Highway Administration*	–
LATITUDE	Average state latitude	*http://dev.maxmind.com/geoip/legacy/codes/state_latlon/*	?
LATXLONG	Latitude times longitude	*http://dev.maxmind.com/geoip/legacy/codes/state_latlon/*	?
LONGITUDE	Average state longitude	*http://dev.maxmind.com/geoip/legacy/codes/state_latlon/*	?
MCMY	Median car model year	*Highway Statistics (various years), Federal Highway Administration, Traffic Safety Facts (various years), National Highway Traffic Safety Administration*	–
MLDA	Minimum legal drinking age	*A Digest of State Alcohol-Highway Safety Related Legislation (various years), Traffic Laws Annotated 1979, Alcohol and Highway Safety Laws: A National Overview of 1980, National Highway Traffic Safety Administration, U.S. Census Bureau*	–
PASSIVE	Percentage of registered vehicles manufactured after 1989 (passive restraint law)	*Traffic Safety Facts (various years), National Highway and Traffic Safety Administration*	–
PCTSUNNY	Percentage of days that are sunny	*http://www.currentresults.com/Weather/US*	+
POPDEN	Population density	*Statistical Abstract of the United States (various years), U.S. Census Bureau*	–
POVERTY	Poverty rate (percentage)	*Statistical Abstract of the United States (various years), U.S. Census Bureau website http://www.census.gov/hhes/poverty/histpov19.html*	+

Table 1. (*Continued*)

Name	Description	Source	Expected Sign
RAIN	Average rainfall in inches	*http://www.currentresults.com/Weather/US*	+
SEATBELT	Indictor for presence of a legislated seat belt law	*U.S. Census Bureau, National Institute on Alcohol Abuse and Alcoholism*	–
SNOW	Average snowfall in inches	*http://www.currentresults.com/Weather/US*	+
SPEED	Maximum posted speed limit, urban Interstate highways	*Highway Statistics (various years), Federal Highway Administration*	+
SUICIDE	Suicide rate (suicides per 100,000 population)	*Statistical Abstract of the United States (various years), U.S. Census Bureau*	+
UNEMPLOY	Unemployment rate (percentage)	*Statistical Abstract of the United States (various years), U.S. Census Bureau*	–
URBANPCT	Percent of population living in urban areas	*Statistical Abstract of the United States (various years), U.S. Census Bureau*	–
YOUNG	Percentage of males (16–24) relative to population of age 16 and over	*State Population Estimates (various years), U.S. Census Bureau http://www.census.gov/population/www/estimates/statepop.html*	+

Cross Sectional – Time Series analysis of traffic fatality rates for 50 states and DC (full sample: 1980–2010; training period: 1980–2005).

precisely locate the observation. LATITUDE differentiates north/south locations. LONGITUDE differentiates east/west locations. LATXLONG further expands the location definition. North/south effects may differ in the east and in the west. We also include RAIN, PCTSUNNY, and SNOW to further differentiate weather conditions. URBANPCT, the percent of the population living in urban areas is another regional variable.

4. ECONOMETRIC METHODS

4.1. Extreme Bounds Analysis

Extreme Bounds Analysis (EBA) was developed by Edward Leamer in a series of articles beginning in 1978 (Leamer, 1978, 1982). It is a methodology of global sensitivity analysis that computes the maximum and minimum values for Bayesian posterior means in the context of linear regression models. Here we use a simple normal-gamma model

$$\begin{aligned} Y_{it} &= X_{it}\beta + \varepsilon_{it} \\ i &= 1, 2, \ldots, 51 \\ t &= 1, 2, \ldots 26 \end{aligned} \tag{1}$$

The prior distributions for this model are specified

$$\begin{aligned} \beta &\sim N(\beta_0, \Sigma_0) \\ \varepsilon &\sim N(0, \pi^{-1}) \\ \pi &\sim \Gamma(\nu, \lambda) \end{aligned} \tag{2}$$

Because our paper focuses attention on variable selection and not on parameter estimation, we do not consider that the ε_{it}'s might exhibit heteroscedasticity or serial correlation (or both). Our selection criteria under EBA is that relatively narrow bounds favor variables that are strongly considered. These extreme values are the upper and lower values that could be obtained when all possible linear combinations of the explanatory variables are considered under all possible model specifications. Clearly these bounds are wide given a proper prior location at zero for sets of variables called "doubtful." These variables are those that a researcher might consider dropping from a regression and thus would be comfortable with a prior mean at zero.

We can visualize EBA using a two-dimensional simplification. The basic model is illustrated in Fig. 1. The likelihood contours implied by the data are shown along with the maximum likelihood estimate, b. In addition, the prior contours implied by the prior location, β_0, and the prior precision, Σ_0^{-1}, are shown. The posterior mean for this sample and prior is a matrix weighted average of the sample and prior values

$$\beta = (\Sigma_0^{-1} + X'X)^{-1}(\Sigma_0^{-1}\beta_0 + X'Xb) \tag{3}$$

The potential set of posterior means for this prior and likelihood are indicated by the relation labeled the information contract curve. The exact

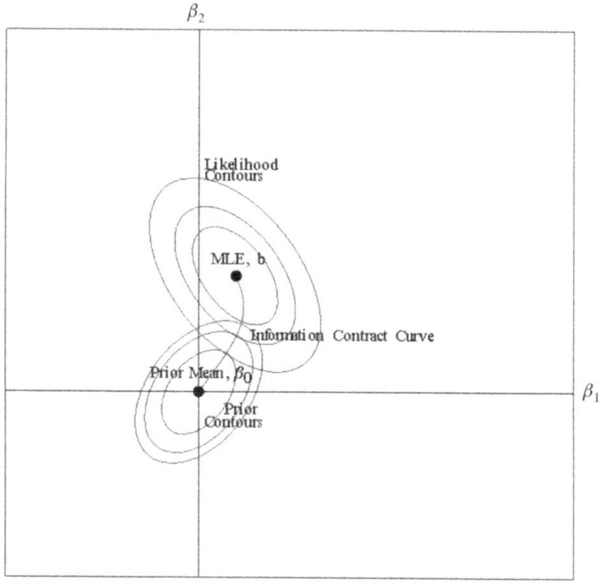

Fig. 1. Likelihood/Prior Contours and Information Contract Curve.

position of the posterior mean along this curve is a function of the relative sample and prior precisions. EBA is illustrated in the context of this model.[8]

The analysis depicted in Fig. 1 requires specifying both a prior mean, β_0, and a prior precision, Σ_0^{-1}. The posterior might encompass a wide range of values even if the prior location is restricted to the origin. In a remarkable result of Chamberlain and Leamer (1976), posterior means fall within a feasible ellipse under minimal assumptions regarding the nature of the precision matrix, Σ_0^{-1}. The feasible ellipse is illustrated in Fig. 2. Nonetheless, as indicated by this figure, the range of potential posterior values encompasses zero. Without further restrictions all variables are necessarily fragile with bounds that do cover zero.

In this paper, we limit our range of values to those that are highly likely, for example, those falling within the 95% confidence ellipsoid. Bounds within the 95% ellipsoid are referred to as being data favored. Fig. 3 illustrates the implication of this restriction for the extreme bounds on the parameter values. In this illustration the extreme bounds on β_2 are positive and

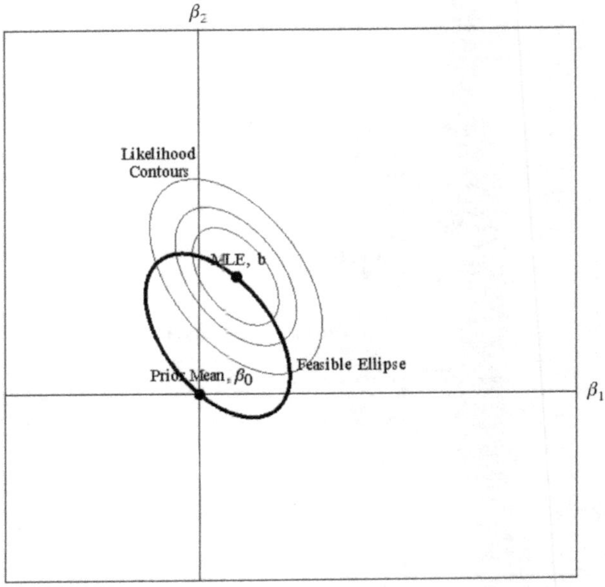

Fig. 2. Feasible Ellipse.

X_2 is not a fragile variable while bounds on β_1 cover zero and thus X_1 is fragile.

Data favored bounds for 11 variables are seen to be non-fragile and are highlighted in Table 2. These include BEER, CELLPOP, ED_COL, LATITUDE, LONGITUDE, LATXLONG, MCMY, POVERTY, SUICIDE, UNEMPLOY, and URBANPCT. All coefficient signs associated with these 11 variables are in agreement with the expected signs as presented in Table 1.

The maximum and minimum values for the regression coefficients in Table 2 may be associated with very implausible models.[9] EBA does not take this into consideration, but Stochastic Search Model Selection and Bayesian Model Averaging do. These procedures are illustrated in the following sections.

4.2. Stochastic Search Variable Selection

Stochastic Search Variable Selection (SSVS) was introduced by George and McCulloch (1993). It is computationally burdensome, and is one of the

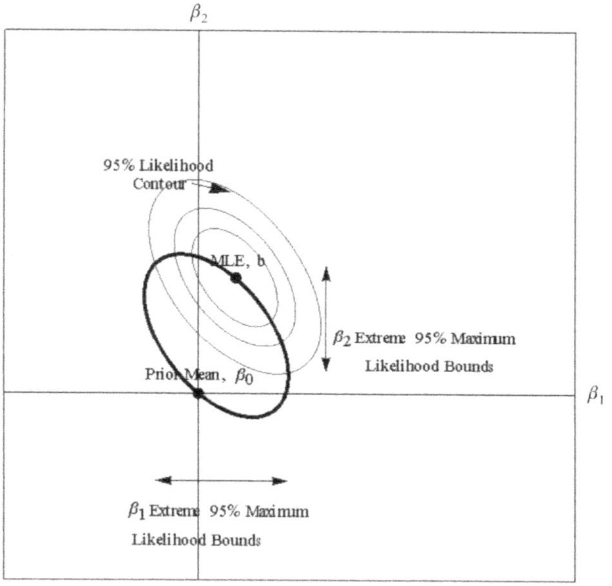

Fig. 3. Extreme Bounds Within 95% Likelihood Contour. All Variables Doubtful.

more recent procedures in Bayesian analysis that takes advantage of the ability to integrate over multidimensional spaces using Markov Chain Monte Carlo (MCMC) methods typically found when dealing with analyses of the posterior density. This is done with a Gibbs sampler in SSVS.

All k of the explanatory variables are included at each iteration of the Markov chain to take advantage of the application of the Gibbs sampler to a hierarchical Bayesian model. The variable selection choice is imposed by means of a latent variable, γ. Each model is represented by a binary vector $\gamma=(\gamma_1, \gamma_2,\ldots, \gamma_k)$. If the variable is to be effectively excluded from the model, then $\gamma_i=0$ and the prior distributions on the slope parameters (β's) are distributed normally with mean zero and small variance. If the variable is to be effectively included in the model, then $\gamma_i=1$ and the prior distributions on the slope parameters are distributed normally with mean zero and large variance.

All k variables are examined in random order at the end of each iteration of the Gibbs sampler to evaluate the marginal impact of effectively including/excluding that variable in the current model.[10] Based on this, a probability of including the variable is computed and the value of γ_i for the

Table 2. Maximum Likelihood and EBA Upper and Lower Bounds Within 95% Confidence Ellipsoids.

	MLE	EBA Upper	EBA Lower
BACLAW	−0.040	0.001	−0.081
BEER	0.127	0.218	0.034
CELLPOP	0.211	0.359	0.060
CRIME	0.062	0.217	−0.095
ED_COL	−0.194	−0.014	−0.371
ED_HS	−0.007	0.164	−0.177
GINI	0.143	0.367	−0.083
INCOME	0.024	0.324	−0.276
INSPECT	0.009	0.038	−0.020
LATITUDE	−0.648	−0.095	−1.193
LATXLONG	−0.902	−0.283	−1.509
LONGITUDE	0.354	0.652	0.052
MCMY	−0.628	−0.370	−0.876
MLDA	0.024	0.081	−0.032
PASSIVE	0.084	0.250	−0.082
PCTSUNNY	0.084	0.220	−0.053
POPDEN	−0.012	0.149	−0.172
POVERTY	0.120	0.235	0.003
RAIN	0.030	0.129	−0.071
SEATBELT	0.013	0.056	−0.029
SNOW	0.058	0.142	−0.027
SPEED	0.028	0.093	−0.037
SUICIDE	0.151	0.291	0.009
UNEMPLOY	−0.119	−0.003	−0.234
URBANPCT	−0.089	−0.015	−0.162
YOUNG	0.061	0.173	−0.052

Cross Sectional – Time Series analysis of traffic fatality rates for 50 states and DC from 1980 to 2005 (training period).
Non-fragile bounds shaded.

next iteration is computed stochastically based on this probability. Initial values of the γ_i are all set at 1 and the initial probabilities of inclusion are set at 0.5. The mean of these iterative values of the probability of inclusion for each variable is given in the table below. A stochastic scheme is implemented to search for the models with the highest posterior densities. In particular, the Gibbs sampler begins with initialized parameters $\gamma^{(0)}$, $\beta^{(0)}$, $\sigma^{2(0)}$ and generates the sequence $\gamma^{(1)}$, $\beta^{(1)}$, $\sigma^{2(1)}$, $\gamma^{(2)}$, $\beta^{(2)}$, $\sigma^{2(2)}$, ... This sequence converges to a posterior distribution which supplies the complete posterior distribution, $P(\beta, \sigma^2, \gamma \mid Y)$. Concurrent with the iterative values of the vector γ are iterative values of the vector $\boldsymbol{p}$, the probability of variable inclusion, enabling us to compute an expected value of the vector $\boldsymbol{\beta}$ for that iteration.

Table 3 summarizes the findings for SSVS for the linear fatality model based on 100,000 iterations.[11] These results utilize standardized data with shaded cells indicating non-fragile bounds as in Table 2. The first column (Mean Coefficient) gives the weighted average for the sequence of slope coefficients, weighted by the probability of inclusion. The second column (Probability Inclusion) is the mean iterative value of the probability of a variable's inclusion in the model. The third and fourth columns are the minimum and maximum values found over the SSVS iterations. Variables for which the min–max interval excludes zero (SSVS non-fragile) are highlighted. These fourteen variables are BEER, CELLPOP, GINI, LATITUDE, LONGITUDE, LATXLONG, MCMY, PASSIVE, POVERTY, SNOW, SPEED, SUICIDE, URBANPCT, and YOUNG.[12]

There are 10 variables with a relatively high probability of inclusion (over 70%) and their coefficients are also shaded in Table 3. With the exception of BACLAW, all of these are non-fragile from an SSVS perspective.

The distribution of the parameter values for the 100,000 iterations is given in Fig. 4. In these figures the heavy vertical line indicates the mean value of the probability of excluding this variable with a scale on the right of the distribution. The black line pictures the distribution of the iterative parameter values. Bimodal distributions occur because the values depend on the inclusion or exclusion of other variables.

4.3. Bayesian Model Averaging

Bayesian Model Averaging (BMA) was addressed extensively by Raftery, Madigan, and Hoetling (1997) following a suggestion by Leamer (1978). In contrast to SSVS which includes all k variables at each iteration, BMA considers all 2^k conceivable discrete models weighting them by their posterior probabilities. By averaging across many model specifications, especially among those with high posterior probability, BMA is able to explicitly account for model uncertainty as it relates to parameter estimation. Although theoretically attractive, the number of potential models is enormous rendering computation of all potential models infeasible. Methods of limiting consideration to a subset of models to be searched given current computing power have made this feasible.[13] Bayesian Information

Table 3. SSVS Results (100,000 Iterations).

Variable Name	Mean Coefficient	Mean Probability Inclusion	SSVS Min	SSVS Max
BACLAW	−0.0283	0.733	−0.0579	0.0000
BEER	0.118	1.000	0.0743	0.165
CELLPOP	0.520	0.919	0.000	0.807
CRIME	0.0145	0.399	−0.0501	0.123
ED_COL	−0.0620	0.344	−0.279	0.000
ED_HS	−0.0280	0.415	−0.207	0.0749
GINI	0.0475	0.346	0.000	0.265
INCOME	−0.0235	0.418	−0.344	0.285
INSPECT	0.00284	0.337	−0.00786	0.0213
LATITUDE	−0.729	1.000	−1.11	−0.403
LATXLONG	−0.928	1.000	−1.32	−0.564
LONGITUDE	0.374	1.000	0.205	0.559
MCMY	−0.624	1.000	−0.839	−0.444
MLDA	0.00618	0.352	−0.0087	0.0482
PASSIVE	0.0469	0.371	0.000	0.272
PCTSUNNY	0.019	0.326	−0.00069	0.128
POPDEN	0.000249	0.414	−0.129	0.101
POVERTY	0.109	0.917	0.000	0.184
RAIN	−0.000704	0.441	−0.0514	0.0663
SEATBELT	0.00420	0.372	−0.0119	0.0315
SNOW	0.0141	0.248	0.00000	0.101
SPEED	0.0107	0.344	0.0016359	0.0632
SUICIDE	0.1635	1.000	0.0687	0.254
UNEMPLOY	−0.0657	0.510	−0.185	0.000
URBANPCT	−0.0821	1.000	−0.121	−0.0458
YOUNG	0.0319	0.397	0.000	0.157

Cross Sectional – Time Series analysis of traffic fatality rates for 50 states and DC from 1980 to 2005 (training period).
High inclusion or non-fragile SSVS bounds are shaded.

Criterion (BIC) priors provide a useful way to quickly discover models with high posterior probability. As presented in Hoetling et al. (1999), BMA provides a straightforward method to summarize the effects of explanatory variables as measured by their regression coefficients as they are manifest in assorted models. We utilize the standard BMA package in *R* which is well documented.[14] Details of BMA and its application to the traffic fatality problem are presented in Table 4.

In this table, there are 16 variables that are always present in the top models with high posterior probability.[15] The coefficients for these are shaded in Table 4. In terms of variable selection, BMA results generally

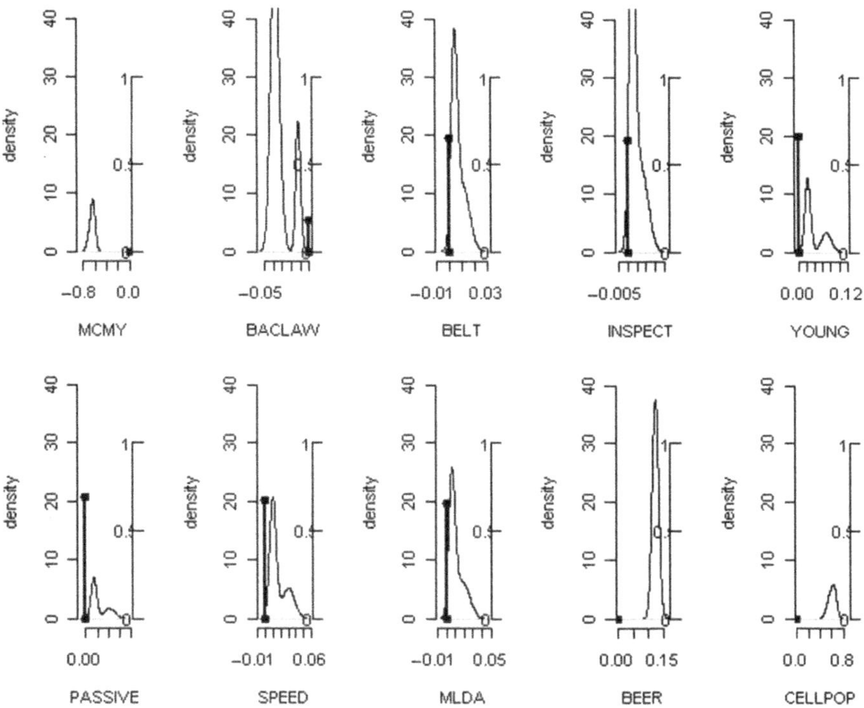

Fig. 4. Density Distributions of Parameter Values in SSVS for 100,000 Iterations (Heavy line indicates mean probability of variable exclusion in next iteration, scale on right).

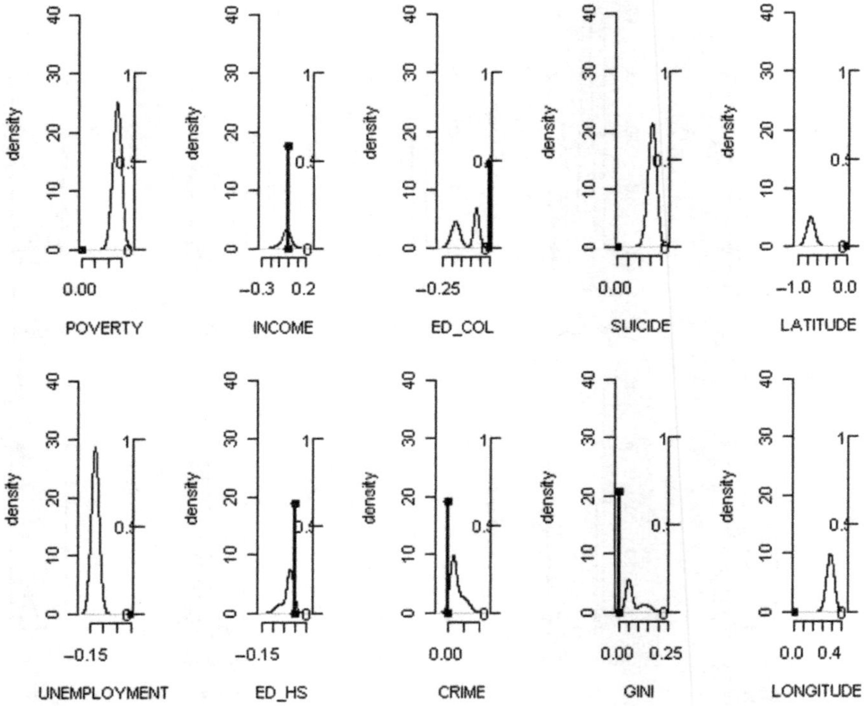

Fig. 4. (*Continued*)

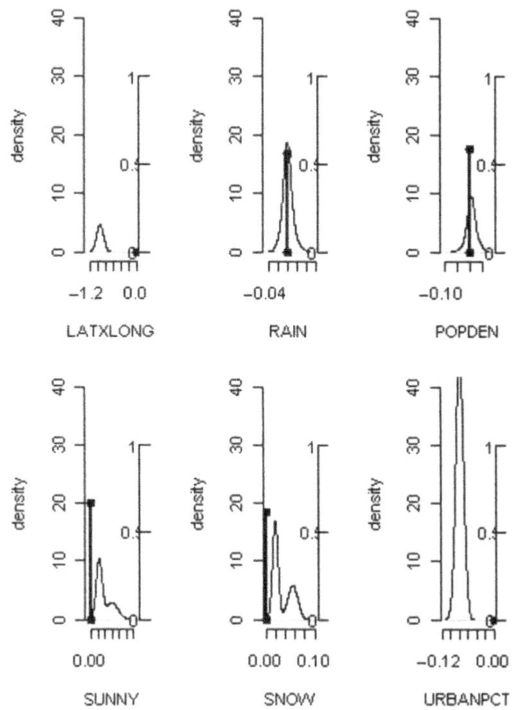

Fig. 4. (*Continued*)

match with SSVS. BMA does choose six more variables than SSVS, namely ED_COL, GINI, PCTSUNNY, SNOW, UNEMPLOY, and YOUNG.

4.4. Bayesian Additive Regression Trees

Bayesian Additive Regression Trees (BART or BayesTree) is a procedure written by Chipman, George, and McCulloch (2009). Their package, available in *R*, was employed here.[16] The procedure is well documented elsewhere and here we introduce basic concepts, their relevance to the current application, and outline the procedure.[17]

BART is designed for out-of-sample forecasting. It is an ensemble method aggregating over a number of semi-independent trees. Each tree is a binary tree model partitioning the data into relatively homogeneous

Table 4. BMA Results.

Variable	Probability of Inclusion	Posterior Mean	Posterior Standard Deviation
BACLAW	1	−0.040	0.007
BEER	1	0.125	0.015
CELLPOP	1	0.213	0.028
CRIME	0.561	0.036	0.036
ED_COL	1	−0.188	0.024
ED_HS	0.008	0.000	0.003
GINI	1	0.146	0.028
INCOME	0.003	0.000	0.003
INSPECT	0.177	0.002	0.004
LATITUDE	1	−0.750	0.084
LATXLONG	1	−0.978	0.100
LONGITUDE	1	0.397	0.048
MCMY	1	−0.558	0.055
MLDA	0.384	0.009	0.013
PASSIVE	0.602	0.047	0.044
PCTSUNNY	0.931	0.060	0.025
POPDEN	0.01	0.000	0.003
POVERTY	1	0.122	0.018
RAIN	0.069	0.002	0.008
SEATBELT	0.085	0.001	0.004
SNOW	1.00	0.053	0.013
SPEED	0.415	0.011	0.015
SUICIDE	1	0.142	0.024
UNEMPLOY	1	−0.125	0.019
URBANPCT	1	−0.083	0.014
YOUNG	0.879	0.051	0.025

Cross Sectional – Time Series analysis of traffic fatality rates for 50 states and DC from 1980 to 2005 (training period).
Probability of inclusion >.7 shaded.

subsets and making forecasts on the basis of the subset in which the observation is contained. The concept of a binary tree is illustrated in Figs. 5(a) and (b). Fig. 5(a) presents a simple tree which explains some vocabulary. All trees start with a root node which contains all the observations in the data set. The data set is bifurcated into two child nodes by means of a splitting rule, here first on the variable MCMY. Observations with MCMY $\leq$ 1985 are put into one child node; observations with MCMY $>$ 1985 are put into the other child node. Subsequently in this diagram, one of the child nodes is split further into two child nodes. This is based on the splitting rule, ED_HS $\leq$ 70 (and ED_HS $>$ 70). This tree has three terminal nodes, illustrated with boxes here, and two internal nodes,

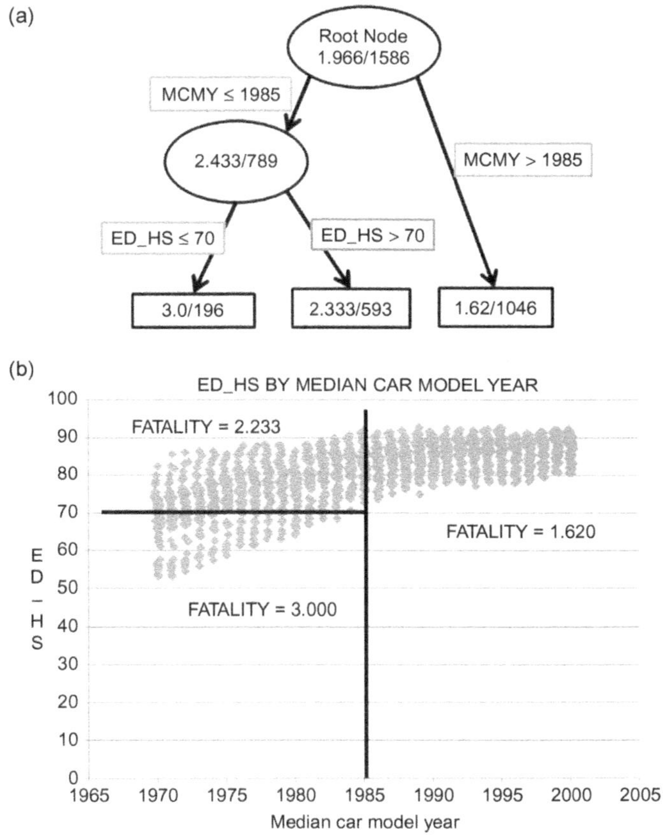

Fig. 5. (a) Simple Tree Model. (b) Scatter Plot Based on Simple Tree Model. Mean Fatality Rates Given in Each Partition.

illustrated with ellipses. The number of terminal nodes is always one more than the number of internal nodes. The splitting rules are given beneath the internal nodes. This tree has depth 2; the splitting employs two variables, MCMY and ED_HS. This partitioning of our data according to the splitting rules given here is shown in a scatter plot in Fig. 5(b). Each observation is contained in one and only one terminal node. A mean traffic fatality rate for each partition in Fig. 5(b) is illustrated.[18]

The basic BART model is

$$y_i = \sum_{j=1}^{m} g(X_{ij}|T^j, M^j) + u_{ij}; u_{ij} \sim N(0, \sigma^2)$$

where i is the observation number ($i = 1, \ldots, n$) and j is the jth tree ($j = 1, \ldots, m$). Here the variable y_i, FATAL, is a variable indicating the fatality rate by location and year. Each tree, j, in the ensemble makes forecasts according to its own tree, T^j, and model, M^j, where M^j defines the parameter values associated with the terminal nodes of T^j. It is a sum of trees model, aggregating the m trees in the ensemble.

The trees are generated independently. Each tree is generated, however, with a boosting algorithm conditional on the other $m-1$ trees, consequently the term semi-independent. Given the m trees generated in any iteration, the residuals are known and a new σ^2 distribution is based on these residuals. An inverse gamma distribution is used for σ^2 and the trees in the next iteration are conditional on the σ^2 drawn from this distribution.

A Markov chain of trees is generated for each tree by means of a stochastic process. Given the existing tree, T^{jk-1}, for forecaster j at iteration $k-1$, a proposal tree, T^*, is generated. The proposal tree is compared to the tree in the last iteration and accepted or rejected in a metropolis algorithm. The generation of the proposal tree as a stochastic process and the acceptance decision are done according to the following steps:

- At iteration k determine the dependent variable, R^{jk}, or the "residuals" for Y conditional on the $m-1$ other trees. $R^{jk} = Y - \sum_{l \neq j} g(X|T^{l\gamma}, M^{l\gamma})$, where $\gamma = k-1$ if $l > j$, and $\gamma = k$ if $l < j$. Determine the previous tree for this tree, T^{jk-1}.
- A decision is made on whether the tree will be split as defined by the probability, $a(1+d)^b$.[19]
- Given a decision to split, a decision is made on the type of split. The types of splits and their associated probabilities are: GROW (.25), PRUNE (.25), CHANGE (.4), SWAP (.1). These are described in Chipman, George, and McCulloch (1998). At the root node there is only one option, GROW. The option CHANGE is feasible only if the tree has depth greater than or equal to 2. For each type of split there are a finite number of choices. GROW will occur at terminal nodes. CHANGE occurs at a pair of internal nodes, one the child of the other.
- The next decision concerns the variable on which the split is made and the splitting rule is again among a finite number of choices. The variables are treated equally. The number of potential splits depends on the variable selected, but for each variable the potential splits are equally likely.

- Given this proposal tree, a posterior distribution is determined for each terminal node based on a "regularization" prior designed to keep individual tree contributions small. Parameters are drawn from the posterior distribution for each terminal node.
- The proposal tree is accepted or rejected by a Metropolis–Hastings algorithm with the probability of accepting T^* equal to α

$$= \min\left(\frac{q(T^{jk-1},T*)}{q(T*,T^{jk-1})}\frac{p(Y|X,T*)p(T*)}{p(Y|X,T^{jk-1})p(T^{jk-1})},1\right)$$

  where $q(T^{jk-1},T^*)$ is the transition probability of going from T^{jk-1} to T^* and $q(T^*,T^{jk-1})$ is the transition probability of going from T^* to T^{jk-1}. The function $q()$ and the probabilities $p(T^*)$ and $p(T^{jk-1})$ are functions of the stochastic process generating the tree. The ratio, $\frac{p(Y|X,T*)}{p(Y|X,T^{jk-1})}$ is a likelihood ratio reflecting the data X and Y, ensuring that the accept/reject decision is a function of the data.
- Acceptance of a tree is dependent on there being a sufficient number of observations in each terminal node of T^*.
- If the tree is accepted $T^{jk} = T^*$, otherwise $T^{jk} = T^{jk-1}$.

The MCMC process is run for a large number of iterations to achieve convergence. The individual trees are not identified. It is possible that trees may be replicated among the set of trees in different iterations. The objective here is not parameter estimation but variable selection. There are two important aspects of the MCMC process: the break-in period and the splitting rules which are described as follows.

4.4.1. Splitting Rules

Before discussion of the performance of our model we look at some of our choices concerning the BART process. First we specify the number of trees. In this example we used 50. For comparison purposes we also used 20 trees. A more thorough investigation might try more or less trees. We also had to specify the parameters of the splitting rule for the tree generating process, $P(\text{split}) = a(1+d)^b$. We selected $a = .95$, the default value. This implies a high likelihood of a split at the root node with a decreasing probability as the depth, d, of the tree increases. The parameter b relates to the bushiness of the tree. First we examined the average number of terminal nodes per tree for the first 50,000 iterations after the break-in period for different values of the parameter b. These are given in Table 5. A value of $b = 1.0$ performs well on root mean squared error (RMSE), correlation

Table 5. Bart Model Performance for 50 and 20 Trees and Power Parameter = *b*.

	BART 50 Trees			OLS
	$b=0.5$	$b=1.0$	$b=1.5$	
RMSE	0.06234	0.019365	0.021627	0.06234
Correlation	0.992584	0.991596	0.989488	0.908594
No. of nodes	5.222615	4.58238	3.87284	
	BART 20 Trees			
	$b=0.5$	$b=1.0$	$b=1.5$	
RMSE	0.021667	0.022594	0.025404	
Correlation	0.989453	0.988517	0.985464	
No. of nodes	8.465195	7.363104	6.018347	

(CORRELATION) with the actual fatality rates for the training data. For comparison purposes, statistics for ordinary least squares are also given. We selected a BART model with 50 trees and $b = 1.0$ for most of the results reported although results are reported for forecasting and tree size for all the BART models. The table confirms that fewer trees and a smaller power parameter lead to bushier trees.

While Table 5 describes the average size of trees, there is substantial variation among the trees in any single iteration. The frequency distribution of tree sizes within the last iteration by the "power" parameter *b* is pictured in Fig. 6. While tree size may vary substantially for any specific forecaster across iterations, the last iteration should be representative for the distribution of tree size in post break-in iterations. In Fig. 6, a 50-tree model is consistent with each forecaster making his decision based on less than 10 variables and greater than or equal to 3. A 20-tree model yields up to 19 leaves.

4.4.2. Break-in Period

The MCMC process is known to converge to a limiting distribution under appropriate conditions and a number of iterations are discarded to insure the process has settled down to this distribution. It is not established, however, when this convergence is reached. We looked at the MCMC history of forecasts for a number of states/years in the training period. In these figures (Figs. 7(a) and (b)) a break-in period of 5,000 iterations was used for the model with 50 trees and $b = 1.0$. Fifty thousand iterations after the break-in period were plotted. The value for each iteration is the sum of the

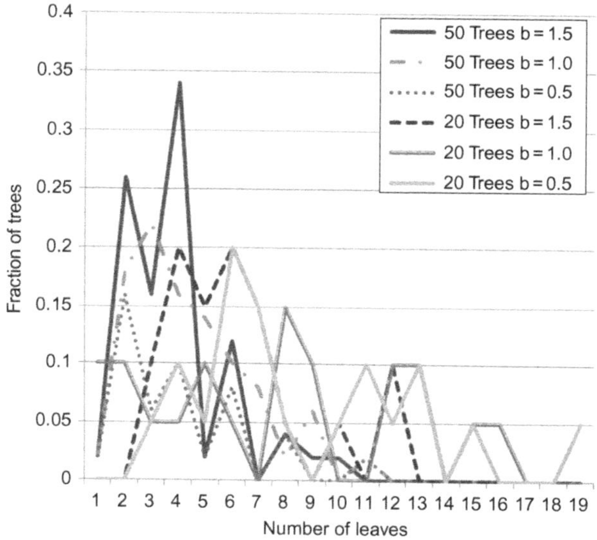

Fig. 6. Distribution of Tree Sizes by Power Parameter b and Number of Trees.

50 trees. The prediction is the average of the 50,000 iterations. The iterations for the selected observations are plotted with a black line. Several state-year observations were selected to see how the process does in different conditions. Although there is variation among the iterations for any observation, the forecasts "mix" well in that there is no functional trend among the iterative forecasts. The MCMC standard error varies among the observations selected, but is relatively uniform within each of the iterations for each state-year observation. It should be noted that all the data are in standardized units, so that California has negative values and Arkansas has positive values.

4.4.3. Variable Choice

An interesting comparison is the variable choice among the trees. This is illustrated in Fig. 8 for 50 trees showing a standard box-and-whisker plot for variable use among the trees in 50,000 post burn-in iterations. The vertical axis gives the average number of trees employing each variable over 50,000 iterations. A value of 50 would indicate a variable used by every tree (a value larger than 50 would indicate that at least one tree split on some variable twice). No such variable exists.

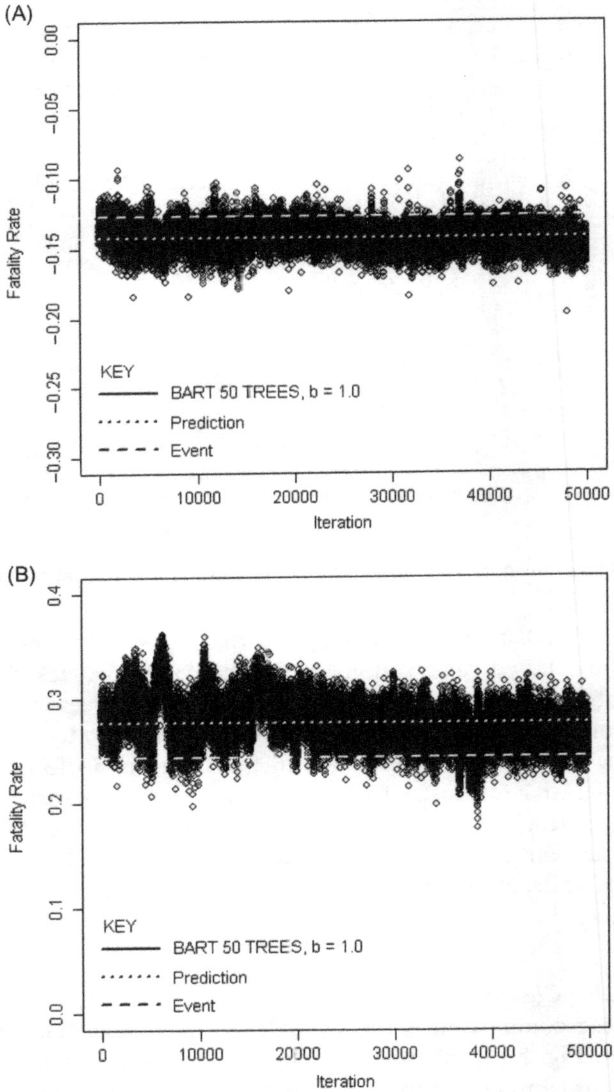

Fig. 7. (a) MCMC History: CALIFORNIA 2005. (b) MCMC History: ARKANSAS 1980.

Although no variable is always used, MCMY is used by almost 25 trees in at least one iteration. The top 13 choices include MCMY, MLDA, BEER, CELLPOP, ED_HS, SUICIDE, CRIME, LATITUDE, LONGITUDE, LAXLONG, PCTSUNNY, RAIN, and URBANPCT. In regards to variable selection, the one new choice that has not been picked up in any of the three previous procedures is RAIN. There is one variable that was chosen in the three other procedures, but not selected as a top choice in BART. This is POVERTY.

4.5. Forecasting

BART and BMA are both techniques designed for forecasting. We wanted to compare the forecasting performance of the techniques with the other methods considered here. EBA does not apply to forecasts, but OLS forecasts are obtained. BMA forecasts are similarly produced by applying the mean parameter values to the test data. Both SSVS and BART are iterative processes. In SSVS each iteration produces a model in the iterative MCMC process. This model is applied to the test data yielding 100,000 iterative forecasts. The model forecast is the mean of these forecasts. BART is also

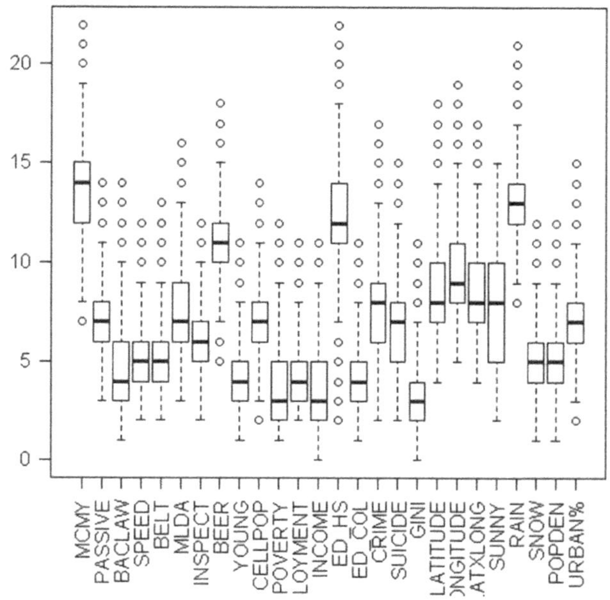

Fig. 8. Variable Use for 50,000 Post Burn-in Draws for 50 Trees, $b = 1.0$.

Table 6. Root Mean Square Forecasting Error for Test Period 2005–2010.

Statistical Technique	Number of Trees	Power Parameter	RMSE
OLS			0.0712
SVSS			0.0642
BMA			0.0575
BART	50	$b = 0.5$	0.0652
BART	50	$b = 1.0$	0.0550
BART	50	$b = 1.5$	0.0602
BART	20	$b = 0.5$	0.0623
BART	20	$b = 1.0$	0.0598
BART	20	$b = 1.5$	0.0592

an iterative process. Each iteration produces an ensemble of trees. That ensemble of trees generated with the training data can be applied to the test data yielding an iterative forecast. The model forecast is the mean of the 50,000 iterative forecasts.

The root mean square forecasting error is a means of comparing the forecasting of the models. These are reported in Table 6.

It should be noted that the best performing method is BART with 50 trees and $b = 1.0$, which was also the best performing method with training data. BMA, the other technique designed for forecasting, performs second best among the techniques evaluated.

5. CONCLUDING COMMENTS

As Bradly Efron noted, how to go about choosing variables to include in a regression is a challenging endeavor.[20] With an increasing amount of data being available to analyze, a way to account for model uncertainty is as important an issue as parameter uncertainty in doing applied econometrics. In this paper we have looked at four Bayesian procedures, none of which assume that any given model is true. Table 7 provides a brief summary of our results.

We first note that there are several variables that are among the favorite for all procedures. The first is MCMY. As noted above, motor vehicle fatality rates have been declining since 1972. We would expect this to continue over time as cars become safer with the modernization of the fleet. More specifically, Blattenberger et al. (2013) estimate a 64% decline in

Table 7. EBA, BMA, SSVS, and BART Comparisons.

Variable	EBA Non-Fragile Bounds	BMA High Inclusion Probability	SSVS Non-Fragile Bounds	SSVS High Inclusion Probability	BART High Use (Top 13)
BACLAW		x		x	
BEER	x	x	x	x	x
CELLPOP	x	x	x	x	x
CRIME					x
ED_COL	x	x			
ED_HS					x
GINI		x	x		
INSPECT					
LATITUDE	x	x	x	x	x
LATxLONG	x	x	x	x	x
LONGITUDE	x	x	x	x	x
MCMY	x	x	x	x	x
MLDA					x
PASSIVE			x		
PCTSUNNY		x			x
POPDEN					
POVERTY	x	x	x	x	
RAIN					x
REALINC					
SEATBELT					
SNOW		x	x		
SPEED			x		
SUICIDE	x	x	x	x	x
UNEMPLOY	x	x			
URBANPCT	x	x	x	x	x
YOUNG		x	x		

Cross Sectional – Time Series analysis of traffic fatality rates for 50 states and DC from 1980 to 2010.

fatality rates when a state modernizes its fleet by seven years. The efficacy of government funded rebates and/or the reduction of taxes on newly purchased vehicles may be worthy of study as to their abilities to speed the modernization of the fleet.

URBANPCT, LATITUDE, LONGITUDE, and LATXLONG are also selected variables across all methods. It seems these geographically related variables might capture many of the characteristics of roadways, speed, and access to fast emergency services as well as weather. Specific weather variables, PCTSUNNY, RAIN, and SNOW are occasionally selected (see Table 7).

SUICIDE is among the list of five policy-related variables that always appear.[21] If using a motor vehicle is a means to accomplish an intended death, public health decisions ought to consider this linkage, especially given that motor vehicle deaths and suicide together are the leading causes of death among young persons in the United States.[22] It may prove worthy to consider the effect of posting information on the availability of mental health facilities for those in need of such health care. In addition, policies may be considered that might result in a reduction of aggressive driving.

BEER is an important policy-related variable that shows up in all five of the procedures. Our analyses suggest that states should continue to impose strict laws on drunk driving.

Another important socioeconomic variable that appears in our results (although not always selected by all criteria) is POVERTY. This might reflect disparity among the states in terms of, for example, fleet age, states' infrastructures relative to highway safety, and access to fast emergency care. This is consistent with other studies such as Blattenberger et al. (2013).

CELLPOP is also chosen by all procedures. Its importance as an explanatory variable is consistent with other studies as well.[23] Our findings suggest that states should continue to implement and enforce laws designed to reduce the negative effects of cell phone use while driving. Only 13 states and the District of Columbia have a primary handheld ban on cell phone use by drivers.[24] Extending this ban across all states and banning hands-free devices are worthy of study with respect to reducing motor vehicle fatality rates along with the life-saving effect of stricter police enforcement of those bans and enhanced fine structures.

NOTES

1. See, for example, Key, Pericchi, and Smith (1999) and Clyde (1999).
2. A comprehensive review of literature related to motor vehicle fatalities is provided in Blattenberger et al. (2013). In this section, we highlight some important issues related to our variable choice problem.
3. See, http://en.wikipedia.org/wiki/List_of_motor_vehicle_deaths_in_U.S._by_year for data on crash fatalities.
4. A detailed review is provided in Blattenberger et al. (2013).
5. See Blattenberger et al. (2012).
6. Our method of imputing cell phone subscriptions correlates with the actual data with a correlation coefficient of .9943.
7. The per se law refers to legislation that makes it illegal to drive a vehicle at a blood alcohol level at or above the specified BAC level. BAC is measured in grams per deciliter.

8. Mathematical developments are found in Leamer (1982).

9. In this paper, we do report data-favored bounds as opposed to global bounds which are very wide.

10. The current model begins with the model of the last iteration plus each newly included variable or minus each newly excluded variables; each variable is assessed.

11. The first 500 iterations were deleted as a break-in period so there were a total of 99,500 iterations employed in the results reported.

12. In Table 3, zero is rounded and not true zero.

13. Limitation to less than 30 potential explanatory variables enhances feasibility.

14. Raftery et al. (2009).

15. Again, with a probability of inclusion greater than 70%.

16. See Chipman et al. (2009).

17. See Chipman, George, and McCulloch (2010).

18. This partition scores poorly but is only used to illustrate the concepts.

19. We selected $a = .95$, the default value. This implies a high likelihood of a split at the root node with a decreasing probability as the depth of the tree, d, increases. A number of values were compared.

20. In: http://www.insightful.com/lars/NIH_LarsProposal.pdf.

21. The five policy-related variables are: BEER, CELLPOP, MCMY, SUICIDE, and URBANPCT.

22. Centers for Disease Control and Prevention, National Center for Injury Prevention and Control. Web-based Injury Statistics Query and Reporting System (WISQARS): www.cdc.gov/ncipc/wisqars.

23. See Blattenberger et al. (2012, 2013).

24. These states include: California, Connecticut, Delaware, Hawaii, Illinois, Maryland, Nevada, New Jersey, New York, Oregon, Vermont (as of 10/2014), Washington, and West Virginia.

REFERENCES

Blattenberger, G., Fowles, R., Loeb, P. D., & Clarke, Wm. (2012). Understanding the cell phone effect on vehicle fatalities: A Bayesian view. *Applied Economics*, *44*, 1823–1835.

Blattenberger, G., Fowles, R., & Loeb, P. D. (2013). Determinants of motor vehicle crash fatalities using Bayesian model selection methods. *Research in Transportation Economics*, *43*, 112–122.

Centers for Disease Control and Prevention. *National center for injury prevention and control. Web-based Injury Statistics Query and Reporting System (WISQARS)*. Retrieved from http://www.cdc.gov/injury/wisqars/. Accessed on July 7, 2012.

Chamberlain, G., & Leamer, E. E. (1976). Matrix weighted averages and posterior bounds. *Journal of the Royal Statistical Society, Series B*, *38*(1), 73–84.

Chipman, H. A., George, E. I., & McCulloch, R. E. (1998). Bayesian CART model search. *Journal of the American Statistical Association*, *93*(443), 935–948.

Chipman, H., George, E. I., & McCulloch, R. (2009). *Implementation of BART: Bayesian additive regression trees*. Retrieved from http://www.r-project.org

Chipman, H. A., George, E. I., & McCulloch, R. E. (2010). BART: Bayesian additive regression trees. *The Annals of Applied Statistics* 4(1), 266–298.

Clyde, M. A. (1999). Bayesian model averaging and model search strategies. *Bayesian Statistics*, *6*, 157–185.

George, E. I., & McCulloch, R. R. (1993). Variable selection via Gibbs sampling. *Journal of the American Statistical Association*, *88*, 881–889.

Hoetling, J., Madigan, D., Raftery, A., & Volinsky, C. (1999). Bayesian model averaging: A tutorial. *Statistical Science*, *14*(4), 382–417.

Key, J. T., Pericchi, L., & Smith, A. F. M. (1999). Bayesian model choice: What and why? *Bayesian Statistics*, *6*, 343–370.

Leamer, E. E. (1978). *Specification Searches: Ad hoc inference with non-experimental data*. New York, NY: Wiley.

Leamer, E. E. (1982). Sets of posterior means with bounded variance priors. *Econometrica*, *50*(3), 725–736.

Raftery, A., Hoetling, J., Volinsky, C., Painter, I., Yeung, K. Y. (2009). *BMA: Bayesian model averaging. R package version 3.10*. Retrieved from http://CRAN.R-project.org/package=BMA. Accessed on 11 February, 2011.

Raftery, A. E., Madigan, D., & Hoetling, J. A. (1997). Bayesian model averaging for linear regression models. *Journal of the American Statistical Association*, *92*, 179–191.

INTRINSIC PRIORS FOR OBJECTIVE BAYESIAN MODEL SELECTION

Elías Moreno[a] and Luís Raúl Pericchi[b]

[a]*Department of Statistics, University of Granada, Granada, Spain*
[b]*Department of Mathematics, University of Puerto Rico, Rio Pedras, San Juan, Puerto Rico, USA*

ABSTRACT

We put forward the idea that for model selection the intrinsic priors are becoming a center of a cluster of a dominant group of methodologies for objective Bayesian Model Selection.

The intrinsic method and its applications have been developed in the last two decades, and has stimulated closely related methods. The intrinsic methodology can be thought of as the long searched approach for objective Bayesian model selection and hypothesis testing.

In this paper we review the foundations of the intrinsic priors, their general properties, and some of their applications.

Keywords: *g*-prior; intrinsic priors; model selection; sampling properties of Bayes factors

JEL classifications: C11; C12; C18

Bayesian Model Comparison
Advances in Econometrics, Volume 34, 279–300

ISSN: 0731-9053/doi:10.1108/S0731-905320140000034012

1. INTRODUCTION

A fully subjective Bayesian approach to model selection typically leaves the priors under each of the models unrelated to each other, and the objective intrinsic priors for model selection solve this gap. The well-known reference priors are objective priors for estimating the parameters of a given model, and they are often improper. This makes them not suitable for model selection as the resulting marginal distributions of the data are defined up to an arbitrary multiplicative constant. However, they are the starting point for the development of *suitable objective* priors for model selection, the so-called intrinsic priors. The intrinsic methodology provides the intrinsic priors that are related to each other across models.

The general concept of intrinsic priors (Berger & Pericchi, 1996a, 1996b) does not use subjective information but is *intrinsic* to the competing models entertained. Further, the *structural* information, based on which are the particular candidate models being compared, is propagated toward the different models through the *intrinsic prior equations*, the functional equations defining the intrinsic priors.

These equations, however, do not have a unique solution when the starting priors are improper, and a large class of prior distributions is solution to the equations. Further, the marginal distribution of the data as the intrinsic prior varies in the class is typically not robust (Moreno, 1997). Thus, a way for choosing a "reasonable" intrinsic prior for each of the model parameters is required. The intrinsic limiting procedure for testing nested hypotheses, and for some particular non-nested hypotheses, by Moreno, Bertolino, and Racugno (1998) generates the intrinsic priors as the unique limit of proper intrinsic priors.

On the other hand, some other efforts have been made to extend the intrinsic methodology to nonstandard situations. For instance, in Berger and Pericchi (2004) the validity of the intrinsic prior equations to more general sampling schemes, for instance, to censored data, was explored. Although the intrinsic priors methodology has been developed for several problems, there is still plenty of room for developments.

Recently, Bayarri, Berger, Forte, and García-Donato (2012) have proposed a sensible *desiderata* of seven principles to be obeyed by a prior for objective Bayesian model selection. They also propose for the normal linear models the so-called robust priors that obey the seven principles, and the corresponding Bayes factor can be written in closed form. It turns out that the intrinsic priors also obey the seven principles (to be proved elsewhere). The variable selection we obtain when using the robust priors is virtually the same as that obtained when using the intrinsic priors. In general, it seems that no one dominates the other, a conclusion that can be drawn by comparing their Type I and II errors. Moreover, for complex linear models both the robust priors and the intrinsic priors exhibit a small region in the parameter space for which consistency does not hold. The intrinsic priors have the advantage of being defined for general model selection, while the robust prior is only defined for normal models.

The intrinsic prior distributions have interesting general desirable properties. For instance, (i) when comparing two nested models, the intrinsic prior distribution for the parameters of the full model is concentrated around the parameter space of the simpler model, thus obeying the Savage's continuity condition, (ii) they are well calibrated (Pericchi, 2005), (iii) the intrinsic prior for the parameters of the full model, conditional on a fixed null model, is a proper probability distribution, although the unconditional prior is defined up to a multiplicative constant that does not play any role in model selection. Further, sampling properties of the Bayes factors for intrinsic priors have been established and compared with that of the BIC for models with finite dimensional parameter spaces (Casella, Girón, Martinez, & Moreno, 2009), and also for complex models, whose dimension grows with the sample size (Moreno, Girón, & Casella, 2010).

A quite interesting scientific impact of the intrinsic priors has been in the motivation of closely related methods. These are, the expected posterior priors (Perez & Berger, 2002), the integral priors (Cano & Salmerón, 2013), and the Power-Expected-Posterior Priors (Fouskakis, Ntzoufras, & Draper, 2014). All of these share the same basis as the intrinsic priors.

The rest of the article is organized as follows. In Section 2 the elements and foundation of the intrinsic priors, and a summary of their main properties are presented. In Section 3 we give an application to variable selection in normal regression. Section 4 contains concluding remarks and a link to software already developed for variable selection.

2. INTRINSIC PRIORS FOR NESTED MODELS

The use of improper noninformative priors for model comparison yield Bayes factors which depend on an arbitrary multiplicative constant. The problem may be present even in comparisons of models of the same dimension. The casual use of "vague proper priors" does not solve the problem and often aggravates it, giving unbounded bias in favor of one of the candidate models. On the other hand, probability intervals alone should not be used for model comparison. For instance, for models with one-dimensional parameter spaces, the posterior HPD (95%) interval of the posterior distribution of the difference of the parameters, or the ratio, is often used for testing the null hypothesis that the model parameters are equal. However, we note that the null hypothesis is not taken into account in the construction of the HPD (95%) interval, so it cannot be used for taking a decision on it. It was Jeffreys (1961) who recommended to distinguish between estimation and testing, and introduced the Bayes factor as the appropriate tool for testing hypotheses.

An attempt to circumvent the ill definition of the Bayes factors for improper noninformative priors is the intrinsic Bayes factor introduced by Berger and Pericchi (1996a), which is a modification of a partial Bayes factor. They assumed the following Bayesian principle: Testing and model selection methods should correspond, in some sense, to actual Bayes factors, arising from reasonable default prior distributions (Berger & Pericchi, 1996b). Thus, although the intrinsic Bayes factor is not a Bayes factor, it was shown to be asymptotically equivalent to a genuine Bayes factor for the so-called intrinsic priors. Thus, the intrinsic priors arose as a justification of the intrinsic Bayes factor in the light of the above *Bayesian principle*. However, our interest here is on the intrinsic priors and the Bayes factor for intrinsic priors, and not on the intrinsic Bayes factor which is based on real training samples.

For simplicity in deriving the intrinsic priors, we consider that only two alternative sampling models $f_i\ (x|\theta_i)$, $\theta_i \in \Theta_i$, $i=1,2$, are proposed for the observed sample $\mathbf{x}=(x_1,\ldots,x_n)$. Let $\pi_i^N(\theta_i)$, $i=1,2$, denote default improper priors. The Bayesian model defined by the pair $\{f_i(x|\theta_i), \pi_i^N(\theta_i)\}$ is denoted as model M_i. The Bayes factor for comparing model M_2 versus M_1 is given by

$$B_{21}^N(\mathbf{x}) = \frac{m_2^N(\mathbf{x})}{m_1^N(\mathbf{x})} = \frac{\int f_2(\mathbf{x}|\theta_2)\pi_2^N(\theta_2)d\theta_2}{\int f_1(\mathbf{x}|\theta_1)\pi_1^N(\theta_1)d\theta_1}$$

which is defined up to the multiplicative constant c_2/c_1. The arbitrary constants c_1, c_2 are inherited from the improper prior $\pi_i^N(\theta_i)$, $i=1,2$.

To define the intrinsic priors we need to follow two steps. In the first step, the arithmetic intrinsic Bayes factor is defined. In the second, it is shown that asymptotically it is equivalent to an actual Bayes factor, the Bayes factor for the intrinsic priors.

To define the arithmetic intrinsic Bayes factor let us consider the set of subsamples $\mathbf{x}(\ell)$ of the sample $\mathbf{x}$ of minimal size ℓ such that $0<m_i^N(\mathbf{x}(\ell))<\infty$ for $i=1,2$, the number of which is denoted by L. These subsamples are called *training samples*. Any training sample converts the improper prior $\pi_i^N(\theta_i)$ into a proper posterior $\pi_i^N(\theta_i|\mathbf{x}(\ell))$, $i=1,2$. Then, the Bayes factor for the subsample $\mathbf{x}(n-\ell)$, where $\mathbf{x}(\ell)\cup\mathbf{x}(n-\ell)=\mathbf{x}$, and the proper posterior $\pi_i^N(\theta_i|\mathbf{x}(\ell))$ is called a "partial" Bayes factor,

$$B_{21}^N(\mathbf{x}(n-\ell))=\frac{\int f_2(\mathbf{x}(\mathbf{n}-\ell)|\theta_2)\pi_2^N(\theta_i|\mathbf{x}(\ell))d\theta_2}{\int f_1(\mathbf{x}(\mathbf{n}-\ell)|\theta_1)\pi_1^N(\theta_i|\mathbf{x}(\ell))d\theta_1}$$

This partial Bayes factor is a well-defined Bayes factor, and can be written as

$$B_{21}^N(\mathbf{x}(n-\ell))=B_{21}^N(\mathbf{x})B_{12}^N(\mathbf{x}(\ell))$$

Unfortunately, $B_{21}^N(\mathbf{x}(n-\ell)$ is not necessarily stable across training samples, and hence depends on the training sample $\mathbf{x}(\ell)$ we use. To avoid this arbitrariness, Berger and Pericchi (1996a) suggested averaging over all training samples

$$B_{21}^{\mathrm{AIF}}(\mathbf{x})=B_{21}^N(\mathbf{x})\frac{1}{L}\sum_{\ell=1}^{L}B_{12}^N(\mathbf{x}(\ell)) \tag{1}$$

which is called the arithmetic intrinsic Bayes factor. $B_{21}^{\mathrm{AIF}}(\mathbf{x})$ does not depend on a particular training sample and can be seen as a more stable version of $B_{21}^N(\mathbf{x}(n-\ell))$, more so as the sample size n increases. Unfortunately, $B_{21}^{\mathrm{AIF}}(\mathbf{x})$ is not a Bayes factor.

However, when the sampling model $f_1(x|\theta_1)$ is nested in $f_2\ (x|\theta_2)$, that is $\Theta_1\subset\Theta_2$ and $f_2\ (x|\theta_2)=f_1(x|\theta_1)$ for any $\theta_2=\theta_1$, an asymptotic equation provides a Bayes factor for "intrinsic priors" (Berger & Pericchi, 1996a).

Let us consider a sequence of i.i.d. random variables $\{X_1,X_2,\ldots\}$ with distribution $f_2\ (x|\theta_2)$, and impose the asymptotic equality

$$\lim_{n\to\infty}B_{21}^{\mathrm{AIF}}(X_1,\ldots,X_n)=\lim_{k\to\infty}B_{21}(X_1,\ldots,X_n),\ [P_{\theta_2}]$$

where $B_{21}(X_1,\ldots,X_n)$ is the Bayes factor for the intrinsic priors $\pi_1^I(\theta_1)$ and $\pi_2^I(\theta_2)$. Some calculations show that this equation defines the intrinsic prior $\pi_1^I(\theta_1)$ to be any prior in the class (Moreno, 1997)

$$\Gamma_1 = \left\{ \pi_1^I(\theta_1) : \int_{\Theta_1} \pi_1^I(\theta_1) d\theta_1 = 1, \int_{\Theta_2} T(\theta_2) \pi_1^I\left(\psi_1(\theta_2)\right) d\theta_2 = 1 \right\}$$

where

$$T(\theta_2) = \frac{\pi_2^N(\theta_2)}{\pi_1^N(\psi_1(\theta_2))} E_{X(\ell)|\theta_2} B_{12}^N(X(\ell))$$

and $\psi_1(\theta_2) \in \Theta_1$ denotes the limit in probability of the MLE $\hat{\theta}_1(x_1, \ldots, x_n)$ when sampling from $f_2(x|\theta_2)$.

Then, for each $\pi_1^I(\theta_1) \in \Gamma_1$, $\pi_2^I(\theta_2)$ is given by

$$\pi_2^I(\theta_2) = T(\theta_2) \pi_1^I\left(\psi_1(\theta_2)\right)$$

We note that for the case where $\pi_1^N(\theta_1)$ is proper, Γ_1 contains only this prior, that is, $\pi_1^I(\theta_1) = \pi_1^N(\theta_1)$, and then $\pi_2^I(\theta_2)$ is the unique prior

$$\pi_2^I(\theta_2) = \pi_2^N(\theta_2) E_{X(\ell)|\theta_2} B_{12}^N(X(\ell))$$

Thus, in that case the intrinsic priors $\left(\pi_1^I(\theta_1), \pi_2^I(\theta_2)\right)$ are unique and proper.

Example 1. *Assume an exponential density $f(x|\lambda) = \lambda \exp(-\lambda x)$ and we want testing the null $H_0 : \lambda = \lambda_0$ versus the alternative $H_1 : \lambda \in \mathbb{R}^+$. This is equivalent to the model selection problem between model*

$$M_1 : \left\{f(x|\lambda), \pi_1^N(\lambda)\right\} \text{ and } M_2 : \left\{f(x|\lambda), \pi_2^N(\lambda)\right\},$$

where $\pi_1^N(\lambda)$ is the proper distribution $\pi_1^N(\lambda) = 1_{(\lambda=\lambda_0)}(\lambda)$, and $\pi_2^N(\lambda) = c_2/\lambda$ is the improper reference prior utilized for estimation. The minimal training sample size is now $\ell = 1$, and the intrinsic priors are unique and given by

$$\pi_1^I(\lambda) = 1_{(\lambda=\lambda_0)}(\lambda)$$

$$\pi_2^I(\lambda) = \frac{1}{\lambda} E_{x|\lambda}\left[B_{12}^N(x)\right] = \frac{\lambda_0}{(\lambda + \lambda_0)^2}$$

We note that $\pi_2^I(\lambda)$ is a long tailed proper prior. Its median is λ_0.

In the general case where $\pi_1^N(\theta_1)$ is not proper, the class of intrinsic priors above $\left(\pi_1^I(\theta_1), \pi_2^I(\theta_2)\right)$ is a huge class of priors. However, if the conditions $\int \pi^I(\theta_1)d\theta_1 = \int \pi^I(\theta_2)d\theta_2 = 1$ are not imposed, and it is assumed instead that the intrinsic priors are limit of proper intrinsic priors, then it can be shown that the pair of prior distributions

$$\pi_1^I(\theta_1) = \pi_1^N(\theta_1)$$

$$\pi_2^I(\theta_2) = \pi_2^N(\theta_2) E_{X(\ell)|\theta_2} B_{12}^N(X(\ell)) \tag{2}$$

are the unique limit of proper intrinsic priors (Moreno et al., 1998). We remark that although these intrinsic prior distributions are not necessarily proper, the corresponding Bayes factor is well-defined.

It is of interest to write the bidimensional intrinsic prior for θ_1 and θ_2 as

$$\pi^I(\theta_1, \theta_2) = \pi_2^I(\theta_2|\theta_1)\pi_1^I(\theta_1)$$

where the intrinsic prior for θ_2, conditional on θ_1, is given by the proper prior distribution

$$\pi_2^I(\theta_2|\theta_1) = \pi_2^N(\theta_2) E_{X(\ell)|\theta_2} \frac{f_1(X(\ell)|\theta_1)}{m_2^N(X(\ell))} \tag{3}$$

The following example illustrates the marginal and conditional intrinsic priors.

Example 2. *Suppose the normal model $N(x|\mu,\sigma^2)$ with unknown variance σ^2, and we want testing $H_1{:}\mu=0$ versus $H_1{:}\mu\neq 0$. This is equivalent to the model selection problem between model*

$$M_1 : \left\{N\left(x|0,\sigma^2\right), \pi_1^N(\sigma)\right\} \text{ and } M_2 : \left\{N\left(x|\mu,\sigma^2\right), \pi_2^N(\mu,\sigma)\right\}$$

where $\pi_1^N(\sigma) = c_1/\sigma^2 1_{(0,\infty)}(\sigma)$ and $\pi_2^N(\mu,\sigma) = c_2/\sigma^2 1_{(-\infty,\infty)\times(0,\infty)}(\mu,\sigma)$ are the improper Jeffreys' priors, and c_1, c_2 arbitrary positive constants. From formula (3) it can be seen that the marginal and conditional intrinsic priors are given by

$$\pi_1^I(\sigma) = \frac{c_1}{\sigma^2}$$

$$\pi_2^I(\mu,\sigma) = \frac{c_1}{\sigma^2}\frac{1-\exp\left[-\mu^2/\sigma^2\right]}{2\sqrt{\pi}\left(\mu^2/\sigma\right)}$$

and

$$\pi_2^I(\mu|\sigma) = \frac{1-\exp\left[-\mu^2/\sigma^2\right]}{2\sqrt{\pi}\left(\mu^2/\sigma\right)}$$

which is a proper prior.

The intrinsic priors enjoy attractive properties that we briefly summarize.

1. Since $\pi_1^N(\theta_1)$ and $\pi_2^N(\theta_2)$ are objective priors, the intrinsic priors do not require prior elicitation on the side of the user.
2. The conditional intrinsic prior $\pi_2^I(\theta_2|\theta_1)$ is a probability distribution that satisfies the "Savage continuity condition" (Gunel & Dickey, 1974). This condition is a widely accepted requirement in testing problems (Berger, 1994; Berger & Sellke, 1987; Casella & Berger, 1987; Casella & Moreno, 2009; Consonni & La Rocca, 2008; Morris, 1987), and it states that the prior on the parameter of full model has to be "concentrated" around the space of the null model parameter. Thus providing a fairer comparison between the two hypotheses under investigation. We note that, if the null model is a reasonable null hypothesis, it is important to be able to distinguish $f_1(x|\theta_1)$ from close alternatives. We note that putting prior probability on extreme models, far from $f_1(x|\theta_1)$, will discount the more reasonable alternatives (Casella & Moreno, 2009; Consonni, Moreno, & Venturini, 2011).
3. The intrinsic priors are invariant to reparametrizations.
4. If we let the training sample size ℓ vary in the set of integers, we can assess posterior robustness of the test with respect to the prior (Consonni et al., 2011).
5. Closed-form expressions for the Bayes factor for intrinsic priors are not always available, and usually a quite simple low-dimensional integration is required.
6. For nested models with fixed dimensions, testing procedures based on the intrinsic priors are consistent. If the model dimension grows with the sample size, consistency is typically not satisfied for models in a "small" subset of alternative models (see Section 3). These models for which inconsistency appears are "close", in a given metric, to the null model.

7. The intrinsic testing procedure can be implemented using an efficient Monte Carlo technique.

3. VARIABLE SELECTION IN LINEAR MODELS

The normal linear model plays a central role in statistics for its wide range of applications. The formulation of this model is as follows. Let Y represent an observable random variable and $X_1,\ldots, X_k$ a potential set of explanatory regressors related through the normal linear model

$$Y = \alpha_0 + \alpha_1 X_1 + \cdots + \alpha_k X_k + \varepsilon_k, \quad \varepsilon_k \sim N\left(0, \sigma_k^2\right)$$

where the vector of regression coefficients $\boldsymbol{\alpha}_{k+1} = (\alpha_0,\alpha_1,\ldots, \alpha_k)'$ and the variance error σ_k^2 is unknown. For a given data set $(\mathbf{y},\mathbf{X})$, where $\mathbf{y}$ is a vector of dimension n of independent observations of Y, and $\mathbf{X}_{k+1}$ an $n \times (k+1)$ design matrix of full rank, the sampling regression model for this data is the n-dimensional normal distribution $N_n\left(\mathbf{y}|\mathbf{X}_{k+1}\alpha_{k+1},\sigma_k^2\mathbf{I}_n\right)$. The intercept only model $N_n\left(\mathbf{y}|\mu_0,\sigma_0^2\mathbf{I}_n\right)$ will be denoted as M_0.

The first important question to be addressed about the model is whether the dimension can be reduced by deleting the noninfluential regressors. This old problem is known as the model selection problem, and has received many solutions none of them satisfying all the requirements. Here we revise the objective Bayesian solution to this problem based on the Bayes factor for the intrinsic priors, and compare with the popular solution based on Zellner's g-prior and some modifications of it.

Let us consider the class of regression models defined by all possible subsets of regressors of $\{X_1,\ldots, X_k\}$. A generic sampling model M_j is the n-dimensional normal density $N_n\left(\mathbf{y}|\mathbf{X}_{j+1}\beta_{j+1},\sigma_j^2\mathbf{I}_n\right)$, where $\beta_{j+1} = (\beta_0,\beta_1, \ldots, \beta_j)$ is an unknown vector of dimension $j+1$, $\mathbf{X}_{j+1}$ is the $n \times (j+1)$ design matrix containing only j regressors, and σ_j^2 is the unknown variance error. There are $\binom{k}{j}$ such a M_j models, and this subclass of regression models will be denoted as $\mathfrak{M}_j$. We note that any model in $\mathfrak{M}_j$ has the same dimension. The whole class of regression models will be denoted as $\mathfrak{M} = \cup_{j=0}^{k}\mathfrak{M}_j$, the number of which is 2^k. The development in the paper will be clear using this somewhat ambiguous, but simpler, notation.

Based on the sample information (**y**,**X**), the variable selection problem is a decision problem where the decision space $\mathfrak{D}$ is the decision of choosing a model in the class $\mathfrak{M}$ and the states of nature $\mathfrak{N}$ are the models in the class $\mathfrak{M}$. For the 0−1 loss function, the optimal decision is the model having the highest model posterior probability. Since the 0−1 loss function penalizes the decision of choosing the wrong model with one unit, regardless how far this model is from the true one, it is sometimes preferable to rank the models according to their model posterior probabilities. (For other loss functions, see, for instance, Clyde & George, 2004).

Following Scott and Berger (2010) the Bayesian approach "*has the attractive feature that they can automatically adjust for multiple testing without needing to introduce ad-hoc penalties.*" We essentially agree with that, but we realize that the automatic dimension correction of the Bayesian approach dramatically varies as the priors for both the models and the parameters within models vary, and this motivates the need of a careful selection of them.

Given a data set (**y**,**X**) coming from an unknown model M_t in $\mathfrak{M}$, and priors for models and model parameters $\{\pi(\beta_j,\sigma_j|M_j),\pi(M_j),M_j\in\mathfrak{M}\}$, the model posterior probability of the model M_j is given by

$$\Pr(M_j|\mathbf{y},\mathbf{X}) = \frac{B_{j0}(\mathbf{y},\mathbf{X})\pi(M_j)/\pi(M_0)}{1+\sum_{\substack{M_j\in\mathfrak{M}\\ M_j\neq M_0}} B_{j0}(\mathbf{y},\mathbf{X})\pi(M_j)/\pi(M_0)}, M_j\in\mathfrak{M}$$

where the Bayes factor $B_{j0}(\mathbf{Y},\mathbf{X})$ is given by

$$B_{j0}(\mathbf{y},\mathbf{X}) = \frac{\int N_n\left(\mathbf{y}|\mathbf{X}_j\beta_j,\sigma_j^2\mathbf{I}_n\right)\pi(\beta_j,\sigma_j|M_j)d\beta_j d\sigma_j}{\int N_n(\mathbf{y}|\alpha_0,\sigma_0^2\mathbf{I}_n)\pi(\alpha_0,\sigma_0|M_0)d\alpha_0 d\sigma_0}$$

In the next sections we will consider specific priors for model parameters $\pi(\beta_j,\sigma_j|M_j)$ and model priors $\pi(|M_j)$ that will be used throughout the paper.

3.1. Intrinsic Priors

Two type of intrinsic priors have been considered in the literature. One arising when using the fact that any model M_j is nested in the full one M_k, called encompassing from above (Casella & Moreno, 2006), and a second one that uses the fact that the intercept only model M_0 is nested in any model M_j, encompassing from below (Girón, Martínez, Moreno, & Torres, 2006). Although both methods essentially choose the same subset of regressors (Moreno & Girón, 2008), and have the same asymptotic behavior for finite k (Casella et al., 2009), we only consider here the intrinsic priors coming from encompassing from below.

For comparing the null model M_0 versus the alternative M_j using the intrinsic priors, we start with the improper reference prior for the parameters (α_0, σ_0) of the null model M_0, and for the parameters (β_{j+1}, σ_j) of the alternative model M_j, that is

$$\pi^N(\alpha_0, \sigma_0) = \frac{c_0}{\sigma_0} \text{ and } \pi^N(\beta_{j+1}, \sigma_j) = \frac{c_j}{\sigma_j}$$

where c_0 and c_j are arbitrary positive constants. Applying the standard intrinsic method we get the intrinsic proper prior for the parameters (β_j, σ_j), conditional on (α_0, σ_0), as

$$\pi^I(\beta_{j+1}, \sigma_j|\alpha_0, \sigma_0) = N_{j+1}\left(\beta_{j+1}|\overline{\alpha}_0, \left(\sigma_j^2 + \sigma_0^2\right)\mathrm{W}_{j+1}^{-1}\right)\mathrm{HC}^+(\sigma_j|\sigma_0)$$

where $\overline{\alpha}'_0 = (\alpha_0, \mathbf{0}'_j)'$

$$\mathbf{W}_{j+1}^{-1} = \frac{n}{j+2}\left(\mathbf{X}'_{j+1}\mathbf{X}_{j+1}\right)^{-1}$$

and $\mathrm{HC}^+(\sigma_j|\sigma_0) = (2/\pi)\,\sigma_0/\left(\sigma_j^2 + \sigma_0^2\right)$ is the half Cauchy distribution on R^+ with location parameter 0 and scale σ_0.

For comparing model M_0 against M_j the intrinsic priors are the pair $(\pi^N(\alpha_0,\sigma_0), \pi^I(\beta_{j+1},\sigma_j|\alpha_0,\sigma_0))$. We note that the intrinsic priors depend on an arbitrary constant c_0 that cancels out in the Bayes factor for comparing model M_0 against M_j, and hence no tuning hyper-parameters have to be adjusted.

3.2. *Zellner's* g*-priors*

One of the most interesting and used priors for variable selection are the so-called Zellner's g-priors (Zellner, 1986; Zellner & Siow, 1980). It is perplexing that they are so popular for model selection, since Zellner proposed these priors for estimation. A reason that g-priors are one of the most popular choices of prior for hypothesis testing is that the marginal density, $m_k(\mathbf{y})$, is available in closed form and it is given by

$$m_k(\mathbf{y}) = \frac{\Gamma(n/2)}{2\pi^{n/2}(1+g)^{k/2}}\left(\mathbf{y}^t\mathbf{y} - \frac{g}{1+g}\mathbf{y}^t\mathbf{X}(\mathbf{X}^t\mathbf{X})^{-1}\mathbf{X}^t\mathbf{y}\right)^{-n/2}$$

Thus the Bayes factors and posterior model probabilities for comparing any two linear models are available in simple and closed form, which is a perceived advantage.

Following Moreno, Girón, and Casella (2014) we bring a comparative study of the Bayesian procedure obtained with the intrinsic priors for model parameters, and some versions on Zellner's g-priors. These versions include the case of $g=n$, a non-fully Bayes prior, and a mixture of g-priors with respect to a hyper-prior $\pi(g)$ to be specified later. Further, some versions of the independent Bernoulli priors proposed by George and McCulloch (1993) will be considered as priors on the model space.

In the formulation of g-priors for variable selection we use here encompassing from below (versions for encompassing from above are given in Scott & Berger, 2010). Furthermore, it is assumed that the intercept and the variance error are common parameters to all models. This implies that under this assumption the number of parameters involved for comparing model M_0 against M_j is $j+2$, while in the original formulation the number of parameters was $j+4$. It can be argued that the regression coefficients and the variance error typically vary across models, so that the assumption that the intercept and the variance error remain the same across models is not a realistic assumption. For simplicity in writing the regression parameters of model M_j are denoted as $(\alpha,\beta_j)'=(\alpha,\beta_1,\ldots,\beta_j)'$and the variance error as σ^2.

For a sample $(\mathbf{y},\mathbf{X})$, most references to g-priors in the variable selection literature (Berger & Pericchi, 2001; Clyde & George, 2000; Fernández, Ley, & Steel, 2001; George & Foster, 2000; Hansen & Yu, 2001; Liang, Paulo, Molina, Clyde, & Berger, 2008, among others), refer to them as the pair $(\pi^N(\alpha,\sigma), \pi^g(\beta_j|\sigma))$where

$$\pi^N(\alpha,\sigma) = \frac{1}{\sigma}1_{R\times R^+}(\alpha,\sigma)$$

and

$$\pi^g(\beta_j|\sigma) = N_j\left(\beta_j|\mathbf{0}, g\sigma^2\left(\mathbf{X}_j'\mathbf{X}_j\right)^{-1}\right)$$

where g is an unknown positive parameter, and $\mathbf{X}_j$ is the matrix of dimensions $n \times j$ resulting from suppressing the first column in the design matrix $\mathbf{X}_{j+1}$ of the original formulation of the regression model M_j.

The hyper-parameter g plays an important role in the behavior of the Bayes factor. Several values for g have been suggested, although not one of them satisfies all the requirements (Berger & Pericchi, 2001; Clyde, Parmigiani, & Vidakovic, 1998; Fernández et al., 2001; George & Foster, 2000; Hansen & Yu, 2001; Liang et al., 2008). For instance, large g values induce the Lindley–Bartlett paradox (Bartlett, 1957), and a fixed value for g induces inconsistency. Therefore, it seems that g should be dependent on n as the literature recognizes.

We consider here two versions of the g-priors. The first one is obtained for $g = n$, which is justified on the ground that it gives a unit information prior (Kass & Wasserman, 1995; Liang et al., 2008), and also because it provides consistent Bayes factors when the number of regressors is fixed.

A major problem with the first version of g-priors is that they have the "*finite sample inconsistency*," (Berger & Pericchi, 2001). The meaning of this is that if one is interested in comparing M_k against the intercept only model M_0, it can be shown that, as the least squares estimate $\hat{\beta}$ goes to infinity (or more generally if the likelihood ratio statistic of M_0 vs. M_k goes to zero), so that one becomes certain that M_0 is wrong, paradoxically, the Bayes factor B_{0k} for comparing M_0 and M_k goes to the nonzero constant

$$\lim_{\hat{\beta} \to \infty} B_{0k}(\mathbf{y}, \mathbf{X}) = (1+g)^{(k-n)/2}$$

It was this defect that caused Jeffreys (1961) to reject conjugate-priors for model selection. It can be seen that the Bayes factor for the intrinsic priors will go to zero when the evidence is overwhelmingly against M_0.

A second version of the g-priors has been derived (Liang et al., 2008) to avoid the finite sample inconsistency property. A way to avoid this inconsistency is to mix $\pi^g(\beta_j|\sigma)$ with respect to an appropriate hyper-prior distribution for g. The g-prior mixture

$$\pi^{Mix}(\beta_j|\sigma) = \int_0^\infty N_j\left(\beta_j|0, g\sigma^2\left(\mathbf{X}_j'\mathbf{X}_j\right)^{-1}\right)\pi(g)dg$$

for the inverse Gamma IG($g|1/2,2/n$) mixing distribution

$$\pi(g) = \frac{(n/2)^{1/2}}{\Gamma(1/2)} g^{-3/2} \exp\left(-\frac{n}{2g}\right)$$

avoids the finite sample inconsistency (Liang et al., 2008; Scott & Berger, 2010).

3.3. Model Priors

The natural default prior on the class of models $\mathfrak{M}$ is the uniform prior, but it has been seen (Moreno et al., 2014) that it is not necessarily a good prior. A generalization of the uniform prior is the parametric class called independent Bernoulli priors (George & McCulloch, 1993; Scott & Berger, 2010). Under these priors the probability of model M_j is given by $p^j(1-p)^{k-j}$, $0 \le p \le 1$. If $p = 1/2$ the uniform prior is obtained.

The justification of the Bernoulli prior is that all the regressor are put in an equal footing, and all have a priori the same probability p of inclusion.

If we take for p a uniform prior we obtain a hierarchical uniform prior for the models. For the decomposition $\mathfrak{M} = \cup_{j=0}^{k} \mathfrak{M}_j$ we have that $\pi^U(M_j|\mathfrak{M}_j) = \binom{k}{j}^{-1}$, and $\pi^U(\mathfrak{M}_j) = 1/(k+1)$, $j = 0,1,\ldots, k$.

3.4. Bayes Factors: Asymptotic

For the data set (**y**,**X**) we bring here the expressions of the Bayes factors for the g-priors when $g = n$, for the mixture of g-priors, and for the intrinsic priors.

1. The Bayes factor B_{j0} for comparing M_j versus M_0 for the g-prior with $g = n$ turns out to be

$$B_{j0} = (1+n)^{(n-j-1)/2}\left(1 + n\mathscr{B}_{j0}\right)^{-(n-1)/2} \tag{4}$$

2. The Bayes factor B_{j0}^{Mix} for the mixture of g-priors is given by

$$B_{j0}^{Mix} = \frac{(n/2)^{1/2}}{\Gamma(1/2)} \int_0^\infty \frac{(1+g)^{(n-j-1)/2}}{\left(g\mathcal{B}_{j0}+1\right)^{(n-1)/2}} g^{-3/2} \exp\left(-\frac{n}{2g}\right) dg \tag{5}$$

where the integral on R^+ does not have an explicit expression but needs numerical integration.

3. The Bayes factor B_{j0}^{IP} for the intrinsic priors is

$$B_{j0}^{\text{IP}} = \frac{2}{\pi}(j+1)^{(j-1)/2} \int_0^{\pi/2} \frac{\sin^{j-1}\varphi \ \left(n+(j+1)\sin^2\varphi\right)^{(n-j)/2}}{\left(n\mathcal{B}_{j0}+(j+1)\sin^2\varphi\right)^{(n-1)/2}} \, d\varphi \tag{6}$$

where the integral on the interval $(0,\pi/2)$ does not have an explicit expression and needs to be evaluated numerically.

The three Bayes factors above depend on the data through the same sufficient statistic $\mathcal{B}_{j0}$, which is the ratio of the square sum of the residuals of models M_j and M_0, that is

$$\mathcal{B}_{j0} = \frac{\mathbf{y}'(\mathbf{I}-\mathbf{H}_j)\mathbf{y}}{\mathbf{y}'\left(\mathbf{I}-\frac{1}{n}\mathbf{1}_n\mathbf{1}'_n\right)\mathbf{y}} \tag{7}$$

where $\mathbf{H}_j$ is the hat matrix associated to X_j.

We observe that the three Bayes factors exhibit a quite different dimensional correction. This suggests that for specific values of n and j, the results obtained from them might be very different, a point that we consider in the next section.

The behavior of these Bayes factors and their associated Bayesian variable selection procedures for moderate and large sample sizes has recently been analyzed by Moreno et al. (2014). Since nowadays large number of regressors appears in many applications they also considered the asymptotic case when the number of regressors grows as the sample size grows.

It was seen that when the number of regressors is fixed and the sample size is large the Bayesian variable selection procedures based on the Bayes factor for the g-prior for $g=n$, for the mixtures of g-priors, and for the intrinsic priors are all asymptotically equivalent.

However, for the case where the dimension of the models grows with the sample size, the behavior of the Bayesian procedure strongly depends on the Bayes factor and the model priors. In this latter context the behavior of the procedure based on the g-prior for $g=n$ is inconsistent, while for the mixture of the g-priors and the intrinsic priors are consistent for the hierarchical uniform model prior and for the uniform or the Bernoulli model prior their consistency depends on the speed the number of regressors grows with the sample size.

3.5. *Small Sample Comparisons*

It is surprising that the two versions of the g-priors give procedures that have undesirable sampling behavior for finite sample sizes and number of regressors. We explore here how these procedures behave for small or moderate values of n, with $j = O(n)$. We note that, for a uniform prior on the class of models, the ranking of the models produced by the posterior probabilities of the models in the whole space $\mathfrak{M}$ is exactly the same as that produced by the Bayes factors B_{j0}, B_{j0}^{Mix}, and B_{j0}^{IP}. It is convenient to compare them in a probabilistic scale, that is, to compare the model posterior model probabilities $P(M_j|\mathbf{y},\mathbf{X})$ arising from the Bayes factors B_{j0}, B_{j0}^{Mix}, and B_{j0}^{IP}.

The Bayesian decision rule when comparing M_j versus M_0 is to reject the null M_0 when the posterior probability of M_j is greater than 1/2. We recall that for any of these Bayes factors, the posterior probability $\Pr(M_j|\mathbf{y},\mathbf{X})$ of model M_j in the space of models $\{M_0, M_j\}$ depends on the data $(\mathbf{y},\mathbf{X})$ through the same statistic $\mathcal{B}_{j0}$. Therefore, the critical regions for comparing the null M_0 and the alternative M_j for each of these Bayesian procedures are

$$R_{j0}^{(g=n)} = \{\mathcal{B}_{j0} : \Pr(M_j|B_{j0}) \geq 1/2\}$$

$$R_{j0}^{Mix} = \left\{\mathcal{B}_{j0} : \Pr\left(M_j|B_{j0}^{Mix}\right) \geq 1/2\right\}$$

and

$$R_{j0}^{\mathrm{IP}} = \left\{\mathcal{B}_{j0} : \Pr\left(M_j|B_{j0}^{\mathrm{IP}}\right) \geq 1/2\right\}$$

These posterior probabilities are monotone increasing functions of $\mathcal{B}_{j0}$, so that the critical regions are intervals in (0,1). Let us bring here the distribution of $\mathcal{B}_{j0}$ under the null M_0 and the alternative M_j, which permits finding the size and power of the test.

Theorem 3. *(i) When sampling from M_0 the distribution of $\mathscr{B}_{j0}$ is the Beta distribution*

$$\mathscr{B}_{j0} \sim \text{Beta}\left(\mathscr{B}_{j0} | \frac{n-j-1}{2}, \frac{j}{2} \right), \quad [M_0]$$

(ii). *When sampling from M_j the distribution of $\mathscr{B}_{j0}$ is the noncentral Beta distribution*

$$\mathscr{B}_{j0} \sim \text{Beta}\left(\mathscr{B}_{j0} | \frac{n-j-1}{2}, \frac{j}{2}; 0, \delta_{j0} \right), \quad [M_j]$$

where

$$\delta_{j0} = \frac{1}{2\sigma_j^2} \beta_j' \mathbf{X}'_j \left(\mathbf{H}_j - \frac{1}{n} \mathbf{1}_n \mathbf{1}'_n \right) \mathbf{X}_j \beta_j$$

is a pseudo-distance between model M_j and M_0.

Proof. *See Theorem 1 in Casella et al. (2009).* ■

From this theorem we can compute the exact values of the probability of the Type I and II errors as a function of the model dimension j and the sample size n. Fig. 1 shows the Type I error probabilities of the decision rules associated to the regions $R_{j0}^{(g=n)}, R_{j0}^{Mix}$, and R_{j0}^{IP} for $j = n/3$ and sample sizes $n \leq 100$.

From Fig. 1 it follows that the Type I error probabilities of the procedures based on g-priors are very close to each other, and much smaller than those of the procedures based on intrinsic priors. We note that, for $n \geq 30$, these probabilities for the g-priors are very close to zero. This behavior is not shared by the procedure based on the intrinsic priors.

The power function of the decision rules $R_{j0}^{(g=n)}$, R_{j0}^{Mix}, and R_{j0}^{IP}, for $n/j = 3$, $\delta_{j0=1}$, and samples size $n \leq 100$, are displayed in Fig. 2.

From Fig. 2 we observe that the power of the procedure based on intrinsic priors is much larger than that based on the g-priors. This is the price the procedures based on g-priors pay for the small Type I error probabilities they have. Further, the power for the intrinsic priors and the mixture

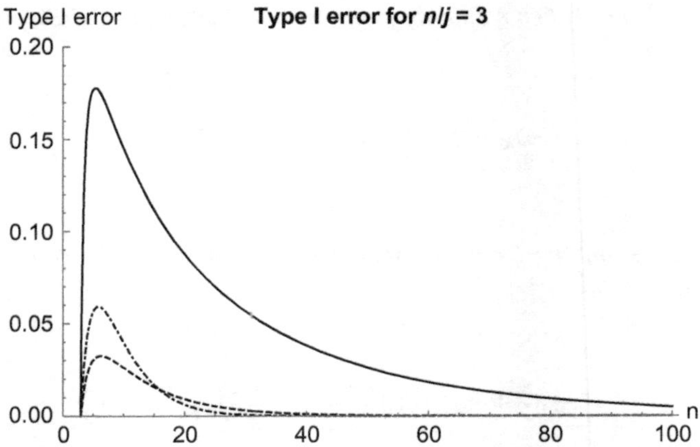

Fig. 1. Type I Error Probabilities for the Intrinsic Procedure (solid line), the g-Prior with $g=n$ (dot-dashed line), and the Mixture of g-Priors (dashed).

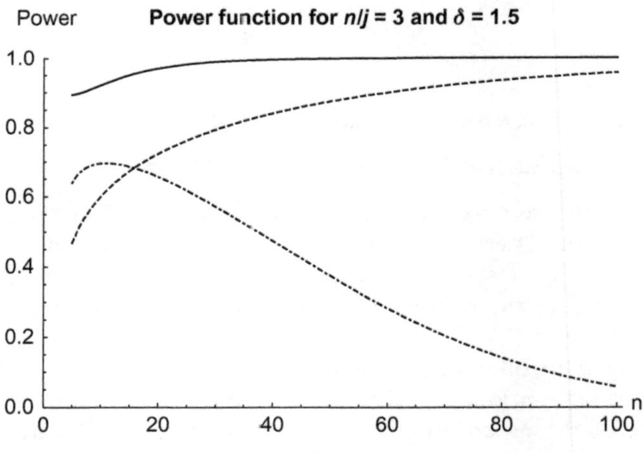

Fig. 2. Power for the Intrinsic Priors (solid line), for g-Priors for $g=n$ (dot-dashed line), and for the Mixture of g-Priors (dashed).

of g-priors increase to one as the sample size n and the model dimension j increase at the same pace, that is, $n/j=r\geq 1$, but the power for the g-prior with $g=n$ increases as n increases up to a certain n and then decreases, which is a surprisingly unreasonable behavior.

On the other hand, it can be seen that the power of the three procedures increases as the pseudo-distance δ_{j0} between model Mj and the null M_0 increases, which is sensible. For a small or moderate δ_{j0} the power of the procedures based on g-priors is much smaller than the power of the intrinsic procedure.

The practical implications of this analysis is that for moderate sample sizes the Bayesian procedure based on the of g-priors are strongly biased toward the null model. Figs. 1 and 2 indicate how unbalanced are the Type I and II error probabilities for the Bayesian procedures based on g-priors compared with that based on the intrinsic priors.

4. CONCLUDING REMARKS

A summary of the definition and properties of the intrinsic priors for model selection have been given, with particular emphasis in variable selection in regression. In the latter case, some popular Bayesian procedures for variable selection have been compared on the basis of their asymptotic and small sample behavior. Two of them use subjective g-priors, and the other the objective intrinsic priors, a choice that we recommend.

When the number of regressors is fixed, or the rate of growing is low, the three Bayesian procedures are consistent; in fact, they all are asymptotically equivalent to the Schwarz approximation (BIC). However, this is not the case when the number of regressors grows with n, an interesting scenario for some applications. The implications are that for models with a large number of regressors, the Bayesian procedure for the g-prior with the choice $g = n$ for comparing the null M_0 versus the alternative M_j, is not consistent. Therefore, this Bayes factor should be ruled out. Further, the Bayes factors for the mixture of g-priors and the intrinsic priors are consistent when the number of regressors k grows with rates $k = O(n^b)$, $b < 1$, and for $k = O(n)$ they are consistent except for a small region of alternative models around the null. Nevertheless, the inconsistency region of the Bayes factor for the mixture of the g-priors contains the inconsistency region of the Bayes factor for the intrinsic priors.

On the other hand, for complex models, the prior on the model space plays an important role in the asymptotic behavior of the Bayesian model selection procedures. The Bernoulli model prior conditional on the hyperparameter p, and the mixture of Bernoulli priors with the uniform prior on p as the mixing distribution, have been considered in Moreno et al. (2014), where the posterior model consistency of the associated Bayesian procedures was analyzed.

We have also presented a small sample size analysis of three Bayesian procedures, which gives out surprising results yielding the conclusion that the intrinsic priors outperform the other two for the g-prior for $g=n$, and the mixture of g-priors. The Type I error of the Bayesian procedures for the latter two priors are unforeseen small, and the consequence is a small power too. The poor behavior of these tests is not shared by the Bayesian procedure for the intrinsic priors which exhibit a more balance Type I and II errors.

5. COMPUTATIONAL IMPLEMENTATION

For variable selection in normal regression and probit models with intrinsic priors, the package "varSelectIP" is available in the CRAN repository in R. Authors: A. Womack, V. Gopal, L. León-Novelo, and G. Casella.

REFERENCES

Bartlett, M. (1957). A comment on D. V. Lindley's statistical paradox. *Biometrika*, *44*, 533–534.

Bayarri, M. J., Berger, J., Forte, A., & García-Donato, G. (2012). Criteria for Bayesian model choice with application to variable selection. *The Annals of Statistics*, *40*, 1550–1577.

Berger, J. O. (1994). An overview of robust Bayesian analysis. *Test*, *3*, 5–124.

Berger, J. O., & Pericchi, L. R. (1996a). The intrinsic Bayes factor for linear models (with discussion). In J. M. Bernardo , J. O. Berger, A. P. Dawid & A. F. M. Smith (Eds.), *Bayesian Statistics 5* (pp. 25–44). Oxford: Oxford University Press.

Berger, J. O., & Pericchi, L. R. (1996b). The intrinsic Bayes factor for model selection and prediction. *Journal of the American Statistical Association*, *91*, 109–122.

Berger, J. O., & Pericchi, L. R. (2001). Objective Bayesian methods for model selection: Introduction and comparison (with discussion). In P. Lahiri (Ed.), *IMS model selection* (Vol. 38, pp. 135–207). Beachwood, OH: Institute of Mathematical Statistics.

Berger, J. O., & Pericchi, L. R. (2004). Training samples in objective Bayesian model selection. *The Annals of Statistics*, *32*, 841–869.

Berger, J. O., & Sellke, T. (1987). Testing a point null hypothesis: The irreconcilability of p-values and evidence. *Journal of the American Statistical Association*, *82*, 112–122.

Cano, J. A., & Salmerón, D. (2013). Integral priors and constrained imaginary training samples for nested and non-nested Bayesian model comparison. *Bayesian Analysis*, *8*, 361–380.

Casella, G., & Berger, R. L. (1987). Reconciling Bayesian and frequentist evidence in the one-sided testing problem. *Journal of the American Statistical Association*, *82*, 106–111.

Casella, G., Girón, F. J., Martinez, M. L., & Moreno, E. (2009). Consistency of Bayesian procedures for variable selection. *The Annals of Statistics*, *37*, 1207–1228.
Casella, G., & Moreno, E. (2006). Objective Bayesian variable selection. *Journal of the American Statistical Association*, *101*, 157–167.
Casella, G., & Moreno, E. (2009). Assessing robustness of intrinsic tests of independence in two-way contingency tables. *Journal of the American Statistical Association*, *104*, 1261–1271.
Clyde, M., & George, E. (2000). Flexible empirical Bayes estimation for wavelets. *Journal of the Royal Statistical Society, Series B*, *62*, 681–698.
Clyde, M., & George, E. (2004). Model uncertainty. *Statistical Science*, *19*, 81–94.
Clyde, M., Parmigiani, G., & Vidakovic, B. (1998). Multiple shrinkage and subset selection in wavelets. *Biometrika*, *85*, 391–401.
Consonni, G., & La Rocca, L. (2008). Tests based on intrinsic priors for the equality of two correlated proportions. *Journal of the American Statistical Association*, *103*, 1260–1269.
Consonni, G., Moreno, E., & Venturini, S. (2011). Testing Hardy-Weinberg equilibrium: An objective Bayesian analysis. *Statistics in Medicine*, *30*(1), 62–74.
Fernández, C., Ley, E., & Steel, M. F. (2001). Benchmark priors for Bayesian model averaging. *Journal of Econometrics*, *100*, 381–427.
Fouskakis, D., Ntzoufras, I., & Draper, D. (2014). Power-expected-posterior priors for variable selection in gaussian linear models. *Bayesian Analysis*. Published Online: September 25, 2014. doi:10.1214/14-BA887
George, E., & Foster, D. P. (2000). Calibration and empirical Bayes variable selection. *Biometrika*, *87*, 731–747.
George, E., & McCulloch, R. E. (1993). Variable selection via Gibbs sampling. *Journal of the American Statistical Association*, *88*, 881–889.
Girón, F. J., Martínez, M. L., Moreno, E., & Torres, F. (2006). Objective testing procedures in linear models: Calibration of the p-values. *Scandinavian Journal of Statistics*, *33*, 765–784.
Gunel, E., & Dickey, G. E. (1974). Bayes factors for independence in contingency tables. *Biometrika*, *61*, 545–557.
Hansen, M. H., & Yu, B. (2001). Model selection and the principle of minimum description length. *Journal of the American Statistical Association*, *96*, 746–774.
Jeffreys, H. (1961). *Theory of probability* (3rd ed.). London: Oxford University Press.
Kass, R. E., & Wasserman, L. (1995). The selection of prior distributions by formal rules. *Journal of the American Statistical Association*, *91*, 1343–1370.
Liang, F., Paulo, R., Molina, G., Clyde, M., & Berger, J. (2008). Mixtures of g-priors for Bayesian variable selection. *Journal of the American Statistical Association*, *103*, 410–423.
Moreno, E. (1997). Bayes factors for intrinsic and fractional priors in nested models. In Y. Dodge (Ed.), *Bayesian robustness* (Vol. 31, pp. 257–270). IMS Lectures Notes–Monograph Series. Hayward, CA: Institute of Mathematical Statistics.
Moreno, E., Bertolino, F., & Racugno, W. (1998). An intrinsic limiting procedure for model selection and hypothesis testing. *Journal of the American Statistical Association*, *93*, 1451–1460.
Moreno, E., & Girón, F. J. (2008). Comparison of Bayesian objective procedures for variable selection in linear regression. *Test*, *17*, 472–490.

Moreno, E., Girón, F. J., & Casella, G. (2010). Consistency of objective Bayes factors as the model dimension grows. *The Annals of Statistics*, *38*, 1937–1952.

Moreno, E., Girón, F. J., & Casella, G. (2014). Posterior model consistency in variable selection as the model dimension grows. *Statistical Sciences*.

Morris, C. N. (1987). Discussion of 'testing a point null hypothesis: The irreconcilability of p-values and evidence'. *Journal of the American Statistical Association*, *82*, 131–133.

Perez, J. M., & Berger, J. O. (2002). Expected-posterior prior distributions for model selection. *Biometrika*, *89*, 491–511.

Pericchi, L. R. (2005). Model selection and hypothesis testing based on objective probabilities and Bayes factors. In D. Dey & C. R. Rao (Eds.), *Handbook of statistics* (Vol. 25, pp. 115–149). Amsterdam: North-Holland.

Scott, J. O., & Berger, J. O. (2010). Bayes and empirical-Bayes multiplicity adjustment in the variable-selection problem. *The Annals of Statistics*, *38*, 2587–2619.

Zellner, A. (1986). On assessing prior distributions and Bayesian regression analysis with g-prior distributions. In P. K. Goel & A. Zellner (Eds.), *Bayesian inference and decision techniques: Essays in honor of Bruno de Finetti* (pp. 233–243). Amsterdam: North-Holland.

Zellner, A., & Siow, A. (1980). Posterior odds ratios for selected regression hypotheses. In J. M. Bernardo, M. H. DeGroot, D. V. Lindley, & A. F. M. Smith (Eds.), *Bayesian statistics*. Proceedings of the first Valencia international meeting held in Valencia (Spain) (pp. 585–603). Valencia: Valencia University Press.

DEMAND ESTIMATION WITH HIGH-DIMENSIONAL PRODUCT CHARACTERISTICS

Benjamin J. Gillen[a], Matthew Shum[a] and Hyungsik Roger Moon[b]

[a]*Department of Economics, California Institute of Technology, Pasadena, CA, USA*
[b]*Department of Economics, University of Southern California, LA, USA*

ABSTRACT

Structural models of demand founded on the classic work of Berry, Levinsohn, and Pakes (1995) link variation in aggregate market shares for a product to the influence of product attributes on heterogeneous consumer tastes. We consider implementing these models in settings with complicated products where consumer preferences for product attributes are sparse, that is, where a small proportion of a high-dimensional product characteristics influence consumer tastes. We propose a multistep estimator to efficiently perform uniform inference. Our estimator employs a penalized pre-estimation model specification stage to consistently estimate nonlinear features of the BLP model. We then perform selection via a Triple-LASSO for explanatory controls, treatment

Bayesian Model Comparison
Advances in Econometrics, Volume 34, 301–323

ISSN: 0731-9053/doi:10.1108/S0731-905320140000034020

selection controls, and instrument selection. After selecting variables, we use an unpenalized GMM estimator for inference. Monte Carlo simulations verify the performance of these estimators.

Keywords: BLP demand model; high-dimensional product characteristics; LASSO; Post-LASSO; shrinkage estimation

JEL classifications: L15, C01, C26, C55

1. INTRODUCTION

A structural model of behavior in which heterogeneous individuals independently make consumption choices according to their own preferences lays the foundation for most empirical investigations into how a product's market share responds to changes in its price. In markets selling complex products, such as cellular phones or automobiles, an individual's preferences could relate to a high-dimensional set of attributes describing these products. Despite this complexity, researchers typically assume that the dominant influences on consumers' decisions can be captured by a relatively small set of "important" characteristics implicitly selected as part of the empirical study. As such, a tractable random-utility model for preferences is predicated by a variable selection exercise that defines which observable attributes should be included and which can be ignored. Indeed, in a truly high-dimensional setting where the dimension of the attributes exceeds the number of product–market observations, such a model selection exercise is necessary simply to identify the estimator in finite samples. In the absence of a principled approach to variable selection, the accuracy of counterfactual inference in post-model selection estimators suffers from inefficiency and can be severely distorted by bias.

In this paper, we consider an automated approach to variable selection for a Berry et al. (1995) (BLP) GMM estimator. In particular, we investigate how variable selection affects post-selection inference for the expected sensitivity of a product's market share to changes in its price. We frame the inference problem as an exercise in estimating the expectation of an endogenous, heterogeneous, treatment effect in the presence of a high-dimensional set of controls and a high-dimensional set of instruments. We propose a set of assumptions to characterize "sparsity" in this context, which ensures that the true model of preference parsimoniously depends on only a few important attributes. We then adopt standard machine learning approaches to variable selection through penalized regression, leveraging the literature on inference in high-dimensional problems to characterize the properties of our estimator.

Driven by computational gains and algorithmic innovations in operations research by computer scientists, researchers are developing new tools for investigating the role of model selection in inference. One cornerstone of these innovations is the LASSO introduced by Tibshirani (1996), which applies penalty to the absolute magnitude of coefficient magnitudes to the sum of squared residuals in a least squares regression model. Subsequent development led by Fan and Li (2001) and Fan and Peng (2004), proposed an adaptive penalization strategy that relaxes the penalty for "important" coefficients to prevent the penalty from distorting the asymptotic distribution of estimates for selected variables. This adaptive feature allows an estimator to achieve "oracle" efficiency in which the asymptotic distribution for the selected variables matches that of a true model that imposes zero-restrictions as if they were known a priori. Huang, Horowitz, and Ma (2008) characterized this property for BRIDGE estimators and Caner (2009) introduced a BRIDGE-style estimator for GMM estimators. Other important extensions for these results beyond linear models are provided by Huang, Horowitz, and Wei (2010), who consider inference in additive nonparametric settings using a grouped LASSO, and Caner and Zhang (2014), who present an elastic net for high-dimensional models identified by a set of moment conditions. The results from Caner and Zhang (2014) are particularly important in the current application, as they allow us to recover the oracle property in a penalized BLP estimator.

As sparsity-based estimation techniques became more broadly applied, new research led by Leeb and Pötscher (2005) began a deeper investigation into the distributional properties of these estimators, identifying contexts in which inference may be distorted despite the oracle property.[1] With Belloni, Chen, Chernozhukov, and Hansen (2012), econometricians started exploring multistep techniques to address the uniform convergence problems associated with post-variable selection inference by adopting a semiparametric perspective focused on characterizing specific treatment effects (Belloni, Chernozhukov, & Wei, 2013; Belloni, Chernozhukov, & Hansen, 2014; Farrell, 2013) or characterizing a finite set of parameteric counterfactuals (Hansen & Kozbur, 2013; Kozbur, 2013). These techniques allow uniform inference on the effects of interest not by exploiting the oracle property, but rather by adjusting the model selection algorithm to "immunize" inference to selection. By using multiple stages of selection, the sampling distribution of post-selection hypothesis test statistics are robust to the misspecification errors that are inevitable in finite-samples. In linear models, this robustness is established by showing the approximation error between a mis-specified model and a properly specified model disappears asymptotically. Though our model is inherently nonlinear, the high-dimensional aspects of the problem enter through a linear component. As such, given a consistent first-stage estimator for the model, we can perform the variable selection and analyze its effects on the sampling distribution of parameter estimates in a linear context.

Though our investigation centers on a specific structural model widely adopted in applied work, the asymptotic properties of our estimator are little different from those analyzed by Belloni et al. (2013b) and Farrell (2013). We follow a multistep procedure to identify a set of variables to control for observable variation in market shares along with a set of "amelioration" controls. This amelioration set immunizes the estimator, ensuring uniform convergence in the dimensions of the parameter space for which we wish to perform inference. To select these variables within a nonlinear model, we use a penalized GMM pre-estimator for the BLP objective function that consistently recovers latent utility shocks in the model. We then perform the following three linear-LASSO regressions:

- Explanatory Controls: linear-LASSO of latent expected utilities on product attributes.
- Treatment Controls: linear-LASSO of prices on product attributes.
- Relevant Instruments: linear-LASSO regression of prices on the selected product attributes, cost shifters, and BLP Instruments.

Finally, we include all variables and instruments which have nonzero attributes from these three LASSO regressions in an unpenalized selected BLP GMM estimator. This approach follows the algorithm laid out in Belloni et al. (2013b), but extends their methodology to a nonlinear GMM setting.

We verify the performance in a set Monte Carlo simulations that require us to confront the computational challenges inherent in estimating structural demand models. We employ the MPEC algorithm from Dubé, Fox, and Su (2012) to optimize the BLP GMM estimator's objective function. An appealing feature of the penalized GMM pre-estimator is that the penalty does not complicate the Hessian of the objective function or the gradients of the equilibrium constraints.

The next section will present the Berry et al. (1995) structural model for demand, the regularity restrictions we impose on the data generating process, and our sparsity assumption on consumer preferences. After that,

Section 3 presents the Three-Stage, Triple-LASSO estimation strategy we use to attain uniform inference. The results of our simulation study are reported in Section 4, before we close with some commentary on possible refinements for our analysis and directions for further research.

2. THE STRUCTURAL MODEL, DATA GENERATING PROCESS, AND SPARSITY

We now detail the structural features of the data generating process and formally state the sampling restrictions we impose on the two. We consider the standard Berry et al. (1995) structural model of demand estimation. The regularity conditions we require are fairly standard for GMM estimators and fully compatible with the structural model. The most novel assumption we introduce relates to our application of sparsity in the consumer preference model.

2.1. BLP Structural Model

The data generating process builds from a set of heterogeneous consumers, with an individual indexed by i, who choose among J products, $j = 1,\ldots, J$ and an outside good $j = 0$ over $t = 1,\ldots,T$ markets. Each product is described by a vector of K characteristics x_{jt} and its price p_{jt}. The latent utility for consumer i from consuming product j in market t is assumed to be linear:

$$u_{ijt} = x'_{jt}\beta_x + p_{jt}\beta_{ip} + \xi_{jt} + \epsilon_{ijt} \tag{1}$$

Implicit in this formulation of utility, we restrict consumers to share uniform tastes for product attributes (β_x) but allow for their sensitivity to prices to be heterogeneous, with $\beta_{ip} \sim N\left(\beta_{0p}, \sigma_p^2\right)$. As usual, the utility formulation also allows for exogenous product–market specific shocks (ξ_{jt}) that provide identifying variation from aggregate market shares. Individual utility's unobserved idiosyncratic component (ϵ_{ijt}) takes the usual Type-I Extreme Value distribution, leading to the logit choice probabilities

$$Pr\{y_{ijt} = j\} = \frac{\exp\left\{x'_{jt}\beta_x + p_{jt}\beta_{ip} + \xi_{jt}\right\}}{1 + \sum_{r=1}^{J} \exp\{x'_{rt}\beta_x + p_{rt}\beta_{ip} + \xi_{rt}\}} \tag{2}$$

To aggregate individual choices into total market shares, we integrate over the individual heterogeneity to translate individual choice probabilities into expected market shares for product j in market t

$$s_{jt} = \int \frac{\exp\left\{x'_{jt}\beta_x + p_{jt}b + \xi_{jt}\right\}}{1 + \sum_{r=1}^{J} \exp\left\{x'_{rt}\beta_x + p_{rt}b + \xi_{rt}\right\}} dF_{\beta_{ip}}\left(b; \beta_{0p}, \sigma_p^2\right) \tag{3}$$

In practice, the integral is typically calculated using pseudo Monte-Carlo methods, however, naïve approaches to numerical integration can be inefficient. As illustrated in applications that approximating complex likelihoods by Heiss and Winschel (2008) and Heiss (2010), Sparse Grid integration can improve both the computation time and the accuracy of numerical integrals, especially in high-dimensional problems. Recent work by Skrainka and Judd (2011) implements Sparse Grid integration for computing the integral in Eq. (3), greatly reducing processing time for estimating the model's objective function.

Identifying the parameters of the model, $\theta = \{\beta_x, \beta_{0p}, \sigma_p\}$ requires linking observed market shares to the latent utility model. A central contribution in Berry et al. (1995) illustrates that, for a given value of θ, a contraction mapping can iteratively solve for the latent product-market shocks (ξ_{jt} (θ; X, Z, p, s)), which can then characterize the mean utility,

$$\delta_{jt}(\theta; X, Z, p, s) = x'_{jt}\beta_x + p_{jt}\beta_{0p} + \xi_{jt}(\theta; X, p, s) \tag{4}$$

Having recovered these latent utility shocks, we can then utilize a set of L exogenous instruments, denoted by z_{jt}, to identify the parameters via the exclusion restriction $E[\xi_{jt}|z_{jt}] = 0$. We let the matrices X, p, s, ξ, δ, and Z collect all observations for the characteristics, price, market shares, utility shocks, latent utilities, and instruments, respectively. We define W as an $L \times L$ weighting matrix and then follow the usual approach of interacting instruments with residuals to define the standard GMM loss function.

$$Q(\theta) = \xi(\theta; X, Z, p, s)' ZW\, Z' \xi(\theta; X, p, s) \tag{5}$$

The instruments themselves are typically taken to include product attributes, observable cost-shifters for producers (typically embedded in a Bertrand competition model), and the attributes of other products competing in a given market. In a classic reference, Nevo (2000) recommends setting the weighting matrix, W, equal to the inverse of the variance–covariance matrix of the instruments, a common implementation. We will return to discuss instrument selection and weighting after defining sparsity for the model in the next section.

2.2. Sampling Assumptions for the Data Generating Process

We begin by presenting a set of concrete sampling assumptions that are sufficient to ensure the penalized GMM estimator we use in the first stage regression provides a consistent estimator for the parameters in the model. The notation and assumptions here are essentially presenting a simplified set of those maintained in Caner and Zhang (2014), with some adaptations to accommodate traditional variable definitions in the BLP model.

Assumption 1. (Sampling Properties for DGP)

1. $K/T \to 0$ as $T \to \infty$. $\theta_0 \in \Theta_K$, Θ_K *is a compact subset of* R^K, *with a compact limit set* $\Theta_\infty := \lim_{T\to\infty}\Theta_K$.
2. *Our moment conditions apply to the* $L \geq K$ *instruments, growing so that* $L/T \to 0$

$$E\left[g_{ljt}(\theta)\right] = E\left[z_{ljt}\xi_{jt}(\theta)\right], l = 1,\ldots,L$$

   Each moment condition $g_{ljt}(\theta)$ *is continuously differentiable in* θ *and the vector of moment functions for a given product–market is denoted* $g_{jt}(\theta)$.
3. *The following uniform law of large numbers as the number of markets* T *becomes large*:

$$\sup_{p\in\{1,\ldots,K\}}\ \sup_{\theta\in\Theta_p}\left\|\frac{1}{JT}\sum_{j,t=1}^{J,T}\left(g_{jt}(\theta) - Eg_{jt}(\theta)\right)\right\| \to_p 0$$

4. *Define the population moment* $m_{lT}(\theta) = E\left[\frac{1}{JT}\sum_{j,t=1}^{J,T} g_{ljt}(\theta)\right]$, *then:*
   (a) $m_{lt}(\theta) \to m_l(\theta)$ *uniformly in* θ *and* K,
   (b) $m_{lt}(\theta)$ *is continuously differentiable and* $m_l(\theta) = 0$ *is continuous in* θ, *and,*
   (c) $m_{lt}(\theta_0) = 0$ *and* $m_{lt}(\theta) \neq 0$, $\forall\theta \neq \theta_0$.

5. *Define the* $L \times K$ *matrix* $\hat{G}_T(\theta) = \frac{1}{T}\sum_{j,t=1}^{J,T} \frac{\partial g_{jt}(\theta)}{\partial \theta'}$.
   (a) *As* T *becomes large, the following uniform law of large numbers holds uniformly in* K *and a neighborhood* N *of* θ_0

$$\left|\left|\hat{G}_T(\theta) - G(\theta)\right|\right|_2^2 \rightarrow_p 0$$

   (b) $G(\theta)$ *is continuous in* θ, *with* $G(\theta_0)$ *having full column rank* K.
6. W_T *is a positive definite matrix and* $||W_T - W||_2^2 \rightarrow_p 0$, *with* W *a symmetric, positive definite, and finite matrix.*
7. *The expected outer product of the score:* $\Omega = \lim_{T \to \infty}(JT)^{-1} \sum_{j,t=1}^{J,T} E\left[g_{jt}(\theta)g_{jt}(\theta)'\right]$ *is a positive, finite matrix*
8. *The minimal and maximal eigenvalues of* $\Sigma = G(\beta_0)'\Omega^{-1}G(\beta_0)$, *denoted* $\underline{e}$ *and* $\overline{e}$, *are bounded between finite constants* $0 < b \leq \underline{e} \leq \overline{e} \leq B < \infty$.

These assumptions are fairly standard in the literature on GMM estimation with some accommodation to reflect the growing number of parameters K and the growing number of instruments L. Importantly for our study, these assumptions allow us to apply Caner and Zhang's (2014) Consistency Theorem 1 to the penalized GMM estimator we propose in the next section. Caner and Zhang (2014) provide additional discussion of these assumptions and their relation to Newey and Windmeijer's (2009) investigation of weak instruments and Zou and Zhang's (2009) study of the adaptive elastic net.

2.3. Sparsity in the Structural Model

With a sufficiently large dimensional set of product attributes, in particular if K is greater than the number of observations JT, the parameter vector θ will not be identified in finite samples. Sparsity provides a reasonable approach to this problem by assuming all but a relatively small number of parameters are restricted to be zero. We follow the customary investigation of the asymptotic distribution for our estimator as the number of markets T becomes large.[2] As such, we are generally considering the number of competing products, J, to be fixed, and the number of markets $T \rightarrow \infty$. To reflect the influence of the high-dimensional feature space, we recall Assumption 1.1, which lets the number of characteristics $K \rightarrow \infty$, while requiring the ratio $K/T \rightarrow 0$. However, we require further restrictions on the number of characteristics assigned that are assigned nonzero influence in consumer utilities:

Assumption 2. (Sparse Consumer Preferences over Product Characteristics)

Consumers' preferences are sparse in the set of product characteristics. That is

a. *All but* $k_T = o(T^{1/4})$ *of the entries in* β_x *are restricted to be equal to zero*
b. *The remaining* k_T *coefficients' true absolute values are all of an order greater than* $T^{-1/2}$.

This formulation of sparsity can be referred to as exact sparsity to indicate that the true data generating process satisfies the zero restrictions exactly. Caner and Zhang's (2014) consistency proof for the elastic net relies on the exact sparsity assumption, and so we maintain it in its present form.[3] This form of sparsity stands in contrast to the notions of approximate sparsity utilized by Belloni et al. (2012, 2013a). The approximate sparsity formulation of the model is naturally motivated as a semiparametric approximation of the relevant conditional expectations, which we intend to consider the extension to approximate sparsity in future work. In an applied context, the exact sparsity condition is, admittedly, somewhat stronger than we would like from a structural modeling perspective. Exact sparsity also appears strictly stronger than necessary for our estimation strategy, as the exact sparsity restriction is required only for consistency in the first-stage regression. The inferential properties of the second-stage estimator are established indirectly using robustness through double-selection and optimal instruments rather than a direct application of the oracle property.

Note that we do not assume the econometrician knows which attributes have zero coefficients. Instead, the objective of the variable selection step is to identify these attributes and impose the correct zero restrictions on the irrelevant attributes. If the post-variable selection estimator almost surely selects the right zero restrictions, then it has the oracle property and its asymptotic distribution will be the same as that of an estimator for a model where the econometrician knows the correct zero restrictions a priori.

Given sparsity in consumer preferences, the oracle property is an appealing objective for efficient estimation. However, the variable selection step itself may risk introducing an omitted variable bias by excluding features of the model whose effect on utility may correlate with variation in price. One of the key intuitions on high-dimensional inference for treatment effects in

Belloni et al. (2012,2013a) addresses this potential omitted variable bias through an "amelioration" set of variables that augments the controls selected to explain variation in the outcome with additional controls that annihilate endogenous variation in the treatment variable. Belloni et al. (2013b) link this amelioration set to the optimal instruments in a Neyman (1959) c(α) test of composite hypotheses. As our interest centers on the response of market shares to manipulable changes in price, our objective is to immunize our inference on outcomes with respect to variation in price. As such, we focus on a multistep selection algorithm that selects not only product attributes that explain market share, but also attributes that could be driving causal variation in relative product prices.

Even outside of the sparse setting, the large number of instruments reflected in competing product attributes introduces a many-moment bias to the GMM estimator commonly encountered in dynamic panel models as highlighted in Newey and Windmeijer (2009). One approach to address the many moment issue would be to adopt a continuous updating GMM estimator, for example, with weak instruments Hansen and Kozbur (2013) propose a JIVE estimator for high-dimensional GMM problems. Another approach follows Reynaert and Verboven (2013), who introduce a two-step estimator that uses a consistent pre-estimate for the model to calculate the Chamberlain (1987) optimal instruments for the BLP problem. The Reynaert and Verboven (2013) approach is particularly well-suited to the current application, as the first-step estimator we use for variable selection can also be used to characterize the optimal instruments for those selected variables.

3. THE TRIPLE LASSO ESTIMATOR

Having established the structural model, sampling properties, and sparsity assumed above, we now describe the algorithm we use to compute our estimator in detail.

3.1. Penalized First-Stage GMM Estimator

Given the assumptions above, a penalized GMM estimator for the BLP objective function can provide a set of consistent estimates for the latent features of the structural model. Denoting the L1 norm of a vector by $||x||_1$

we simply perturb the objective function and define the first-stage parameter estimates, implied utility shocks, and mean utilities, respectively as

$$\begin{aligned} \tilde{\theta} &= \arg \min_{\{\theta \in \Theta\}} \xi(\theta; X, Z, p, s)' Z W Z' \xi(\theta; X, p, s) + \lambda \|\theta\|_1 \\ \tilde{\xi} &= \xi\left(\tilde{\theta}; X, p, s\right) \\ \tilde{\delta} &= X\tilde{\beta}_x + p\tilde{\beta}_{0p} + \tilde{\xi} \end{aligned} \tag{6}$$

If the penalty term λ goes to zero at rate $1/T$, this penalized GMM estimator will consistently select the important features of the model. We assume this in the following.

Assumption 3. (Asymptotically Diminishing Penalty in First Stage Regression)

The penalty term in the first-stage estimator is asymptotically negligible. In particular, $\lambda/T \rightarrow 0$

As the penalty is not adaptive, that is, the penalty is not tuned to each parameter based on a pre-estimation step, the resulting estimates for θ will not have the oracle property. In particular, the penalty will lead estimates for the selected attributes to be shrunk toward zero. Nonetheless, the recovered utility shocks $\xi(\theta;s)$ and mean utilities $\delta(\theta;s)$ from the estimator retain the parametric rates of consistency necessary for the next selection step.

The L1 LASSO penalty is often coupled with an L2 Ridge penalty term in the elastic net to improve the stability of parameter estimates in the presence of collinearity. We opted for a simpler penalty in this application, as a quadratic penalty for the coefficients would complicate the Hessian for the objective function. By simply adopting the LASSO penalty, the Hessian remains unchanged and the Jacobian is only slightly modified, so the Dubé et al. (2012) algorithm remains well-adapted to the problem. The drawback to this approach, however, is that it will require a second step where we estimate a post-selection unpenalized estimator excluding those variables that are not selected in the utility model. We leave the development of a single-step, oracle efficient estimator as a promising candidate for future work.

Nevo (2000) proposes using the inverse of the cross-product of the instruments for the weighting matrix, so $W = (Z'Z)^{-1}$. With a large number of instruments in finite-samples, this matrix may be poorly scaled since $Z'Z$ will be rank-deficient. For this reason, we apply an asymptotically negligible regularization to the weighting matrix so that $W = (Z'Z + \psi I)^{-1}$. A more formal regularization approach could follow Carrasco's (2012) proposed regularization for the many instruments problem.

3.2. A Triple-LASSO for Attribute and Instrument Selection

We now use the results from the first-stage GMM estimator to identify which variables and instruments to include in a second-stage, unpenalized GMM estimator. In an analogy to standard implementations for Feasible Generalized Least Squares estimators, the consistency of $\tilde{\delta}$ and $\tilde{\xi}$ allows us to perform selection in a relatively simple, linear, auxiliary regression. As such, we characterize and implement this selection step using the LASSO for linear models, though an alternative consistent model selection device would yield similar sampling properties.

In the LASSO formulation, each of these estimators includes some form of L1 penalty denoted by $\lambda_{\{\cdot\}}$ that is allowed to grow with the sample size, but we need to impose restrictions so that it has a diminishing impact relative to the sample size asympotically.

Assumption 4. (Asymptotically Diminishing Penalty in Selection)
The penalty terms incorporated in the auxiliary selection estimators are all asymptotically negligible. In particular

$$1.\ \lambda_x/T \to 0 \quad 2.\ \lambda_p/T \to 0 \quad 3.\ \lambda_{zp}/T \to 0$$

These penalties present tuning parameters that are difficult to characterize beyond their magnitude relative to the information contained in the sample to ensure that the penalty retains influence relative to sampling error. In our simulation exercises, we selected this value using a block cross-validation technique, which performed well.

3.2.1. Explanatory Control Variables

A primary consideration in selecting attributes for demand estimation is to include those features that drive mean utility. We identify these attributes by a LASSO regression of the recovered mean utilities, $\tilde{\delta}$, on product attributes.

$$\hat{\gamma}_x = \arg\min \sum_{jt} \left(\tilde{\delta}_{jt} - x\prime_{jt}\gamma_x\right)^2 - \lambda_x \|\gamma_x\|_1 \tag{7}$$

Note that, even without the penalty term, the LASSO regression would asymptotically estimate a zero parameter for those characteristics that are irrelevant to demand estimation.

3.2.2. Treatment-Selection Control Variables

In addition to those attributes which explain mean utilities, we also include a set of variables to control for attribute-driven variation in price. These treatment-selected attributes are identified by the penalized LASSO regression

$$\hat{\gamma}_p = \arg\min \sum_{jt} \left(p_{jt} - x'_{jt}\gamma_p\right)^2 - \lambda_p \|\gamma_p\|_1 \tag{8}$$

This selection step incorporates the double LASSO proposed in Belloni et al. (2013a). As noted in Assumption 4.2, the penalty term λ_p becomes small relative to the sample size and selection is consistent.

3.2.3. The Optimal Instrument Attributes

The Chamberlain (1987) optimal instruments for the BLP estimator are discussed in Reynaert and Verboven (2013). As is usual in IV estimators, the optimal instruments for the exogenous control variables are the control variables themselves and the optimal instrument for the endogenous variable is its exogenous conditional expectation. To identify this set of instruments, we include the selected attributes identified in the previous two sections, along with a set of instruments for explaining variation in price and its impact on utility.

$$\hat{\zeta}_p = \arg\min \sum_{jt} \left(p_{jt} - z'_{jt}\zeta_p\right)^2 - \lambda_{zp} \|\zeta_p\|_1 \tag{9}$$

We could further refine our instruments by following Reynaert and Verboven's (2013) characterization of the optimal instrument for the variance in consumers' response to price as the expected derivative of latent utilities with respect to σ_p. We conjecture that the optimal instrument for σ_p could be constructed from the first-stage penalized regression estimates, but have yet to establish this result formally. This extension is not trivial, as it requires additional smoothness assumptions to ensure the expectation of these derivatives can be consistently estimated by our first-stage model.

3.3. The Unpenalized Post-Triple Selection Estimator

Our final estimates for the parameters in the demand system come from an unpenalized GMM estimator of the BLP model using the selected attributes and instruments described above. The attributes associated with the parameters whose estimates in either of the two control selection regressions 7 and 8 identify the set of product attributes to include in the demand model. These attributes, along with the instruments that have nonzero coefficients in regression 9, then form the set of selected instruments with which to identify that model.

Define X^* and Z^* as the respective matrices of selected controls and instruments and let $\Theta^* \subset \Theta$ as the subset of parameter space where the unselected attributes' corresponding coefficients are fixed at zero. Then then Post-Triple LASSO BLP estimator is given by

$$\begin{aligned} \theta^* &= \arg\min_{\theta \in \Theta^*} Q^*(\theta) \\ Q^*(\theta) &= \xi(\theta; X^*, Z^*, p, s)' Z^* W Z^{*\prime} \xi(\theta; X^*, Z^*, p, s) \end{aligned} \tag{10}$$

The post-triple LASSO BLP estimator satisfies the conditions for uniformly consistent inference on β_{0p} established in Belloni et al. (2013b). Uniform consistency and uniform asymptotic normality for σ_0 is not assured, however, as inference could be affected by an Andrews (2001) problem in a region of parameter space near the boundary value of $\sigma_0 = 0$.

4. MONTE CARLO SIMULATIONS

We verify the proposed estimator's performance in a number of simulation experiments. Here, we describe our simulation experiments, present the set of estimators we evaluate, and characterize the finite-sample distributions of the estimates for the price-related parameters β_{0p} and σ_p^2.

4.1. The Simulated Data Generating Process

To describe the data generating process used in our simulations, we begin by generating product attributes that are jointly normally distributed with a fixed covariance matrix as follows:

$$X_{jt} \sim N(0, \Sigma_X); \quad \Sigma_{X,\{i,j\}} = 0.5^{|i-j|}/16 \tag{11}$$

The scaling of the covariance matrix allows the variation in utilities due to product attributes in our utility model to roughly match the scale of variation in prices. In our simulations, only the first four attributes of X_{jt} are important to the model. All other attributes vary independently of consumer demand behavior and variation in price. A representative consumer's latent utility is then given by

$$\begin{aligned} u_{ijt} &= [4, 4, 2, 2, 0, \ldots, 0]x_{jt} - \beta_{ip}p_{jt} + \xi_{jt} + \epsilon_{ijt} \\ \epsilon_{ijt} &\sim \text{Type II Extreme Value} \\ \xi_{jt} &\sim N(0, 1/4); \beta_{ip} \sim N(5, 0.1) \end{aligned} \tag{12}$$

Price is endogenous to the system, correlated with the important elements of X_{jt} and ξ_{jt}, but its effect is identified by some exogenous variation in the product markup ν_{jt} that is identifiable by its projection onto a single exogenous instrument z_{jt}.

$$\begin{aligned} p_{jt} &= \xi_{jt} + [1, 1, 0, 0, 0, \ldots, 0]x_{jt} + \nu_{jt}; \nu_{jt} \sim U[0, 1] \\ z_{jt} &= \nu_{jt} + [1, 1, 1, 1, 0, \ldots, 0]x_{jt} + \eta_{jt}; \eta_{jt} \sim N(0, 1) \end{aligned} \tag{13}$$

We assume all shocks in all equations are i.i.d. across products and markets. We simulate 20 markets with 10 products, making an expected number of 100 observations, and evaluate the estimators with $K \in \{5, 20, 50, 100, 200\}$.

4.2. The Fitted Estimators

As a benchmark, we evaluate the sampling distribution of the post-Triple LASSO BLP estimator relative to that of an estimator that drops all variables with a zero coefficient from the model. The Oracle Benchmark Estimator is estimated using an unpenalized BLP estimator optimized using the MPEC algorithm imposing the zero restrictions on attributes as if they were known a priori. By exploiting this information to maximum effect for perfect variable selection, the Oracle Benchmark Estimator provides the best possible estimate in the model.

A standard, unpenalized BLP estimator provides another reference point for our proposed estimator's performance. In practice, this estimator is only identified for finite samples when $K < T$, so we can't estimate it for all of our simulation specifications. However, this standard provides a useful device for identifying the loss of efficiency by not exploiting sparsity in those samples for which it is identified.

We also consider two estimators where model selection is not done robustly. The first non-robust estimator comes from the output of our first-stage LASSO-Penalized GMM Estimator. This estimator will not have the oracle property, as the penalty is shrinking its coefficients toward zero and, in so doing, introducing a finite-sample bias. However, we'd also expect a failure of uniform inference, characterized by a breakdown in the normality of the estimator's sampling distribution.

The second non-robust estimator performs only a single-round of attribute selection, selecting the model that best explains latent utility. We call this estimator the Post-Single LASSO BLP estimator. This estimator will also suffer from a failure of uniformly consistent inference, delivering a non-normal sampling distribution for the parameter estimates.

4.3. The Sample Distribution of Estimators

We consider the simulated mean, bias, standard error, and root mean square error for each of the five estimators from the previous subsection as reported in Table 1. We find the performance lines up nicely with the theoretical properties. For low-dimensional problems, the unpenalized GMM estimator, penalized GMM estimator, and the Post-Triple LASSO estimator all perform quite well, delivering estimates that compare favorably to the Oracle estimator. In these cases, the only striking result comes from the Post-Single LASSO estimator, which is badly distorted by a selection-induced omitted variable bias. As the number of nuisance attributes becomes large, the importance of consistent selection and an unpenalized second-stage regression become apparent. After $K = 100$, the unpenalized GMM estimator is not identified in finite samples, giving rise to highly unstable parameter estimates. Further, the penalized GMM estimator begins to suffer from its bias toward zero, even though its root mean square error is still quite reasonable until $K = 200$. Importantly, the Post-Triple LASSO estimator's performance doesn't deteriorate that badly relative to the Oracle Benchmark even when the number of parameters is twice the expected number of observations.

Table 1. Sampling Properties of Estimated Price Parameters.

	Simulated Bias for β_{0p}					Simulated Bias for σ_0				
	$K=5$	$K=20$	$K=50$	$K=100$	$K=200$	$K=5$	$K=20$	$K=50$	$K=100$	$K=200$
Oracle Benchmark	0.00	0.04	(0.07)	0.06	0.16	0.05	0.05	0.06	0.08	0.08
Unpenalized GMM	0.03	0.12	0.08	–	–	0.00	0.00	0.00	–	–
Penalized GMM	0.04	0.12	0.08	0.41	1.18	0.04	0.04	0.04	0.02	0.03
Post-Single LASSO	0.65	0.51	0.32	0.18	0.93	0.07	0.07	0.08	0.08	0.08
Post-Triple LASSO	0.06	0.12	(0.04)	0.07	0.12	0.05	0.05	0.06	0.08	0.08
	Simulated Standard Error for β_{0p}					Simulated Standard Error for σ_0				
Oracle Benchmark	0.46	0.42	0.65	0.29	0.65	0.04	0.04	0.03	0.02	0.01
Unpenalized GMM	0.68	0.33	0.93	–	–	0.00	0.00	0.00	–	–
Penalized GMM	0.68	0.33	0.90	0.44	0.62	0.04	0.04	0.04	0.08	0.06
Post-Single LASSO	1.68	1.20	1.21	1.84	1.98	0.02	0.02	0.02	0.01	0.00
Post-Triple LASSO	0.61	0.83	0.53	0.62	0.71	0.04	0.04	0.03	0.02	0.01
	Simulated Mean Square Error for β_{0p}					Simulated Mean Square Error for σ_0				
Oracle Benchmark	0.46	0.42	0.65	0.29	0.67	0.06	0.06	0.07	0.08	0.08
Unpenalized GMM	0.68	0.35	0.93	–	–	0.00	0.00	0.00	–	–
Penalized GMM	0.69	0.35	0.90	0.60	1.33	0.06	0.06	0.05	0.08	0.07
Post-Single LASSO	1.80	1.30	1.25	1.85	2.19	0.08	0.08	0.08	0.08	0.08
Post-Triple LASSO	0.61	0.84	0.54	0.62	0.72	0.06	0.06	0.07	0.08	0.08

The table reports simulated sampling bias, standard error, and root mean square error for five different approaches to model selection for demand estimation. The data generating process is presented in Section 4.1 and the presented results are computed from a set of 500 simulation runs for $J=10$ products competing in $T=20$ markets, giving an expected count of 100 observations. The Oracle Benchmark provides the best infeasible estimator, based on perfect model selection a priori. The unpenalized GMM estimator is the usual BLP estimator based on Eq. 5. The Penalized GMM estimator comes from the first-stage LASSO-BLP estimation characterized by Eq. 6. The Post-Single LASSO estimator includes only those variables selected via Eq. 7. The Post-Triple LASSO uses our full algorithm of triple-selection followed by unpenalized estimation of the selected model, as characterized by Eq. 10.

Next, we evaluate the LASSO selection procedure itself to ensure the estimator is including the correct attributes in the model and check the frequency with which incorrect attributes are included. Table 2 reports these statistics, illustrating that the Penalized GMM and Post-Triple LASSO estimators effectively include the important attributes in the model. It also suggests the presence of an omitted variable bias in the Post-Single LASSO estimator, which regularly omits at least one important variable. We also see that the Triple-LASSO is fairly effective at filtering out irrelevant attributes, including on average only one to two additional attributes beyond the important controls. Finally, we note that, although the Penalized GMM estimator is influencing parameter values, it does assign a number of null attributes significant coefficients that are not selected by the Triple-LASSO algorithm. This lack of sharp selection suggests a call for more aggressive penalization, but this enhanced penalty would further bias the estimates. We conjecture that these issues could be jointly addressed by deriving an adaptively penalized estimator for this problem, a possibility we leave for further research.

Finally, we consider the full sampling distribution of the price-related parameters. Fig. 1 presents kernel density plots of the estimates for β_{0p}

Table 2. Model Selection Accuracy and Efficiency.

Panel A Frequency of Selecting Important Attributes					
	$K=5$	$K=20$	$K=50$	$K=100$	$K=200$
Oracle Benchmark	100%	100%	100%	100%	100%
Unpenalized GMM	100%	100%	100%	100%	100%
Penalized GMM	100%	100%	100%	100%	100%
Post-Single LASSO	75%	75%	75%	75%	39%
Post-Triple LASSO	100%	100%	100%	100%	88%
Panel B: Average Number of Null Attributes Selected					
Oracle Benchmark	–	–	–	–	–
Unpenalized GMM	3.76	17.78	46.75	97.15	198.87
Penalized GMM	3.76	17.77	46.73	93.81	192.13
Post-Single LASSO	0.02	0.01	0.02	0.03	1.09
Post-Triple LASSO	0.86	1.15	1.25	1.34	2.24

The table reports on the effectiveness of our proposed model selection algorithms. Panel A reports the frequency with which the model selected the four "important" attributes assigned nonzero coefficients in the simulated data generating process. Panel B reports the average number of additional, "null" attributes which are included in the selected model. The data generating process and estimators match those reported in Table 1.

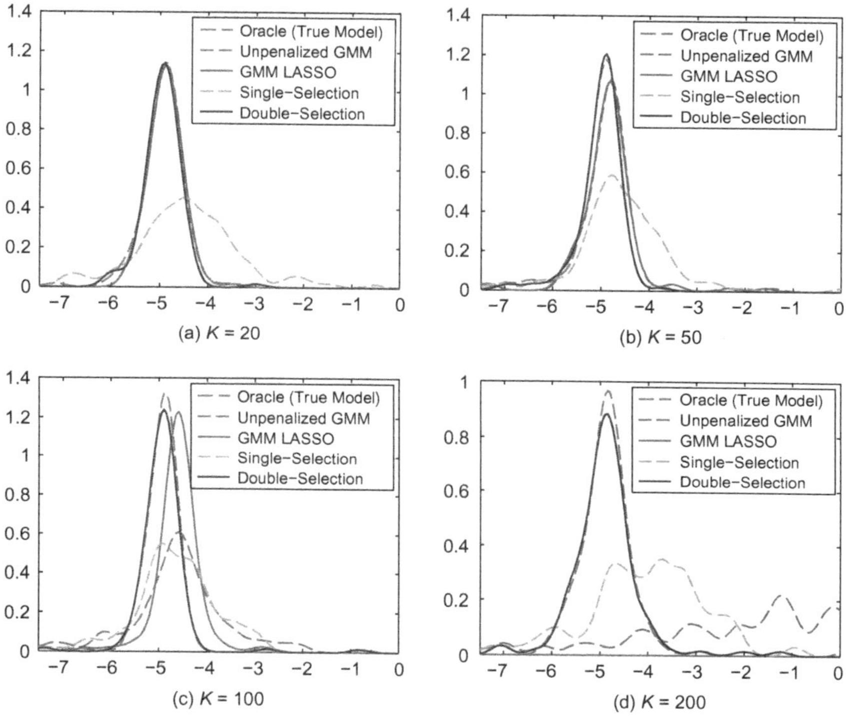

Fig. 1. Distribution of Simulated Price Coefficient β_{0p}. *Note*: The figure plots kernel density estimates of the simulated sampling distribution for our demand estimation models as we increase the number of irrelevant attributes in the model. The data generating process and estimators match those reported in Table 1.

for the proposed estimators. We again see the pattern indicated in Table 1. In a low-dimensional context, all the estimators except the variable-omitting Post-Single LASSO estimator perform relatively well. As the number of null features grows, the bias becomes more severe for the penalized GMM estimator because the penalty term grows relative to the magnitude of the GMM objective function. We also see the unpenalized GMM estimator begin to lose efficiency with a relatively large number of parameters. However, even with a high-dimensional parameter space, the Post-Triple LASSO estimator performs comparably to the Oracle benchmark.

5. TODAY'S CONCLUSIONS AND TOMORROW'S DIRECTIONS

Our analysis provides a proposed method for selecting attributes from a high-dimensional feature space for estimating consumer demand models. We adopt sufficiently strong conditions to show that our estimator achieves uniform, $\sqrt{n}$-consistent inference. Further, we have verified this performance and the importance of robust selection in our simulation study.

Though our estimator has appealing finite-sample properties, the conditions under which it was derived are stronger than strictly necessary. A natural next step would more closely inspect the features of our first-stage estimator that may allow for a more efficient estimation strategy. We are particularly interested in an adaptive penalization strategy that controls the bias introduced by the LASSO estimator by penalizing each parameter based on the results of a first-stage estimator. This approach would extend the adaptive elastic net proposed by Caner and Zhang (2014), however it would require extending their results to a setting with many irrelevant instruments. There are some promising innovations in the work by Fan and Liao (2012), Hansen and Kozbur (2013), and Cheng and Liao (2013) for models with an exponential number of irrelevant instruments that could provide theoretical foundations for the exercise. This exercise may be more amenable to an empirical likelihood approach, following the specification of Conlon (2013) for the demand estimation problem and the asymptotic analysis of Tang and Leng (2010).

Another potential refinement of our approach would consider directly modeling the optimal instruments for the random coefficients. As noted by Belloni et al. (2013b), there is a tight link between the robustness properties of a Post-Double LASSO and the optimal instruments identifying hypothesis tests for the relevant parameters. Using our first-stage regression results, we could apply the algorithms discussed in Reynaert and Verboven (2013) to include only the optimal instruments for our problem rather than the full set of instruments selected in the third iteration of the LASSO.

Finally, it would be interesting to explore the feasibility of selecting the attributes for which consumers display heterogeneous preferences. In our current exercise, the coefficients for product attributes were fixed in the population and the price effect was known a priori to be heterogeneous. Can we effectively test for multiple random coefficients and identify which attributes should be assigned these random coefficients? How does selection interact with unobserved heterogeneous effects? Evaluating this question is complicated by the non-negativity of the variance parameter for the random coefficients, which introduces an Andrews (2001) problem.

At a foundational level, we would also like to motivate the sparsity assumption from a model of consumer behavior featuring either limited attention or satisficing. This structural foundation could be more readily established in the context of an approximate formulation of sparsity. That is, even though consumers focus on only a few attributes, we would rather not assume they jointly coordinate on which attributes to consider. It would be more natural to directly model the approximation error induced by model selection to the consumers' true utility representation. Such a formulation would also lead to a model of consumer demand with complex products that more closely reflects the empirical environment.

NOTES

1. Post-model selection estimators suffer from a failure of uniform convergence in parameter space. Leeb and Pötscher (2005) show that this failure occurs in the regions of parameter space where the parameter's value is local to zero, or of order smaller than $n^{-1/2}$, distorting parametric-rate inference against local alternatives. Though the oracle property leads to estimators that perform well pointwise and can be effectively used to test against nonlocal alternatives, this distortion indicates a practical limitation of oracle efficient estimators. Indeed, a number of papers, including Leeb and Pötscher (2006, 2008), cast doubt on the feasibility of uniformly consistent inference based on the oracle property.

2. In a novel analysis of highly competitive markets, Armstrong (2012) characterizes the effect of a large number of products in the marketplace. As our interest centers on learning about demand behavior for a small number of complex products, the traditional many-market asymptotics provide the relevant limiting behavior for our estimator.

3. Caner and Zhang (2014) present an adaptive elastic net estimator in which the number of nonzero coefficients can be allowed to grow at a rate of $T^{1/3}$, so Assumption 2(a) is slightly stronger than it needs to be. However, the sharper 1/4-th root rate of growth in the nonzero parameter vector is consistent with a nonadaptive penalty allowing us to retain consistency in a first-step estimator. Assumption 2 (b) is often made weaker or stated in terms of more primitive objects relating to the penalty of the estimator. However, the presentation of Assumption 2(b) here highlights the importance of the discontinuity that allows us to exclude those variables whose effects are of order less than $T^{-1/2}$.

ACKNOWLEDGMENT

We are grateful to David Brownstone, Martin Burda, Garland Durham, Jeremy Fox, Stefan Holderlein, Ivan Jeliazkov, Dale Poirier, Guillame

Weisang, Frank Windmeijer, and seminar participants at the Advances in Econometrics Conference on Bayesian Model Comparison and the California Institute of Technology for helpful comments. We owe special thanks to Alexander Charles Smith for important insights early in developing the project.

REFERENCES

Andrews, D. W. K. (2001). Testing when a parameter is on the boundary of the maintained hypothesis. *Econometrica*, *69*(3), 683–734.

Armstrong, T. (2012). *Large market asymptotics for differentiated product demand estimators with economic models of supply*. Technical report, Working Paper. Department of Economics, Yale University, New Haven, CT.

Belloni, A., Chen, D., Chernozhukov, V., & Hansen, C. (2012). Sparse models and methods for optimal instruments with an application to eminent domain. *Econometrica*, *80*(6), 2369–2429.

Belloni, A., Chernozhukov, V., & Hansen, C. (2014). Inference on treatment effects after selection amongst high-dimensional controls. *Review of Economic Studies*, *81*(2), 608–650

Belloni, A., Chernozhukov, V., & Wei, Y. (2013). Honest confidence regions for logistic regression with a large number of controls. Working Paper. Department of Economics, MIT, Cambridge, MA.

Berry, S., Levinsohn, J., & Pakes, A. (1995). Automobile prices in market equilibrium. *Econometrica*, *63*(4), 841–890 ISSN 00129682. doi:10.2307/2171802. Retrieved from http://dx.doi.org/10.2307/2171802

Caner, M. (2009). LASSO-type GMM estimator. *Econometric Theory*, *25*(1), 270–290.

Caner, M., & Zhang, H. H. (2014). Adaptive elastic net for generalized methods of moments. *Journal of Business & Economic Statistics*, *32*(1), 30–47.

Carrasco, M. (2012). A regularization approach to the many instruments problem. *Journal of Econometrics*, *170*(2), 383–398.

Chamberlain, G. (1987). Asymptotic efficiency in estimation with conditional moment restrictions. *Journal of Econometrics*, *34*(3), 305–334.

Cheng, X., & Liao, Z. (2013). Select the valid and relevant moments: An information based LASSO for GMM with many moments. Working Paper. Department of Economics, University of Pennsylvania, Philadelphia, PA.

Conlon, C. T. (2013). The empirical likelihood MPEC approach to demand estimation. Working Paper. Department of Economics, Columbia University, New York, NY.

Dubé, J.-P., Fox, J., & Su, C.-L. (2012). Improving the numerical performance of static and dynamic aggregate discrete choice random coefficients demand estimation. *Econometrica*, *80*(5), 2231–2267.

Fan, J., & Li, R. (2001). Variable selection via nonconcave penalized likelihood and its oracle properties. *Journal of the American Statistical Association*, *96*(456), 1348–1360.

Fan, J., & Liao, Y. (2012). *Endogeneity in ultrahigh dimension*. Working Paper. Benheim Center of Finance, Princeton University, Princeton, NJ.

Fan, J., & Peng, H. (2004). Nonconcave penalized likelihood with a diverging number of parameters. *The Annals of Statistics*, *32*(3)928–961.

Farrell, M. H. (2013). Robust inference on average treatment effects with possibly more covariates than observations. Working Paper. Booth School of Business, University of Chicago, Chicago, IL.

Hansen, C., & Kozbur, D. (2013). *Instrumental variables estimation with many weak instruments using regularized JIVE*. Working Paper. Booth School of Business, University of Chicago, Chicago, IL

Heiss, F. (2010). The panel probit model: Adaptive integration on sparse grids. In W. Greene & R. C. Hill (Eds.), *Maximum simulated likelihood methods and applications* (Vol. 26, pp. 41–64). Advances in Econometrics. Bingley, UK: Emerald Group Publishing Limited.

Heiss, F., & Winschel, V. (2008). Likelihood approximation by numerical integration on sparse grids. *Journal of Econometrics*, *144*, 62–80.

Huang, J., Horowitz, J. L., & Ma, S. (2008). Asymptotic properties of bridge estimators in sparse high-dimensional regression models. *The Annals of Statistics*, *36*(2)587–613.

Huang, J., Horowitz, J. L., & Wei, F. (2010). Variable selection in nonparametric additive models. *Annals of Statistics*, *38*(4), 2282.

Kozbur, D. (2013). *Inference in additively separable models with a high dimensional component*. Working Paper. Booth School of Business, University of Chicago, Chicago, IL.

Leeb, H., & Pötscher, B. M. (2005). Model selection and inference: Facts and fiction. *Econometric Theory*, *21*(01)21–59.

Leeb, H., & Pötscher, B. M. (2006). Can one estimate the conditional distribution of post-model-selection estimators? *The Annals of Statistics*, *34*, 2554–2591.

Leeb, H., & Pötscher, B. M. (2008). Sparse estimators and the oracle property, or the return of Hodges estimator. *Journal of Econometrics*, *142*(1), 201–211.

Nevo, A. (2000). A practitioner's guide to estimation of random-coefficients logit models of demand. *Journal of Economics & Management Strategy*, *9*(4), 513–548.

Newey, W. K., & Windmeijer, F. (2009). Generalized method of moments with many weak moment conditions. *Econometrica*, *77*(3), 687–719.

Neyman, J. (1959). Optimal asymptotic tests of composite statistical hypotheses. *Probability and Statistics*, *57*, 213.

Reynaert, M., & Verboven, F. (2013). Improving the performance of random coefficients demand models: The role of optimal instruments. *Journal of Econometrics*, *179*(1), 83–98.

Skrainka, B. S., & Judd, K. L. (2011). High performance quadrature rules: How numerical integration affects a popular model of product differentiation. Working Paper. Hoover Institution, Stanford University, Palo Alto, CA.

Tang, C. Y., & Leng, C. (2010). Penalized high-dimensional empirical likelihood. *Biometrika*, *97*(4), 905–920.

Tibshirani, R. (1996). Regression shrinkage and selection via the lasso. *Journal of the Royal Statistical Society: Series B (Methodological)*, *58*, 267–288.

Zou, H., & Zhang, H. H. (2009). On the adaptive elastic-net with a diverging number of parameters. *Annals of Statistics*, *37*(4), 1733.

COPULA ANALYSIS OF CORRELATED COUNTS

Esther Hee Lee
EViews, IHS Global Inc., Irvine, CA, USA

ABSTRACT

Copula modeling enables the analysis of multivariate count data that has previously required imposition of potentially undesirable correlation restrictions or has limited attention to models with only a few outcomes. This article presents a method for analyzing correlated counts that is appealing because it retains well-known marginal distributions for each response while simultaneously allowing for flexible correlations among the outcomes. The proposed framework extends the applicability of the method to settings with high-dimensional outcomes and provides an efficient simulation method to generate the correlation matrix in a single step. Another open problem that is tackled is that of model comparison. In particular, the article presents techniques for estimating marginal likelihoods and Bayes factors in copula models. The methodology is implemented in a study of the joint behavior of four categories of US technology patents. The results reveal that patent counts exhibit high levels of correlation among categories and that joint modeling is crucial for eliciting the interactions among these variables.

Bayesian Model Comparison
Advances in Econometrics, Volume 34, 325–348

ISSN: 0731-9053/doi:10.1108/S0731-905320140000034021

Keywords: Copula; count data; marginal likelihood; Markov chain Monte Carlo (MCMC)

JEL classifications: C11; C25; C52

1. INTRODUCTION

The analysis of multivariate nonlinear models requires careful modeling of the correlation among responses. Unlike linear settings, where inference may be possible even if such correlation is ignored, nonlinear models that do not account for correlation may be severely misspecified. One area of research where incorporating a flexible and interpretable correlation structure in empirical models has been a particularly difficult task is the analysis of multivariate count data (see, e.g., Cameron & Trivedi, 1998; Winkelmann, 2008). As a consequence, models for multivariate counts have either sacrificed generality or have been specified with potentially undesirable correlation restrictions imposed for the sake of retaining computational tractability. For example, multivariate Poisson (MVP) models are often constructed by including a common Poisson process that enters every outcome in the model, but this approach can only produce nonnegative correlations (Karlis, 2003; Kocherlakota & Kocherlakota, 1992). Both positive and negative correlations can be obtained in the Poisson-lognormal model (Aitchison & Ho, 1989; Tunaru, 2002), however, this model imposes a high degree of overdispersion and the implied correlations are typically very small. In general, the analysis of multivariate counts has been difficult because of the lack of suitable discrete multivariate distributions that can support complex correlation structures (for a discussion, see Chib & Winkelmann, 2001).

To address these problems, this paper approaches the analysis of correlated counts using a Gaussian copula model. A copula is a function that represents the joint distribution in terms of its marginals (Sklar, 1959), and hence can be used to *couple* any discrete and/or continuous distributions. An appealing feature of copula modeling is that it can be used to retain well-known parametric families for the marginal distributions even though they may not be easily extendable to multivariate settings.

To define a copula, one should note that for any arbitrary cdf $F_j(\cdot)$, $j = 1,\ldots, q$, if u_j is uniform on the interval [0, 1], then $y_j = F_j^{-1}(u_j)$ is distributed according to $F_j(\cdot)$. The mapping from u_j to y_j is one-to-one when y_j is continuous and many-to-one when y_j is discrete. The multivariate function C on the unit hypercube $[0, 1]^q$

$$C(u_1, \ldots, u_q) = F(F_1^{-1}(u_1), \ldots, F_q^{-1}(u_q)) = F(y_1, \ldots, y_q)$$

is called a copula if it is a continuous distribution function and each marginal is a uniform distribution on [0, 1]. It defines a correspondence between the marginal distributions F_j and the joint distribution F. Although many types of copulas exist (see Clayton, 1978; Demarta & McNeil, 2005; Frank, 1979; Genest & Rivest, 1993; Gumbel, 1960; Huard, Evin & Favre, 2006; Zimmer & Trivedi, 2006), Gaussian copulas are a natural choice when moving beyond the bivariate case. By taking $y_j = F_j^{-1}(u_j)$ for $j = 1,\ldots, q$ and $\mathbf{u} = (u_1, \ldots, u_q)'$ on the unit intervals, this paper focuses on the Gaussian copula $\mathbb{C}$

$$\mathbb{C}(\boldsymbol{u}|\boldsymbol{\Omega}) = \Phi_q\{\Phi^{-1}(F_1(y_1)), \ldots, \Phi^{-1}(F_q(y_q))|\boldsymbol{\Omega}\}$$

where Φ represents a univariate standard normal cdf and Φ_q is the q-variate cdf for a latent vector $z_i = (z_{i1}, \ldots, z_{iq})'$, $z_i \sim N(0,\boldsymbol{\Omega})$, where $\boldsymbol{\Omega}$ is in correlation form. Note that $u_j = \Phi(z_{ij})$, and the dependence among $\mathbf{y}_i = (y_{i1}, \ldots, y_{iq})'$ is created through the correlation of the latent variables z_i.

To illustrate the main idea, suppose that the jth count y_{ij} $(j = 1,\ldots, q)$ takes its values according to the latent variable z_{ij}

$$y_{ij} = k \times 1\gamma_{\mathrm{j,k-1}} < \mathrm{z_{ij}} \leq \gamma_{\mathrm{j,k}}\}, \quad k = 1,\ldots, q$$

where cutpoints $\gamma_{j,k}$ determine the discretization of the data. Fig. 1 displays how a Gaussian copula connects the univariate Poisson probability $\Pr(Y = k|\lambda)$ with $\ln(\lambda) = x'\beta$ to the standard normal pdf and cdf ($\phi(\cdot)$ and $\Phi(\cdot)$), respectively, in the following relation

$$\Pr(Y = k|\lambda) = \int_{\gamma_{k-1}}^{\gamma_k} \phi(z|0, 1)dz = \Phi(\gamma_{k-1}) - \Phi(\gamma_k)$$

Note that the cutpoints $\gamma_k = \Phi^{-1}(F(Y = k|\lambda))$ are uniquely defined as a function of λ and β. It shows that by using the latent variables, each of the marginals F_j is transformed to the standard normal distribution.

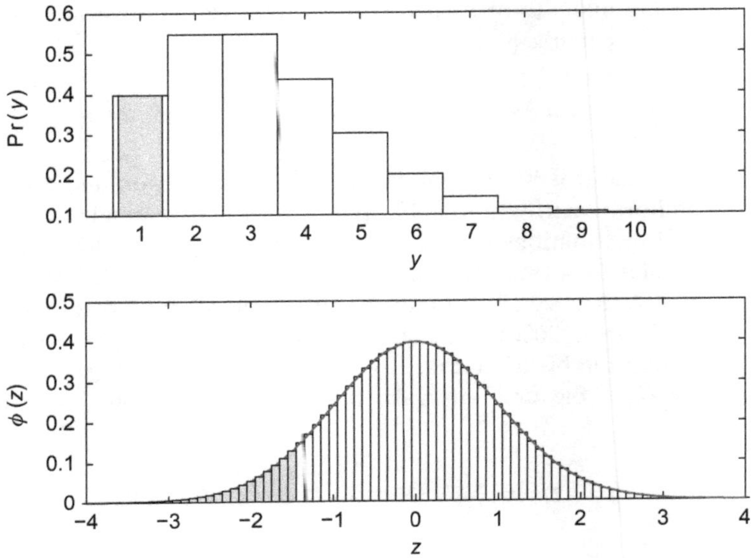

Fig. 1. The Univariate Gaussian Copula with The Poisson Marginal. The Shaded Areas of Both Panels Represent the Same Probability.

Even though the copula is Gaussian, the data $\mathbf{y}_i$ $(i=1,\ldots,n)$ can be non-Gaussian and can include count, censored, ordered, and any other type for which F_j is known, or any combination of discrete and continuous marginals. For the case of count data, the individual marginal distributions F_j $(j=1,\ldots,q)$ can be modeled as Poisson, negative binomial, or Poisson-lognormal, whereby formal model comparison tests can be used to provide useful information about the presence and degree of over dispersion that may be inherent in the data. The multivariate probit model can be considered a simple example of a Gaussian copula model with univariate Probit marginals. Since the Gaussian copula provides a useful and general way of modeling correlation between arbitrary data generating processes, it is particularly attractive when no natural or flexible multivariate distribution exists for the data being considered, but the dependence among $(y_1,\ldots,y_q)$ is of interest.

While a Gaussian copula has long existed in the statistics literature (for extensive theoretical discussion on copulas see Joe, 1997; Nelson, 1999; Trivedi & Zimmer, 2005), its applications in econometrics are relatively recent. Even though the term copula is not explicitly used, by using a

Gaussian copula, van Ophem (1999) considers correlation of bivariate counts when two Poisson marginals have different parameters. Song (2000) considers the Gaussian copula to provide a class of multivariate discrete and continuous models, and estimates the parameters for the continuous marginals in the context of maximum likelihood estimation. Pitt, Chan, and Kohn (2006) show a general estimation method for the multivariate Gaussian copula model when the marginals are either discrete or continuous, but estimate the components of $\mathbf{\Omega}$ one element at a time.

This paper makes the following contributions. First, it extends the applicability of the copula approach to settings with high-dimensional outcomes. For those settings, the paper provides an efficient sampling algorithm for $\mathbf{\Omega}$ in which all elements are sampled jointly and mitigates the computational inefficiencies and poor mixing problems that may arise in one-at-a-time samplers. By employing the parameter expanded reparameterization and Metropolis–Hastings (PX-RPMH) algorithm introduced by Liu and Daniels (2006), this paper provides an alternative approach to simulate all components of the correlation matrix $\mathbf{\Omega}$ in a single step by first drawing a covariance matrix from an inverse Wishart distribution and then transforming it into a correlation matrix that is accepted with a specific MH acceptance probability.

Second, this paper presents techniques for marginal likelihood estimation. Bayesian model selection has been an open issue in copula models because marginal likelihood computation has been complicated by the presence of posterior full-conditional densities of nonstandard form. For this reason, earlier work has used methods for model selection based on information criteria such as Akaike information criterion (AIC), Schwarz information criterion (SIC), and deviance information criterion (DIC) (for details see dos Santos Silva & Lopes, 2008; Huard et al., 2006; Spiegelhalter, Best, Carlin & van der Linde, 2002). This paper shows how to extend the approach in Chib and Jeliazkov (2001) to the context of PX-RPMH sampling to provide a simple and reliable estimate of the marginal likelihood.

Finally, the proposed techniques are used to study the joint behavior of patent counts in the United States. While copulas have been applied to model value-at-risk problems (Cherubini, Luciano & Vecchiato 2004; Embrechts & McNeil, 2003; Embrechts, McNeil & Straumann, 2002), insurance decisions (Frees, Carriere & Valdez 1996; Klugman & Parsa, 1999), and health care analysis (Prieger, 2002; Smith, 2003; Zimmer & Trivedi, 2006), patent data applications have not yet been pursued despite their importance. However, joint modeling is crucial for eliciting substitution and complementarity patterns among different patent categories. Due to the difficulty of analyzing statistical models that account for the interactions among patent counts, such models have been generally difficult to estimate. This study shows that the copula approach can be effective in examining issues regarding the joint behavior of the four categories of patents and features of their conditional distribution such as overdispersion.

The remainder of the paper is organized as follows. Section 2 presents the formal framework behind count modeling with negative binomial marginal distributions, which offer a generalization over Poisson models and allow for different degrees of overdispersion for the individual data series. Section 3 presents the proposed Markov chain Monte Carlo (MCMC) estimation algorithm. Building upon the proposed estimation algorithm, Section 4 presents techniques for marginal likelihood estimation in copula models. The results of a simulation study are presented in Section 5, while an empirical application to patent analysis is reported in 6. Concluding remarks are presented in Section 7.

2. MODEL DESCRIPTION

To illustrate the Gaussian copula framework for count data problems, we consider n observations $\mathbf{y}_1, \ldots, \mathbf{y}_n$ where each is of dimension q. Specifically, we consider a Gaussian copula model in which the marginal distributions F_j $(j = 1, \ldots, q)$ are negative binomial

$$y_{ij} \sim \text{NegBin}(\lambda_{ij}, \delta_j), \quad i = 1, \ldots, n$$

with mean λ_{ij} and variance $\lambda_{ij}(1+\delta_j)$. The mean is parameterized as $\lambda_{ij} = \exp(x_{ij}'\beta_j)$ to allow dependence on covariates, and $\boldsymbol{\delta} = (\delta_1, \ldots, \delta_q)'$ is an overdispersion parameter. For subsequent use, let $\lambda_i = (\lambda_{i1}, \ldots, \lambda_{iq})'$ and $\beta = (\beta_1, \ldots, \beta_q)'$.

From the Gaussian copula representation, the likelihood contribution can be derived as

$$f(\mathbf{y}_i | \boldsymbol{\beta}, \boldsymbol{\delta}, \boldsymbol{\Omega}) = \int_{B_{iq}} \cdots \int_{B_{i1}} \phi_q(z_i | \mathbf{0}, \boldsymbol{\Omega}) dz_i \tag{1}$$

where the cutpoints defining the region of integration $B_{ij} = (\gamma_{i,j-1}, \gamma_{i,j}]$ are specified as a function of the parameters β_j and δ_j

$$\begin{aligned} \gamma_{i,j-1} &= \Phi^{-1}(F_j(y_{ij} - 1|\beta_j, \delta_j, z_{ij})) \\ \gamma_{i,j} &= \Phi^{-1}(F_j(y_{ij}|\beta_j, \delta_j, z_{ij})) \end{aligned} \tag{2}$$

where F_j $(j = 1,\ldots, q)$ denotes the jth negative binomial marginal distribution. In this setting, we need to note that the likelihood contribution of the ith subject in Eq. (1) has no closed form solution so that simulation methods become indispensable. These methods are discussed next.

3. ESTIMATION OF THE MODEL

Given the data $\boldsymbol{y} = (\boldsymbol{y}_1, \ldots, \boldsymbol{y}_n)'$, a joint posterior distribution of the parameters $(\boldsymbol{\beta}, \boldsymbol{\delta})$ and the latent data $z = (z_1, \ldots, z_n)'$ can be expressed as

$$\pi(\boldsymbol{\beta}, \boldsymbol{\delta}, z|\boldsymbol{y}) \propto \pi(\boldsymbol{\beta}|\beta_0, B_0)\pi(\boldsymbol{\delta}|\delta_0, D_0)\pi(\boldsymbol{\Omega}|\nu_0, R_0) \prod_{i=1}^{n} \phi_q(z_i|\boldsymbol{0}, \boldsymbol{\Omega})\Pr(\boldsymbol{y}_i|z_i, \boldsymbol{\beta}, \boldsymbol{\delta}, \boldsymbol{\Omega})$$

where the priors are assumed to be $\boldsymbol{\beta} \sim N(\beta_0, B_0)$, $\boldsymbol{\delta} \sim \mathrm{IG}(\nu_0, d_0)$, and the inverse Wishart on $\Omega \sim \mathrm{IW}(\Sigma|\nu_0, R_0)$ (the choice of the prior on Ω will be discussed later in detail).

The main idea of the simulation approach is to construct the following MCMC sampling scheme

$$[z_{ij}|y_{ij}, \theta_j, z_{i\backslash j}, \Omega]; \quad [\theta_j|y_{ij}, z_{i\backslash j}, \Omega]; \quad [\Omega|z]$$

where $\theta_j = (\beta_{j'}, \delta_j)'$ and $z_{\cdot\backslash j} = (z_{\cdot 1}, \ldots, z_{\cdot j-1}, z_{\cdot j+1}, \ldots, z_{\cdot q})$.

3.1. *Sampling* z_{ij}

We first discuss the generation of z_{ij} from a univariate normal distribution $\phi(z_{ij}|\mu_{ij}, \Omega_{ij}^2)$ truncated between γ_{j-1} and γ_j that are defined in Eq. (2). The conditional mean μ_{ij} and variance Ω_{ij}^2 of z_{ij} given $z_{i\backslash j}$ are given by

$$\begin{aligned} \mu_{ij} &= \Omega_{j\backslash j}\Omega_{\backslash j\backslash j}^{-1} z_{i\backslash j} \\ \Omega_{ij}^{2} &= \Omega_{jj} - \Omega_{j\backslash j}\Omega_{\backslash j\backslash j}^{-1}\Omega_{\backslash jj} \end{aligned} \tag{3}$$

where $\Omega_{j\backslash j}$, $\Omega_{\backslash jj}$, and $\Omega_{\backslash j\backslash j}$ consist of all elements of $\boldsymbol{\Omega}$ that are not in $\Omega_{\cdot j}$, $\Omega_{j\cdot}$, and Ω_{jj}, respectively.

3.2. *Sampling θ_j*

To fix notation, define $\theta_{\backslash j} = (\theta'_{-j}, \theta'_{+j})'$ where $\theta_{-j} = (\theta_1, \ldots, \theta_{j-1})'$ denotes the blocks up to $j-1$ and $\theta_{+j} = (\theta_{j+1}, \ldots, \theta_q)'$ denotes the blocks beyond j. The MH algorithm proceeds by sampling a vector of $(q+1)$ parameters from the density proportional to

$$\pi(\theta_j | y_{\cdot j}, z_{\cdot\backslash j}, \theta_{\backslash j}, \Omega) \propto \pi(\theta_j | y_{\cdot j}, z_{\cdot\backslash j}, \Omega) \propto \pi(\theta_j | \theta_{\backslash j}) \prod_{i=1}^{n} f(y_{ij} | z_{i\backslash j}, \theta_j, \Omega)$$

where

$$f(y_{ij} | z_{i\backslash j}, \theta_j, \Omega) = \int \phi(z_{ij} | z_{i\backslash j}, \Omega) f(y_{ij} | z_{ij}, \theta_j) dz_{ij}$$

By using the cutpoints γ_{ij} in Eq. (2), we can easily see that

$$\Pr(Y_{ij} = y_{ij} | z_{i\backslash j}, \theta_j, \Omega) = \Phi\left(\frac{\gamma_{ij} - \mu_{ij}}{\Omega_{ij}}\right) - \Phi\left(\frac{\gamma_{ij-1} - \mu_{ij}}{\Omega_{ij}}\right)$$

where μ_{ij} and Ω_{ij} are specified in Eq. (3). Also, to ensure numerical stability in the calculation of $F_j(\cdot)$, $\Phi(\cdot)$, and $\Phi(\cdot)^{-1}$, at each step the algorithm verifies that $\epsilon < \Phi^{-1}(F_j(\cdot)) < 1 - \epsilon$, where $\epsilon = 10^{-6}$.

Suppose that θ_j is the current value and $\theta_j^{\dagger}$ is the generated value from the proposal. The MH probability of move from θ_j to $\theta_j^{\dagger}$ is

$$\alpha_j(\theta_j, \theta_j^{\dagger} | y_{\cdot j}, z_{\cdot\backslash j}, \Omega) = \min\left\{1, , \frac{\pi(\theta_j^{\dagger} | y_{\cdot j}, \theta^{\dagger}_{-j}, \theta_{+j}, z_{\cdot\backslash j}, \Omega)}{\pi(\theta_j | y_{\cdot j}, \theta^{\dagger}_{-j}, \theta_{+j}, z_{\cdot\backslash j}, \Omega)} \frac{f_T(\theta_j | \hat{\theta}_j, V, \nu)}{f_T(\theta_j^{\dagger} | \hat{\theta}_j, V, \nu)}\right\} \tag{4}$$

The proposal density is a multivariate t distribution $f_T(\theta_j|\hat{\theta}_j, V, \nu)$ with the modal value $\hat{\theta}_j$ and inverse of the negative Hessian V at the mode of $\ln f(\theta_j|y_{\cdot j}, z_{\cdot\backslash j}, \theta_{\backslash j}, \Omega)$, and degrees of freedom ν. We need to note that the proposal density needs to dominate the true density in the tails (for more details see Chib & Jeliazkov, 2001).

3.3. Sampling of Ω

The parameter-expanded reparameterization and Metropolis–Hastings (PX-RPMH) algorithm proposed in Liu and Daniels (2006) was employed to generate $\mathbf{\Omega}$ given z from the density proportional to

$$\pi(\mathbf{\Omega}|\mathbf{y}, z) \propto \pi(\mathbf{\Omega}) \prod_{i=1}^{n} f(\mathbf{y}_i|\theta, \mathbf{\Omega})$$

where $\boldsymbol{\theta} = (\theta_1, ..., \theta_q)'$. Sampling a correlation matrix $\mathbf{\Omega}$ can be problematic due to the two constraints: positive definiteness and fixed diagonal elements. However, the PX-RPMH algorithm circumvents these problems by simulating a positive definite *covariance* matrix $\mathbf{\Sigma}$ from the inverse Wishart distribution and modifies it to be a *correlation* matrix $\mathbf{\Omega}$. Specifically, it relates $\mathbf{\Sigma}$ to $\mathbf{\Omega}$ in the following fashion

$$\mathbf{\Omega} = D^{-1} \Sigma D^{-1}$$

by using an expansion parameter $D = \text{diag}\left(\sqrt{\sigma_{11}}, ..., \sqrt{\sigma_{qq}}\right)$. Since $\mathbf{\Omega}$ has fewer parameters than Σ, an additional constraint

$$z_i = D^{-1} z_i^* \text{ where } \sum_{i=1}^{n} z_{ij}^{*2} = n, \quad j = 1, ..., q \tag{5}$$

is required to make the transformation from $\mathbf{\Sigma}$ to $\mathbf{\Omega}$ a one-to-one mapping. The Jacobian of the transformation is given by $J_{\Sigma \to \Omega} = |D^{-1}|^{n+q+1}$.

Regarding the selection of the prior for $\mathbf{\Sigma}$, Zhang, Boscardin, and Belin (2006) proposed a parameter-extended Wishart prior which is a joint Wishart distribution for $(\mathbf{\Omega}, D)$. Liu (2001) considered a Jeffrey's prior on $\mathbf{\Omega}$. Chib and Greenberg (1998) used a multivariate truncated normal distribution as the prior distribution for $\mathbf{\Omega}$ in a multivariate probit model. Pitt et al. (2006) put the covariance selection prior, introduced by Wong, Carter, and Kohn (2003), which allows the off-diagonals of its inverse to be identically zero. Liu and Daniels (2006) employed the uniform prior of Barnard, McCulloch, and Meng (2000).

However, the paper employs an inverse Wishart prior $f_{\mathrm{IW}}(\boldsymbol{\Sigma}|\nu_0, R_0)$ which has been widely used in Bayesian estimation. A proposal value $\boldsymbol{\Sigma}^{\dagger}$ is also obtained from the inverse Wishart proposal density. We need to note that the Wishart family for the unrestricted covariance matrix $\boldsymbol{\Sigma}$ captures important features of the restricted correlation matrix $\boldsymbol{\Omega}$. The following is a summary of the PX-RPMH algorithm using the Wishart family.

PX-RPMH Algorithm: Given the current value $\boldsymbol{\Sigma}$ and Eq. (5),

- PX step: Sample $\boldsymbol{\Sigma} \sim \mathrm{IW}_q(\nu, S)$ where $\nu = \nu_0 + n$ and $S = \left[R_0^{-1} + \sum_{i=1}^{n} z_i^* z_i^{*\prime}\right]^{-1}$.
- MH step: Move to the new value $\boldsymbol{\Sigma}^{\dagger}$ with the probability $\alpha(\boldsymbol{\Sigma}, \boldsymbol{\Sigma}^{\dagger}|z)$

$$\alpha(\boldsymbol{\Sigma}, \boldsymbol{\Sigma}^{\dagger}|y, z) = \min\left\{1, \frac{f_{\mathrm{IW}}(\boldsymbol{\Sigma}^{\dagger}|\nu_0, R_0)\prod_{i=1}^{n} f(y_i|\boldsymbol{\theta}, \boldsymbol{\Sigma}^{\dagger})}{f_{\mathrm{IW}}(\boldsymbol{\Sigma}|\nu_0, R_0)\prod_{i=1}^{n} f(y_i|\boldsymbol{\theta}, \boldsymbol{\Sigma})} \frac{f_{\mathrm{IW}}(\boldsymbol{\Sigma}|\nu, S)}{f_{\mathrm{IW}}(\boldsymbol{\Sigma}^{\dagger}|\nu, S)}\right\} \tag{6}$$

  and stay at the old value $\boldsymbol{\Sigma}$ with probability $1 - \alpha(\boldsymbol{\Sigma}, \boldsymbol{\Sigma}^{\dagger}|z)$.
- RP step: Transform $\boldsymbol{\Sigma}^{\dagger}$ into $\boldsymbol{\Omega}^{\dagger}$ by $\boldsymbol{\Omega}^{\dagger} = D^{\dagger\,-1}\boldsymbol{\Sigma}^{\dagger}D^{\dagger\,-1}$.

4. MARGINAL LIKELIHOOD COMPUTATION

The analysis of the marginal likelihood has been an open issue in the copula modeling because the appropriate model specification is rarely known. Given the data $\boldsymbol{y}$, the marginal likelihood is the integral of the likelihood $f(\boldsymbol{y}|\Theta)$ with respect to the prior density $\pi(\Theta)$ of the parameters Θ

$$m(\boldsymbol{y}) = \int f(\boldsymbol{y}|\Theta)\pi(\Theta)d\Theta$$

On the basis of the marginal likelihood, one can proceed Bayesian model choice by pairwise comparisons of any two models $\mathcal{M}_A$ and $\mathcal{M}_B$ through their posterior odds ratio

$$\frac{\Pr(\mathcal{M}_A|y)}{\Pr(\mathcal{M}_B|y)} = \frac{\Pr(\mathcal{M}_A)}{\Pr(\mathcal{M}_B)} \mathrm{B}_{A,B}$$

where the first fraction on the right-hand side is known as the prior odds and the second as the Bayes factor

$$\mathrm{B}_{A,B} = \frac{m(y|\mathcal{M}_A)}{m(y|\mathcal{M}_B)}$$

which is the ratio of marginal likelihoods of the two models.

Based on the recognition that the marginal likelihood $m(y)$ is a normalizing constant of the posterior density, Chib (1995) provides an expression

$$m(y) = \frac{f(y|\Theta^*)\pi(\Theta^*)}{\pi(\Theta^*|y)}$$

which is referred to as the basic marginal likelihood identity. Note that this density can be evaluated at any point Θ^*, however, for estimation efficiency this paper uses the posterior mean of parameters. Evaluating the right-hand side of this identity at any appropriate point Θ^* and taking logarithms one can obtain the expression

$$\ln m(y) = \ln f(y|\Theta^*) + \ln\pi(\Theta^*) - \ln\pi(\Theta^*|y)$$

Since $f(y|\Theta^*)$ and $\pi(\Theta^*)$ are either available directly or by an alternative calculation, calculation of the marginal likelihood is reduced to finding an estimate of the posterior ordinate $\pi(\Theta^*|y)$ at a single point Θ^*.

In the context of the paper, since the parameter vector Θ consists of $\boldsymbol{\theta} = (\theta_1, \ldots, \theta_q)$ and Σ, the posterior ordinate at Θ^* can be estimated by the decomposition

$$\pi(\Theta^*|y) = \pi(\theta^*|y, \Sigma^*)\pi(\Sigma^*|y)$$

Due to the approach by Chib and Jeliazkov (2001), estimation of the first ordinate is decomposed by $\pi(\boldsymbol{\theta}^*|y, \Sigma^*) = \prod_{j=1}^{q} \pi(\theta_j^*|y_{\cdot j}, z_{\cdot j}, \Sigma^*, \theta_{-j}^*)$, and each of which is estimated by adapting Eq. (4)

$$\hat{\pi}(\theta_j^*|y_{\cdot j}, z_{\cdot \backslash j}, \Sigma^*, \theta_{-j}^*) = M^{-1}\sum\nolimits_{g=1}^{M} \alpha_j(\theta_j^{(g)}, \theta_j^*|y_{\cdot j}, \theta_{-j}^*, \theta_{+j}, z_{\backslash j}, \Sigma^*) f_T(\theta_j| \hat{\theta}_j, V, \nu) \times \frac{1}{J^{-1}\sum\nolimits_{g=1}^{J} \alpha_j(\theta_j^*, \theta_j^{(j)}|y_{\cdot j}, \theta_{-j}^*, \theta_{+j}, z_{\backslash j}, \Sigma^*)}$$

where the draws $\{\theta_j^{(g)}\}$ are from the conditional posterior $\pi(\theta_j, \theta_{+j}|y_{\cdot j}, z_{\cdot \backslash j}, \Sigma, \theta_{-j}^*)$ and the draws $\{\theta_j^{(j)}\}$ are from the conditional product measure $\pi(\theta_{+j}|y_{\cdot j}, z_{\cdot \backslash j}, \Sigma, \theta_{-j}^*, \theta_j^*) f_T(\theta_j^*|\hat{\theta}_j, V, \nu)$. Estimation of the second ordinate $\pi(\Sigma^*|\mathbf{y})$ is done by adapting Eq. (6)

$$\hat{\pi}(\Sigma^*|\mathbf{y}) = \hat{\pi}(\boldsymbol{\sigma}^*|\mathbf{y}) = \frac{M^{-1}\sum\nolimits_{g=1}^{M} \alpha(\boldsymbol{\sigma}^{(g)}, \boldsymbol{\sigma}^*|\mathbf{y}, z^{(g)}) f_{\text{IW}}(\boldsymbol{\sigma}^{(g)}|\nu, S)}{J^{-1}\sum\nolimits_{g=1}^{J} \alpha(\boldsymbol{\sigma}^*, \boldsymbol{\sigma}^{(j)}|\mathbf{y}, z^{(j)})}$$

where $\boldsymbol{\sigma} = \{\sigma_{jk} : j \geq k\}$ $(1 \leq j, k \leq q)$ denotes the half-vectorized $q(q+1)/2$ dimensional column vector. The draws $\{\boldsymbol{\sigma}^{(g)}, z^{(g)}\}$ are from $\pi(\boldsymbol{\sigma}, z|y)$ and the draws $\{\boldsymbol{\sigma}^{(j)}, z^{(j)}\}$ are from $\pi(\boldsymbol{\sigma}, z|\mathbf{y}) f_{\text{IW}}(\boldsymbol{\sigma}^*|\nu, S)$.

5. SIMULATION STUDY

This paper conducts two simulation studies by using a Gaussian copula approach with negative binomial marginals. The first study delivers insight into the advantages of copula modeling in deriving the dependence structure between observations $\mathbf{y}_i = (y_{i1}, \ldots, y_{in})'$. Specifically, the Gaussian copula relates the correlations imposed by the latent variables $z_i \sim N(\mathbf{0}, \Omega)$ to the correlations among $\mathbf{y}_i$. The second study illustrates the ability of the proposed methodology to estimate parameters in the marginal distributions and the elements of the correlation matrix Ω.

For illustrative purposes, this paper provides the implied correlations between bivariate counts $\mathbf{y}_i = (y_{i1}, y_{i2})'$ under several parameter settings. Specifically, the off-diagonal element Ω_{12} of the correlation matrix $\Omega_{2\times 2}$ is set to be $\pm 0.3, \pm 0.6$, or ± 0.9. The elements of the mean λ_i are fixed to be 0.1ι, 1ι, or 10ι $(\iota = 1_{2\times 1})$, and for obtaining a positive overdispersion parameter all elements of δ are set to be 5 and/or 10. This setting would lead to the generalization of the Gaussian copula in modeling the multivariate dependence only with the correlations from the latent normal distribution.

Table 1 shows that the values of $\text{Corr}(y_{\cdot 1}, y_{\cdot 2}) = \rho_{12}$, captured by the Gaussian copula, largely correspond to the values of $\text{Corr}(z_{\cdot 1}, z_{\cdot 2}) = \Omega_{12}$.

Table 1. Dependence Structures ($\mathrm{Corr}(y_{\cdot 1}, y_{\cdot 2}) = \rho_{12}$) for Bivariate Counts.

Ω_{12}	δ	$\lambda = 0.1\iota$	$\lambda = 1\iota$	$\lambda = 10\iota$
−0.9	5ι	−0.095	−0.624	−0.817
	10ι	−0.096	−0.651	−0.851
−0.6	5ι	−0.093	−0.439	−0.553
	10ι	−0.094	−0.453	−0.572
−0.3	5ι	−0.069	−0.234	−0.278
	10ι	−0.070	−0.240	−0.287
0.3	5ι	0.142	0.260	0.296
	10ι	0.143	0.262	0.300
0.6	5ι	0.348	0.538	0.593
	10ι	0.345	0.535	0.596
0.9	5ι	0.695	0.837	0.896
	10ι	0.690	0.833	0.895

When $\lambda_i = 10\iota$ and $\delta_j = 10$, we see that the values of ρ_{12} closely match the values of Ω_{12}. The cases with the positive values of ρ_{12} exhibit relatively close matches, compared to the cases with the negative values of ρ_{12}.

The simulation study demonstrates two advantages of the copula model. One advantage is to have a rich correlation structure between counts due to removing the restriction of positive correlations. Kocherlakota and Kocherlakota (1992) provided a comprehensive discussion of the MVP model, and stated that the relatively simple structure is the reason for its popularity in count data problems. However, for the MVP model to be useful, the correlations between counts should be nonnegative. Consider a vector $\mathbf{y}_i = (y_{i1}, \ldots, y_{iq})'$ with $y_{ij} = z_{ij} + u_i$, where $z_{ij} \sim \mathrm{Poisson}(\lambda_{ij})$ and $u_i \sim \mathrm{Poisson}(\gamma)$ are independent. Because of the convolution property of independent Poisson variables, the marginal distribution of y_{ij} is also a Poisson distribution with parameter $\lambda_{ij} + \gamma$. The correlation and covariance matrices of $\mathbf{y}'_i$ are given by

$$\mathrm{Corr}(y_{ij}, y_{ik}) = \frac{\gamma}{\sqrt{(\lambda_{ij} + \gamma)(\lambda_{ik} + \gamma)}}, \quad j \neq k$$

$$\mathrm{Var}(\mathbf{y}_i) = \Lambda_i + \gamma \mathbf{1}\mathbf{1}'$$

where $\Lambda_i = \mathrm{diag}(\lambda_{i1}, \ldots, \lambda_{iq})$ and $\mathbf{1}$ is a q vector of ones. Since $\gamma \geq 0$ and the denominator of the correlation is the square root of $(\lambda_{ij} + \gamma)(\lambda_{ik} + \gamma)$, we can see that the MVP model allows nonnegative correlations only. It is shown that the correlations of $\mathbf{y}_i$ in the multivariate negative binomial (MVNB) model should be nonnegative (see Winkelmann, 2001, 2008). In Table 1, we see that the copula model, unlike the aforementioned models, seems promising in its ability to explain a wide range of correlation structures by allowing for both negative and positive correlations between counts.

Another advantage of copula modeling is to facilitate the analysis of correlated counts with overdispersion. Empirical studies, for example, Hall, Griliches, and Hausman (1986), and Hausman, Hall, and Griliches (1984), pointed out that overdispersion is a common feature of count data. In order to deal with overdispersion, the MVNB model has been studied as an analogue to the MVP model. Importantly, Aitchison and Ho (1989) showed that a multivariate Poisson-log-normal (MVLN) model allows for overdispsersion in Poisson marginals. Table 1 suggests that the copula model also has this property. We can find the richness of the copula model in its ability to model a flexible range of correlations regardless of variation in overdispersion parameter δ (and the mean value λ_i).

The second simulation study considers a Gaussian copula model with $i = 1, \ldots, 300$ and $j = 1, \ldots, 4$. β_j is set to be a 4×1 vector of 0.3's and a 4×1 vector of covariates x_{ij} is independently generated from the standard normal distribution. For an overdispersion parameter δ, a relatively large number for $\boldsymbol{\delta} = 0.65\iota'$ is used to achieve convergence in optimization. The Gaussian copula constructs the dependence structures among $\mathbf{y}_i = (y_{i1}, \ldots, y_{in})'$ by using latent variables $\mathbf{z}_i$ which are generated from $N(\mathbf{0}, \Omega)$ such that

$$\Omega = \begin{pmatrix} 1.000 & -0.668 & 0.096 & 0.500 \\ -0.668 & 1.000 & 0.223 & -0.294 \\ 0.096 & 0.223 & 1.000 & 0.027 \\ 0.500 & -0.294 & 0.027 & 1.000 \end{pmatrix}$$

We conducted the proposed MCMC simulations with the full conditional distributions of $\boldsymbol{\beta}$, $\boldsymbol{\delta}$, $\{z_i\}$, and $\boldsymbol{\Omega}$ where $\boldsymbol{\theta} \equiv (\boldsymbol{\beta}, \boldsymbol{\delta})$ and $\boldsymbol{\Omega}$ were updated by M−H and PX-RPMH steps, respectively. A chain of 55,000 iterations was generated and the last 50,000 samples were retained. To apply the proposed sampling method we used the following priors: $\boldsymbol{\beta} \sim N(\mathbf{0}, I_4)$, $\boldsymbol{\delta} \sim \text{IG}(0.5I_4, 0.5I_4)$, and $\boldsymbol{\Sigma} \sim \text{IW}(6, I_4)$. The parameters $\boldsymbol{\theta} = (\boldsymbol{\beta}, \boldsymbol{\delta})$ were initialized to be MLE values and the off-diagonal elements of $\boldsymbol{\Omega}$ were initialized at zeros which were far from the true values.

The dependence structures among $\mathbf{y}_i$ in this instance are shown in a correlation matrix $\hat{\Omega}$ as

$$\hat{\Omega} = \begin{pmatrix} 1.000 & -0.020 & 0.047 & 0.277 \\ -0.020 & 1.000 & 0.141 & -0.100 \\ 0.047 & 0.141 & 1.000 & 0.004 \\ 0.277 & -0.100 & 0.004 & 1.000 \end{pmatrix}$$

The maximum likelihood estimates (MLE) from each of the univariate marginal distribution are computed by maximizing the jth likelihood function and these estimates are reported in Table 2. The posterior distributions of the parameters β_{jk} and δ_j $(j, k = 1,\ldots, 4)$ are obtained by applying the proposed MCMC algorithm. Our results for the posterior are summarized in Table 2 and Fig. 2. We see that the estimates for $\boldsymbol{\beta}$ relatively converge to the true values, but the MLE estimates for $\boldsymbol{\delta}$ have notable inconsistency. It is shown that the proposed simulation method produces estimates with smaller variability than MLE. The more interesting finding in this study is obtaining the posterior estimates for $\boldsymbol{\Omega}$ by generating all elements of $\boldsymbol{\Omega}$ in a single step. We see that the estimates for all diagonals converge to the true

Table 2. ML and MCMC Estimates for the Gaussian Copula Model.

Parameter	True	Univariate Model		Multivariate Model		
		MLE	SE	Posterior	NSE	SD
β_{11}	0.3	0.22	0.064	0.26	0.002	0.045
β_{12}	0.3	0.33	0.060	0.32	0.002	0.044
β_{13}	0.3	0.37	0.063	0.25	0.002	0.044
β_{14}	0.3	0.30	0.064	0.32	0.002	0.048
β_{21}	0.3	0.24	0.054	0.31	0.002	0.047
β_{22}	0.3	0.39	0.055	0.30	0.002	0.050
β_{23}	0.3	0.24	0.048	0.26	0.002	0.050
β_{24}	0.3	0.34	0.050	0.32	0.002	0.051
β_{31}	0.3	0.26	0.063	0.26	0.003	0.048
β_{32}	0.3	0.30	0.060	0.23	0.003	0.051
β_{33}	0.3	0.39	0.055	0.31	0.002	0.049
β_{34}	0.3	0.26	0.067	0.24	0.002	0.051
β_{41}	0.3	0.29	0.053	0.28	0.002	0.045
β_{42}	0.3	0.28	0.049	0.32	0.002	0.046
β_{43}	0.3	0.36	0.060	0.35	0.002	0.043
β_{44}	0.3	0.34	0.054	0.24	0.002	0.045
δ_1	0.65	0.98	0.218	0.69	0.005	0.090
δ_2	0.65	2.38	0.770	0.63	0.005	0.092
δ_3	0.65	1.22	0.337	0.70	0.005	0.095
δ_4	0.65	1.73	0.414	0.67	0.004	0.088

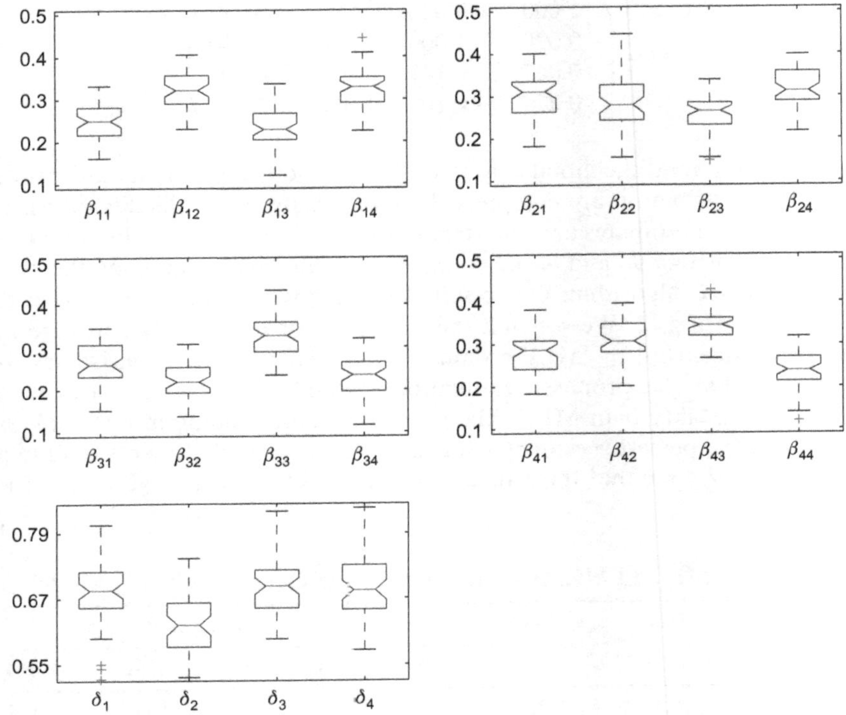

Fig. 2. MCMC Estimates for the Multivariate Gaussian Copula Model.

Table 3. Posterior Estimates for the Free Elements of Correlation Matrix Ω.

	True	Mean	SD	NSE	Lower	Upper
Ω_{21}	−0.668	−0.674	0.040	0.001	−0.753	−0.405
Ω_{31}	0.096	0.145	0.039	0.001	0.078	0.231
Ω_{32}	0.223	0.252	0.038	0.001	0.187	0.335
Ω_{41}	0.500	0.478	0.032	0.001	0.414	0.539
Ω_{42}	−0.294	−0.291	0.039	0.001	−0.366	−0.213
Ω_{43}	0.027	0.048	0.040	0.001	−0.030	0.127

Table 4. Log Marginal Likelihood Estimates.

	Type of Sampling	Marg.	NSE
PX-RPMH with joint uniform	All elements at one time	−399.96	0.0671
PX-RPMH with inverse Wishart	All elements at one time	−399.93	0.0349

value 1, and the estimates for the off-diagonal elements are reported in Table 3.

To assess the analysis of the marginal likelihood computation more formally, we performed the evaluation of the marginal likelihood with two different proposals; one is the joint uniform proposal which is provided in Liu and Daniels (2006)

$$q(\Omega, \Omega^{\dagger}|z) = |\Omega|^{-\frac{q+1}{2}} I\left\{\Omega_{jk} : \Omega_{jk} = 1(j=k), |\Omega_{jk}| < 1(j \neq k)\right\} \prod_{i=1}^{n} f(y_i|\boldsymbol{\theta}, \Omega)$$

and the other is the inverse Wishart proposal. This will show how prior (and proposal) selection on the covariance matrix $\boldsymbol{\Sigma}$ leads to a more efficient estimator of the correlation matrix Ω for the PX-RPMH algorithm. Table 4 presents that the inverse Wishart proposal case produces the value −399.93 with a numerical standard error 0.0349, while the joint uniform proposal case produces the value −399.96 with a numerical standard error 0.0671. Note that the larger numerical standard error of the joint uniform proposal case is a consequence of higher inefficiency factors (for more details see Chib & Jeliazkov, 2001), while the loss in efficiency is small by having the inverse Wishart prior.

6. APPLICATION

This section considers an application to computer patents in the United States. Patent counts are often viewed as a measure of the number of successful outcomes among a large (but unobserved) number of projects, each of which has a small probability of success. By focusing on the correlations among patent counts from different categories, this research shows how different kinds of patent categories are related to each other at the firm-level. Understanding the dependence structure of patent categories enables us to gain insights into the firms' behavior and the complementarity or substitution patterns that may exist in intra-firm R&D investment decisions.

The application makes two contributions to the analysis of patent data. The first is to allow for overdispersion in each patent category. Since the Poisson distribution is widely used for many count data problems in econometrics, it seems to be a natural assumption that patent counts have the Poisson marginals $\boldsymbol{y}_i \sim \text{Po}(\boldsymbol{\lambda}_i)$ with $\ln(\boldsymbol{\lambda}_i) = X_i\boldsymbol{\beta}$ where $\boldsymbol{y}_i$ is the number of patents by firm i and X_i is a vector of explanatory variables, primarily R&D expenditure. However, the patent data exhibit two notable features: relatively large means and heavy upper tails, the latter of which usually leads to the presence of overdispersion. The attempt to mitigate the overdispersion of the patent data has been of interest in several studies. Hausman et al. (1984) and Hall et al. (1986) used a fixed-effect Poisson model that accommodates overdispersion by adding unobserved firm-specific factors in the marginals. More flexible (semi)parametric approaches assume that the Poisson mean is random where the distribution of the mean is referred to a mixing distribution (see Alfò & Trovato, 2004). By assuming that the mixing distribution is a continuous parametric form, Wang, Cockburn, and Puterman (1998) presented a mixed-Poisson regression model in the context of maximum likelihood estimation. Guo and Trivedi (2002) considered semiparametric mixture models by dropping strong parametric assumptions about the mixing distribution. This study shows that the Gaussian copula model with the Negative Binomial distributed marginals also allows for overdispersion.

The second contribution of the application is to offer a new approach to the empirical study of patents. Due to the difficulty of analyzing statistical models that account for substitution and complementarity patterns among multiple patent counts, such models have received little attention in the previous literature. However, since the copula can relate any marginal distributions to a joint distribution, we can account for the joint behavior of patent counts. In this example, we consider the dependence structure of patent counts $\boldsymbol{y}_i$ as the correlation matrix $\boldsymbol{\Omega}$ where the firm's profitability $z_i \sim N(\boldsymbol{0}, \Omega)$ is unobserved. We need to note that, as described in Section 5, the Gaussian copula could sustain a full range of correlations between counts y_i only with the correlation structure imposed by the latent variables z_i.

The sample consists of 498 observations on 254 manufacturing firms during the 1980s from the US Patent & Trademark Office (USPTO) dataset of Hall, Jaffe, and Trajtenberg (2001) which is available on the National Bureau of Economic Research (NBER) website. By recognizing that the propensity of patenting may differ across industries depending on the relative costs of innovation versus imitations, industries with the (relatively) high imitation costs are considered. Specifically, the study contains the four subcategories from a technology category, Computers and Communications, given in Table 5, in which firms have a strong incentive to file patents to protect their intellectual properties. To avoid difficulties by having interindustry variation, this study restricts its attention to firms within computers and instruments industry which is classified in Mairesse and Hall (1996).

The three covariates, introduced in Mairesse and Hall (1996), reflect the characteristics of individual firms (these covariates are described in Table 6). The measure of yearly sales is included. The measure of the labor is the average number of workers at the end of the fiscal year. R&D capital, described in Hall (1990), is constructed using a perpetual inventory with a depreciation rate of 15% per year. Note that to measure the contribution of R&D to firms' productivity this study constructs the sample from the firms that perform and report R&D spending in the 1980s.

The MCMC simulation was run with 5,000 burn-in iterations and 50,000 further iterations for inference to approximate the posterior distribution. A normal prior on β_j, an inverse gamma prior on δ_j, and an inverse Wishart prior on $\mathbf{\Omega}$ are used.

Table 5. US Computers and Communications Patents Classification.

Variable	Category	Patent Measure
COM	21	Number of communications patents
CHS	22	Number of computer hardware and software patents
CP	23	Number of computer peripherals patents
IS	24	Number of information storage patents

Table 6. Description of Firms Characteristics.

Variable	Descriptions	Mean	SD
SALE	Log of real sales (millions)	6.830	7.076
LABOR	Log of full time employments (1,000s)	2.398	2.647
CAPITAL	Log of real R&D capital, end of year (millions)	5.593	5.825

Note that all values are real for the Fiscal year of the data (see Compustat files of Standard and Poor for details).

Estimates of the parameter β_j, $\mathbf{\Omega}$, and δ_j (for $j = 1,\ldots,4$), with the posterior standard deviations given in brackets, are presented in Table 7. The results in Table 7 largely accord with economic theory. Of particular interest is the fact that in all cases the coefficients on ln(CAPITAL) are positive, and for COM, CHS, CP, and IS, they are also economically and statistically significant. This indicates that innovations in these categories are capital-intensive and this implies that the stock of R&D capital can be a key role patenting activity. The results also suggest that the introduction of patents tends to be done by large firms, as measured by the size of the company workforce ln(LABOR). The coefficient on that variable in the communications category is large and statistically significant. Interestingly, counter-intuitively, the coefficients on ln(SALES) in these categories are predominantly negative, and most of them are not statistically significant. These puzzling findings can be explained by the presence of asymmetric information. In particular, firms that do not have steady sales revenue such as start-ups are often cash constrained and one way for such firms to exhibit their research innovations and overall productivity is to register patents. In this case patents serve a dual role: they protect the firms, innovations from infringement and also send a positive signal to potential outside stakeholders. In contrast, firms that have more reliable sources of revenue due to higher sales have lower incentives to patent their innovations and may

Table 7. Posterior Means with Standard Deviations in Brackets for the Covariate Coefficients Associated with Each of the Four Measures Defined in Table 5 and Overdispersion Parameter δ_j.

Parameter	Univariate Model				Multivariate Model			
	COM	CHS	CP	IS	COM	CHS	CP	IS
Constant	0.97	−1.71	−5.83	−2.11	0.92	−1.46	−6.10	−2.03
	(0.847)	(0.784)	(1.196)	(0.763)	(1.132)	(1.002)	(1.642)	(0.631)
ln(SALE)	−0.30	−0.20	0.08	−0.19	−0.29	−0.19	−0.24	−0.18
	(0.210)	(0.201)	(0.313)	(0.193)	(0.240)	(0.258)	(0.410)	(0.130)
ln(LABOR)	0.76	0.35	0.27	0.22	0.77	0.32	0.10	0.22
	(0.187)	(0.167)	(0.265)	(0.171)	(0.257)	(0.201)	(0.451)	(0.148)
ln(CAPITAL)	0.08	0.61	0.72	0.63	0.078	0.59	0.70	0.60
	(0.100)	(0.097)	(0.152)	(0.114)	(0.109)	(0.223)	(0.221)	(0.088)
$\ln(\hat{\delta}_j)$	−0.17	−0.02	−0.56	−0.46	−0.18	0.03	−0.57	−0.51
	(0.085)	(0.086)	(0.135)	(0.104)	(0.100)	(0.086)	(0.128)	(0.109)

Table 8. Posterior Means with Standard Deviations in Brackets for the Correlation Matrix Ω.

Parameter	COM	CHS	CP	IS
$\ln(\hat{\Omega})$	1			
	(0.000)			
	0.070	1		
	(0.072)	(0.000)		
	0.113	0.299	1	
	(0.063)	(0.055)	(0.000)	
	−0.083	0.229	0.135	1
	(0.057)	(0.054)	(0.072)	(0.000)

instead opt to protect their intellectual property in other ways. These considerations are especially relevant in the computers and instruments industry, where patents have short life cycles and can often be circumvented by competitors.

Table 7 illustrates that overdispersion is a common feature across all categories in both the copula and univariate regression models. As a consequence, allowing for overdispersion by considering a negative binomial specification, as opposed to estimating a Poisson model, appears well justified.

Table 8 displays that the estimated $\boldsymbol{\Omega}$ in the Gaussian copula model reveals interesting dependences among patent categories, and supports the case for joint modeling and estimation. Specifically, the estimates suggest that patents in the computer hardware and software category are highly correlated with counts in the computer peripherals and information storage categories, while the correlation between patents in the communications category are relatively mildly correlated with those in the remaining categories. To test for the relevance of the copula correlation structure, we use the likelihood ratio test. The log likelihood of the independent negative binomial model is $L_A = -4050.06$, whereas the log likelihood of the Gaussian copula model is $L_B = -4023.27$, leading to a likelihood ratio test statistic $-2(L_A - L_B) = 53.58$. This test statistic has a χ^2 distribution with 6 degrees of freedom (i.e., the number of off-diagonal elements in Ω) and 5% critical value of 12.59, and therefore the data rejects the independent negative binomial specification.

Lastly, from the Gaussian copula specification the marginal likelihood is found to be −623.07 with a numerical standard error of 0.4382.

7. CONCLUSION

This paper investigates a statistical model based on Gaussian copulas for multiple count variables and presents an efficient Bayesian approach for estimating the model. Due to the difficulty of analyzing statistical models that can account for substitution patterns among multiple count variables, such models have been difficult to estimate. Recent advances in econometrics have made the analysis possible. A Gaussian copula approach gives a tractable way of handling a multivariate regression of count data when the marginal distributions are members of well-known parametric families or are tractable. This approach takes the marginal distributions of the variables given and constructs the covariance structure between the original variables by assuming a multivariate Gaussian distribution for the latent variables. In experiments involving the Gaussian copula model with negative binomial marginals, we illustrate that the proposed MCMC algorithm is efficient in sampling the posterior distribution and estimating the marginal likelihood. By applying the proposed methodology to the analysis of the patent propensity, this study not only supports the case of joint modeling but also shows that the copula approach has many potential empirical applications.

REFERENCES

Aitchison, J., & Ho, C. H. (1989). The multivariate Poisson-log normal distribution. *Biometrika, 76*, 643–653.

Alfò, M., & Trovato, G. (2004). Semiparametric mixture models for multivariate count data, with application. *Econometrics Journal, 7*, 426–454.

Barnard, J., McCulloch, R., & Meng, X.-L. (2000). Modeling covariance matrices in terms of standard deviations and correlations, with application to shrinkage. *Statistica Sinica, 10*, 1281–1311.

Cameron, A. C., & Trivedi, P. K. (1998). *Regression analysis of count data*. Cambridge: Cambridge University Press.

Cherubini, U., Luciano, E., & Vecchiato, W. (2004). *Copula methods in finance*. Chichester: Wiley.

Chib, S. (1995). Marginal likelihood from the Gibbs output. *Journal of the American Statistical Association, 90*, 1313–1321.

Chib, S., & Greenberg, E. (1998). Analysis of multivariate probit models. *Biometrika, 85*, 347–361.

Chib, S., & Jeliazkov, I. (2001). Marginal likelihood from the Metropolis–Hastings output. *Journal of the American Statistical Association, 96*, 270–281.

Chib, S., & Winkelmann, R. (2001). Markov chain Monte Carlo analysis of correlated count data. *Journal of Business and Economic Statistics*, *19*, 428–435.
Clayton, D. G. (1978). A model for association in bivariate life tables and its application in epidemiological studies of familial tendency in chronic disease incidence. *Biometrika*, *65*, 141–151.
Demarta, S., & McNeil, A. J. (2005). The t copula and related copulas. *International Statistical Review*, *73*, 111–129.
dos Santos Silva, R., & Lopes, H. F. (2008). Copula, marginal distributions and model selection: A Bayesian note. *Statistics and Computing*, *18*, 313–320.
Embrechts, P., & McNeil, A. (2003). *Modelling dependence with copulas and applications to risk management*. Amsterdam: Elsevier.
Embrechts, P., McNeil, A., & Straumann, D. (2002). *Correlation and dependence in risk management: Properties and pitfalls*. Cambridge: Cambridge University Press.
Frank, M. J. (1979). On the simultaneous associativity of F(x, y) and x+y−F(x, y). *Aequationes Mathematicae*, *19*, 194–226.
Frees, E. W., Carriere, J. F., & Valdez, E. (1996). Annuity valuation with dependent mortality. *Journal of Risk and Insurance*, *63*, 229–261.
Genest, C., & Rivest, L.-P. (1993). Statistical inference procedures for bivariate Archimedean copulas. *JASA*, *88*, 1034–1043.
Gumbel, E. J. (1960). Distributions des valeurs extrêmes en plusieurs dimensions. *Publications de L'Institute de Statistique*, *9*, 171–173.
Guo, J. Q., & Trivedi, P. K. (2002). Flexible parametric models for long-tailed patent count distributions. *Oxford Bulletin of Economics and Statistics*, *64*, 63–82.
Hall, B. H. (1990). *The manufacturing sector master file: 1959–1987*. NBER Working Paper No. 3366. NBER.
Hall, B. H., Griliches, Z., & Hausman, J. A. (1986). Patents and R&D: Is there a lag. *International Economic Review*, *27*, 265–283.
Hall, B. H., Jaffe, A. B., & Trajtenberg, M. (2001). *The NBER patent citations data file: Lessons, insights, and methodological tools*. NBER Working Paper No. 8498. NBER.
Hausman, J. A., Hall, B. H., & Griliches, Z. (1984). Econometric models for count data with an application to the Patent-R&D relationship. *Econometrica*, *52*, 271–320.
Huard, D., Evin, G., & Favre, A.-C. (2006). Bayesian copula selection. *Computational Statistics & Data Analysis*, *51*, 809–822.
Joe, H. (1997). *Multivariate models and dependence concepts: Monographs on statistics and applied probability*. London: Chapman-Hall.
Karlis, D. (2003). An EM algorithm for multivariate Poisson distribution and related models. *Journal of Applied Statistics*, *30*, 63–77.
Klugman, S. A., & Parsa, R. (1999). Fitting bivariate loss distributions with copulas. *Insurance: Mathematics and Economics*, *24*, 139–148.
Kocherlakota, S., & Kocherlakota, K. (1992). *Bivariate discrete distributions*. New York, NY: Marcel Dekker.
Liu, X., & Daniels, M. J. (2006). A new algorithm for simulating a correlation matrix based on parameter expansion and reparameterization. *Journal of Computational and Graphical Statistics*, *15*, 897–914.
Mairesse, J., & Hall, B. H. (1996). *Estimating the productivity of research and development: An exploration of GMM methods using data on French and United States manufacturing firms*. NBER Working Paper No. 5501. NBER.

Nelson, R. B. (1999). *An introduction to copulas*. New York, NY: Springer.

Pitt, M., Chan, D., & Kohn, R. (2006). Efficient Bayesian inference for Gaussian copula regression models. *Biometrika, 93*, 537–554.

Prieger, J. E. (2002). A flexible parametric selection model for non-normal data with application to health care usage. *Journal of Applied Econometrics, 17*, 367–392.

Sklar, A. (1959). Fonctions de répartition à n dimensions et leurs marges. *Publications del'Institute de Statistique de l'Universite de Paris, 8*, 229–231.

Smith, M. (2003). Modeling selectivity using Archimedean copulas. *Econometrics Journal, 6*, 99–123.

Song, P. X.-K. (2000). Multivariate dispersion models generated from Gaussian copula. *Scandinavian Journal of Statistics, 27*, 305–320.

Spiegelhalter, D. J., Best, N. G., Carlin, B. P., & van der Linde, A. (2002). Bayesian measures of model complexity and fit. *Journal of the Royal Statistical Society: Series B, 64*, 583–639.

Trivedi, P. K., & Zimmer, D. M. (2005). Copula modeling: An introduction for practitioners. *Foundations and Trends in Econometrics, 1*, 1–111.

Tunaru, R. (2002). Hierarchical Bayesian models for multiple count data. *Austrian Journal of Statistics, 31*, 221–229.

van Ophem, H. (1999). A general method to estimate correlated discrete random variables. *Econometric Theory, 15*, 228–237.

Xuefeng, L. (2001). Bayesian Analysis of Multivariate Probit Model. *Journal of Computational and Graphical Statistics, 10*, 75–81.

Wang, P., Cockburn, I. M., & Puterman, M. L. (1998). Analysis of patent data – A mixed-Poisson-regression-model approach. *American Statistical Association Journal of Business & Economic Statistics, 16*, 27–41.

Winkelmann, R. (2001). Seemingly unrelated negative binomial regression. *Oxford Bulletin of Economics and Statistics, 62*, 553–560.

Winkelmann, R. (2008). *Econometric analysis of count data* (5th ed.). Berlin: Springer.

Wong, F., Carter, C., & Kohn, R. (2003). Efficient estimation of covariance selection models. *Biometrika, 90*, 809–830.

Zhang, X., Boscardin, J. W., & Belin, T. R. (2006). Sampling correlation matrices in Bayesian models with correlated latent variables. *Journal of Computational and Graphical Statistics, 15*, 880–896.

Zimmer, D. M., & Trivedi, P. K. (2006). Using trivariate copulas to model sample selection and treatment effects: Application to family health care demand. *Journal of Business and Economic Statistics, 24*, 63–76.